Lecture Notes in Physics

T0171937

Volume 955

The Lecture Notes in Physics

The series Lecture Notes in Physics (LNP), founded in 1969, reports new developments in physics research and teaching-quickly and informally, but with a high quality and the explicit aim to summarize and communicate current knowledge in an accessible way. Books published in this series are conceived as bridging material between advanced graduate textbooks and the forefront of research and to serve three purposes:

- to be a compact and modern up-to-date source of reference on a well-defined topic
- to serve as an accessible introduction to the field to postgraduate students and nonspecialist researchers from related areas
- to be a source of advanced teaching material for specialized seminars, courses and schools

Both monographs and multi-author volumes will be considered for publication. Edited volumes should, however, consist of a very limited number of contributions only. Proceedings will not be considered for LNP.

Volumes published in LNP are disseminated both in print and in electronic formats, the electronic archive being available at springerlink.com. The series content is indexed, abstracted and referenced by many abstracting and information services, bibliographic networks, subscription agencies, library networks, and consortia.

Proposals should be sent to a member of the Editorial Board, or directly to the managing editor at Springer:

Lisa Scalone
Springer Nature
Physics Editorial Department
Tiergartenstrasse 17
69121 Heidelberg, Germany
Lisa.Scalone@springernature.com

More information about this series at http://www.springer.com/series/5304

Jeffrey Linsky

Host Stars and their Effects on Exoplanet Atmospheres

An Introductory Overview

 Springer

Jeffrey Linsky
JILA
University of Colorado and NIST
Boulder, CO, USA

ISSN 0075-8450   ISSN 1616-6361   (electronic)
Lecture Notes in Physics
ISBN 978-3-030-11451-0   ISBN 978-3-030-11452-7   (eBook)
https://doi.org/10.1007/978-3-030-11452-7

This Springer imprint is published by the registered company Springer Nature Switzerland AG.
The registered company address is: Gewerbestrasse 11, 6330 Cham, Switzerland

Preface

The study of exoplanets—their discovery, physical and chemical properties, evolution, climate, possible presence of surface liquid water, and the many requirements for habitability—is an exciting topic for both scientists and the general public who want to know whether we are alone in the universe or whether life forms, perhaps intelligent ones, exist elsewhere. Many ground-based observatories and a cohort of spacecraft, now in orbit or soon to be launched, will address these questions, and the public is waiting in anticipation of the answers. While several thousand exoplanets have been confirmed through transit, radial velocity, direct imaging, and other techniques, there are only a few exoplanets with masses similar to or somewhat larger than the Earth that have well-characterized radii and densities consistent with the existence of rocky surfaces. There exist a much larger number of exoplanets discovered to date that are more massive than the Earth without rocky surfaces that are more like Jupiter or Neptune. The menagerie of exoplanets includes objects not seen in the solar system including super-Earths, mini-Neptunes, and hot-Jupiters with a wide range of surface temperatures.

It is now recognized that the stars at the center of planetary systems, the so-called host stars, control the environment of their siblings. Imagine what the Earth would be like if our host star were not a middle-aged warm and relatively inactive G-type star but rather a much cooler M-type star that emits flares 1000 times stronger than today's Sun. What was the young Earth like when the young Sun emitted 70% less light but far stronger ultraviolet and X-rays with flares orders of magnitude stronger than the largest seen in historical times? To address these and other questions concerning the evolution of exoplanetary atmospheres and their suitability for life forms, one must understand how and at what strength stars emit high-energy radiation and accelerate supersonic magnetized winds. While exoplanet researchers recognize the importance of understanding the output of host stars, many researchers may not be familiar with the results of many observational and theoretical studies of stellar atmospheres, which is a separate field of research.

As an astronomer whose scientific background is the study of stellar atmospheres, in particular their chromospheres, coronae, and winds, I have often been asked to explain or predict the environment of exoplanets created by their host stars so that researchers can simulate and model the properties of exoplanet atmospheres. In response, I have given a series of lectures at universities and research institutes describing the properties of stars including their magnetic fields, their magnetically heated outer regions, their loss of magnetized ions and electrons to space (their stellar wind), and their effects on exoplanet atmospheres. This book is an extension of these lectures.

This book is written for the benefit of researchers, students, and the wider public who have strong interest in and perhaps some background concerning exoplanets to address and hopefully answer the following questions:

- What is the radiation, particle, and magnetic environment of different types of exoplanets at different distances from a variety of host stars, and what was the past environment that has shaped the evolution of exoplanets to the present?
- What is the source, strength, and variability of the radiation, particles, and magnetic fields that constitute the environment of exoplanets?
- How can we measure or predict the characteristics of this environment?
- What effect does the environment have on an exoplanet's atmosphere, in particular, its survival and chemistry?
- How does an exoplanet's environment determine its habitability?

These are questions that are critically important for understanding exoplanets but are usually not posed from the perspective of host stars.

In this book, I will describe the physical processes responsible for the emission of radiation and acceleration of winds of host stars that together control the environment of an exoplanet. I will cover those aspects of a star that are essential for creating this environment and those aspects of an exoplanet that are controlled by its environment. A host star and its exoplanets should not be studied in isolation as they form an integrated system. My approach is to provide a comprehensive overview of this broad topic, which means that I cannot go as deep into many technical aspects as is typically done in narrowly circumscribed technical papers. I include a large list of references to guide those interested in pursuing these questions.

I take this opportunity to thank the many people who have supported me in this long venture. My thanks go first to my wife Lois for her infinite patience and encouragement. I also thank the many people who have allowed me to use their figures in this book. There are many people who have read and commented on draft versions of parts of this book, including Luca Fossati, Kevin France, Moira Jardine, Rachel Osten, Jay Pasachoff, John Pye, Sarah Rugheimer, Evgenya Shkolnik, and Allison Youngblood. I thank them for their insights and also thank Ofer Cohen, Hilary Egan, Moira Jardine, Sarah Peacock, Evgenya Shkolnik, and others for calling my attention to their interesting but not yet published papers.

Boulder, CO, USA   Jeffrey Linsky

Contents

Chapter 1
Why Are Host Stars Important for Understanding Exoplanet Atmospheres?

Like planets in the solar system, exoplanets form, evolve, and interact with their host stars in many ways. Exoplanets form out of protostellar disks, which contain positive ions (H^+, H_2^+, and H_3^+) and other radicals produced when molecules in the disk are photo-dissociated by stellar UV and X-ray photons and then charge-exchanged by protons in the stellar wind. These positive ions become the formation seeds of complex molecules including simple organics.

As exoplanets accrete material and grow to the final size, their atmospheres are subjected to intense UV and X-radiation and high-energy particle bombardment from the young host star. This bombardment has two important effects. One is that the high-energy radiation (X-rays below 10 nm[1] and extreme-UV (EUV) radiation at 10–91.2 nm) photoionizes hydrogen high in the exoplanet's atmosphere where photon energy in excess of 13.5 eV ionizes hydrogen to heat the gas and drive a hydrodynamic wind. Table 1.1 lists the names and wavelength intervals for the different spectral regions used in this book. Whether a planet can retain its atmosphere and the conditions for significant mass loss both depend upon the strength of the host star's high-energy radiation and wind, the distance of the exoplanet from its host star, the gravitational potential of the exoplanet, and the initial chemical composition of the exoplanet's atmosphere. It is critically important to understand how the high-energy stellar flux has evolved from the proto-planet stage to the present time as there is strong evidence that X-ray (see Chap. 5) and EUV (see Chap. 6) emission decline by several orders of magnitude as the stellar-rotation rate and magnetic-field strength decrease from their high values when the star was beginning its evolution along the main sequence (see Chap. 9).

The likelihood of an exoplanet retaining its atmosphere also depends on the properties of the stellar wind. Winds of late-type dwarf stars are likely to be hot, highly ionized, and magnetic supersonic outflows if the Sun is a useful prototype. I

---

[1] In this book I express all wavelengths in nanometers rather than angstroms to follow the physics convention. 1 nm = 10 Å.

© Springer Nature Switzerland AG 2019
J. Linsky, *Host Stars and their Effects on Exoplanet Atmospheres*,
Lecture Notes in Physics 955, https://doi.org/10.1007/978-3-030-11452-7_1

Table 1.1 Spectral regions

X-rays	<10 nm	<100 Å
EUV (extreme ultraviolet)	10–91.2 nm	100–912 Å
LUV (Lyman ultraviolet)	91.2–115 nm	912–1150 Å
FUV (far ultraviolet)	115–180 nm	1150–1800 Å
NUV (near ultraviolet)	180–320 nm	1800–3200 Å
UV (ultraviolet)	115–320 nm	1150–3200 Å
XUV	<91.2 nm	<912 Å
Ultraviolet C	100–280 nm	1000–2800 Å
Ultraviolet B	280–315 nm	2800–3150 Å
Ultraviolet C	315–400 nm	3150–4000 Å

will discuss the basis for this conjecture and how to measure stellar wind properties in Chap. 8.

Stellar winds can erode exoplanet atmospheres by several mechanisms. The simplest is by collisions of stellar wind protons and electrons with neutrals at the top of the exoplanet's atmosphere that impart sufficient energy to eject the target neutrals and molecules from the exoplanet's gravity. Mars likely lost its atmosphere and oceans by a multistep process involving photodissociation of water and other molecules in the exoplanet's atmosphere followed by charge exchange with protons in the stellar wind and pickup of the resulting ions by the stellar wind's magnetic field. The effectiveness of this and related processes depends upon the properties of the stellar wind and whether the planet has a sufficiently strong magnetic field to deflect the stellar wind at distances far from the planet. I describe the physical processes that lead to mass loss and measurements of stellar winds in Chap. 10. Exoplanet magnetic fields have not yet been measured, but I describe possible methods for measurements in Chap. 14.

I describe the formation of stellar far-ultraviolet (FUV 91.2–160 nm) and near-ultraviolet (NUV 160–300 nm) emission in Chap. 4. The FUV emission of the host star and, in particular, the very strong Lyman-α (121.56 nm) radiation (see Chap. 6), photo-dissociate CO_2, H_2O, CH_4, and other molecules in an exoplanet's atmosphere. Photochemical reactions dominate over collisional reactions (equilibrium chemistry) in an exoplanet's upper atmosphere, typically above a pressure of 1 mbar. However, the penetration depth of stellar UV radiation depends on wavelength and the chemical composition of the atmosphere as described in Chap. 11.

Abiotic oxygen chemistry is driven by photodissociation of CO_2 and H_2O that produces atomic oxygen with subsequent reactions leading to O_2 and O_3 in competition with NUV radiation that photodissociates ozone. As a result, the spectral energy distribution of the host star's UV radiation, the relative amounts of FUV and NUV radiation (see Chap. 7), play a key role is determining whether O_2 and O_3 have significant abundances in the atmosphere of a lifeless exoplanet. The detection of O_2 and O_3 in an exoplanet's atmosphere without the detection of other observables could, therefore, be a false detection of photosynthesis.

Global magnetic fields are present in Mercury, Earth, Jupiter, and Saturn. They are known to have been present at an earlier phase for Mars but are not presently measured in Venus. Magnetic fields have not yet been measured on exoplanets despite searches for auroral emission and nonthermal radio emission. Should strong global fields be present on some exoplanets, as indeed appears likely given solar system examples, then the magnetic fields of close-in exoplanets should interact with host star fields, perhaps producing observable star-planet interactions. In Chap. 14, I will explore this topic.

One motivation for writing this book is that researchers who study exoplanet atmospheres and habitability often do not have a background in the physics and phenomenology of host stars that provide the environment in which exoplanets live and evolve. This survey of stellar emission and phenomenology identifies some important ways which stellar high-energy emission and winds modify exoplanet atmospheres. Host stars also control exoplanets in other ways such as gravitationally induced synchronization of rotational and orbital periods. These and other gravitational effects are beyond the scope of this book.

Chapter 2
Stellar Activity–Phenomenology and General Principles

The term "stellar activity" is routinely used to characterize phenomena that are more energetic, brighter, time-variable or otherwise different from the normal behavior of a star. Stellar activity plays a central role in the evolution of exoplanet atmospheres, but the term is rarely defined or tied to underlying causes of active phenomena on stars. In this chapter, I will describe the different types of stellar activity whose underlying cause is heating and particle acceleration by magnetic processes.

2.1 Activity Phenomena and Activity Indicators

Before attempting to define stellar activity, let's first consider the phenomena that are generally called active. Stellar flares, which are rapid enhancements of stellar emission with time scales of seconds to minutes, are observed at X-ray and ultraviolet wavelengths but also in spectral lines in the optical region. The presence of large amounts of 10^6–10^7 K gas in stellar coronae and 10^4–10^5 K gas in stellar chromospheres cannot be explained only by the radiative and convective transport of energy from the star's interior. Another energy source is needed to explain these active phenomena. Similarly, the presence of dark sunspots observed on the solar surface and starspots inferred from photometric variations and Doppler images as dark spots rotate on and off stellar disks cannot be explained by the upward energy flow in stars with radiative/convective atmospheres. The cause of this activity indicator is clearly related to the strong magnetic fields measured in sunspots.

Stellar activity is most easily seen at wavelengths where the background emission from nonactive (quiet) regions on the star is faint. High contrast between active and quiet emission regions is important for detection of activity indicators because active phenomena such as flares likely cover only a small fraction of a star's surface. Stars similar to and cooler than the Sun are faint at X-ray and ultraviolet wavelengths, whereas the warm gas in stellar chromospheres is bright in the ultraviolet and the

© Springer Nature Switzerland AG 2019
J. Linsky, *Host Stars and their Effects on Exoplanet Atmospheres*,
Lecture Notes in Physics 955, https://doi.org/10.1007/978-3-030-11452-7_2

hot gas in stellar coronae is bright at X-ray wavelengths. Chapter 4 will describe the radiation that chromospheres emit at ultraviolet wavelengths and Chap. 5 will describe the X-radiation that the million degree coronae emit.

Many activity indicators are spectral features (emission lines and continua) that are readily observed. These include X-ray emission line and continuum emission, ultraviolet emission lines, and optical emission lines including the Ca II H and K lines, Hα and higher Balmer lines, and the He I 1083.0 nm line. For a listing of the important emission lines that are activity indicators see Table 4.1. Observations of strong magnetic fields, in particular the fraction of the stellar surface covered by strong magnetic fields (the filling factor), are an important activity indicator.

2.2 What Is Stellar Activity?

The term activity is widely used to characterize a variety of phenomena observed in stars, in particular for spectral type F, G, K, and M stars with convective zones. While the term is rarely defined or described in detail, active stellar phenomena play crucial roles in determining the chemical composition of exoplanet atmospheres and even determine whether exoplanets can retain their atmospheres. Here is what I believe characterizes stellar activity:

Conversion of magnetic energy: The energy source for active phenomena is the stellar magnetic field. Our understanding of precisely how the conversion of magnetic energy to heat and accelerated particles actually occurs if far from complete, but it could involve the rapid reconnection of oppositely directed magnetic fields on very small scales or shocks of magnetic waves (e.g., Alfvèn waves or magneto-acoustic waves).

Nonradiative heating: The thermal structure in stellar photospheres is determined by the transfer of radiative energy from stellar interiors to space and by the convective transport of heat from below. Both processes depend primarily on the star's effective temperature and gravity, but the details of the thermal structure depend in part on the sources of opacity and thus on the star's chemical composition. Changes in the thermal structure of a radiative/convective photosphere, especially in the low densities high in the atmosphere requires a magnetic heat source. One quantitative measure of activity is the fractional amount of a star's bolometric luminosity that is converted to radiation in the chromosphere and corona. Stars are called active when, for example, $\log L_x/L_{\mathrm{bol}}$ is in the range 10^{-4} to its saturation level of 10^{-3}. Stars with much smaller levels of $\log L_x/L_{\mathrm{bol}}$, such as the Sun's ratio of 10^{-7} to 10^{-6}, are called inactive.

Inhomogeneous atmospheres: Convection in a stellar atmosphere creates an inhomogeneous thermal structure in the photosphere as a result of rising hot gas and falling cool gas to form the granular structure seen in the Sun. Convection also generates acoustic waves with amplitudes that increase as the waves rise to less dense regions leading to shocks. Shocks can heat the upper layers of a photosphere and produce regions of rising temperature with height called chromospheres. No models of purely acoustic wave heating, however, have been shown to heat the upper atmosphere to coronal temperatures. If the Sun is a sensible guide, stars with convective zones will have localized magnetic fields in addition to global-scale fields, and the localized magnetic fields can produce small-scale structures (e.g., starspots, plages, flares).

Time variable atmospheres: Magnetic heating processes are generally variable on time scales from microseconds to minutes. Rapid heating events called flares are observed at all wavelengths but the largest contrast with quiescent fluxes occurs where the stellar background emission is smallest, generally at X-ray and UV wavelengths. The rise time scale for flares can be seconds to a few minutes and the decay time scales (typically exponential shape) are typically many minutes to several hours. The lifetime of starspots depends on their size and the shear of a star's differential rotation. Small sunspots have lifetimes of about 6 days, whereas giant spots on active stars and spectroscopic binaries can live for up to 10 years before being disrupted by shear motions (Hall and Henry 1994).

All activity indicators are generally believed to be powered by magnetic heating processes. There are strong reasons for this conclusion. Figure 2.1 shows the correlation of X-ray emission produced by heating in stellar coronae with the total unsigned magnetic flux Φ in the underlying photosphere. Although the precise ways

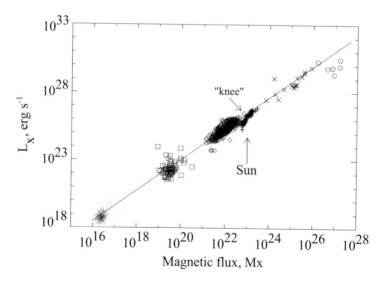

Fig. 2.1 X-ray spectral radiance L_x vs the total unsigned magnetic flux (Maxwells units) in the solar/stellar photosphere. The data consist of quiet Sun (dots), solar X-ray bright points (squares), solar active regions (diamonds), solar disk averages (pluses), G, K and M dwarfs (crosses), and T Tauri stars (circles). The line is the power-law fit to all of the data with L_x proportional to Φ^p, where $p = 1.13 \pm 0.05$. The "knee" indicates a change in power-law index that may be real or instrumental. Figure from Pevtsov et al. (2003). Reproduced by permission of the AAS

in which magnetic fields heat plasma remain an unsolved problem, the correlation of X-ray emission with magnetic flux over eight orders of magnitude in both quantities for both the Sun and stars is highly suggestive of causation. Chapter 3 describes our present understanding of magnetic fields in stars and how they can be measured.

A second reason follows from the discovery and explanation of minimum fluxes (often called basal fluxes) of emission lines in inactive stars. Following previous studies (e.g., Oranje and Zwaan 1985), Schrijver (1987) showed that there is a tight power-law correlation of the coronal X-ray flux with the Ca II and Mg II emission when the basal flux of the chromospheric emission lines is subtracted. The resulting power-law fits of the excess Ca II and Mg II emission, that is the observed emission minus the basal flux) with X-ray emission are independent of stellar effective temperature, luminosity class, and binarity. For an extended discussion of the correlations among activity indicators see Chap. 9. Although the basal flux could be powered by very weak magnetic heating, Schrijver (1987) argued that this is unlikely given the lack of correlation the basal flux with coronal emission and the dependence of the basal flux only on classical stellar parameters (effective temperature and gravity). Instead, he speculated that the basal flux seen in chromospheric lines measures purely acoustic wave heating with no magnetic component. Schröder et al. (2012) noted that during the Sun's deep magnetic minimum in 2009–2010 when there were no plages (active regions) on the solar disk, the Ca II flux level was consistent with the basal flux level of the least active G-type stars. However, Schröder et al. (2012) argued that the weak X-ray emission seen in solar coronal holes and the basal X-ray flux seen in inactive stars could result from a very low level of magnetic heating.

Pérez Martínez et al. (2014) showed that radiative equilibrium profiles of the Ca II lines computed by the state-of-the-art PHOENIX model photosphere code are the same as the observed basal flux in these lines, and that the use of excess rather than measured fluxes also reduces the scatter of the Ca II emission about the basal flux line. Figure 2.2 shows the excellent match of the synthetic spectrum computed by the PHOENIX code and the observed spectrum of the K2 V star HD 10361 near and below the Ca II K line emission feature.

The basal flux levels for the Ca II and Mg II line depend only on a star's effective temperature and not its gravity. Pérez Martínez et al. (2014) found that the basal surface flux for the Ca II H and K lines[1] in their sample of 25 G, K, and M dwarf and giant stars fits the relation $\log F_{\text{CaIIH+K}}^{\text{basal}} = (7.05 \pm 0.31) \log T_{\text{eff}} - (20.86 \pm 1.15)$ with no evidence for a dependence on stellar gravity. This relation is consistent with earlier fits, for example a power-law index of 8 (Strassmeier et al. 1994) or 7.7 (Pasquini et al. 2000) for the Ca II K line. In their analysis of 177 G, K, and M giants and supergiants, Pérez Martìnez et al. (2011) found for the Mg II h and k lines that the basal surface flux is $\log F_{\text{MgIIh+k}}^{\text{basal}} = (7.33 \pm 0.47) \log T_{\text{eff}} - (21.75 \pm 1.72)$, with an increase in basal flux with decreasing stellar gravity less than a factor of two and likely within systematic errors despite a factor of 10^4 range in stellar

---

[1]Pasachoff and Suer (2010) present the history of this notation.

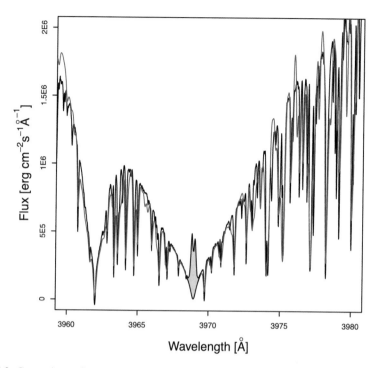

Fig. 2.2 Comparison of the synthetic radiative equilibrium spectrum (red) near the Ca II K line with the observed spectrum (black) for the K2 V star HD 10361. The model photosphere parameters are $T_{eff} = 4800$ K and $\log g = 4.5$. The gray shaded region is the chromospheric flux of the K line after subtraction of the radiative equilibrium model photosphere. Figure from Pérez Martínez et al. (2014). Reproduced from MNRAS by permission of the Oxford University Press

gravity (cf. Cardini 2005). This absence of a significant dependence on stellar gravity is surprising but may result from the near cancellation of the greater acoustic wave energy in the photospheres and greater H^- radiative damping in the extended chromospheres of giant stars as suggested by Pérez Martínez et al. (2011). Note that the basal flux levels for the Ca II and Mg II are approximately equal.

Radiative hydrodynamic simulations are revealing the complex physical processes occurring in a nonmagnetic chromospheres. The one-dimensional radiative hydrodynamical calculations of Carlsson and Stein (1995) describe the propagation of acoustic waves generated in the solar photosphere that propagate upwards into shocks in the chromosphere. The high temperatures in these intermittent shocks can explain the time-averaged basal emission seen in the Ca II and Mg II lines formed in the lower chromosphere, even though the spatially averaged mean temperature in the chromosphere may not increase with height. The three-dimensional radiative hydrodynamic calculations of Wedemeyer et al. (2004) present a picture of the solar chromosphere in which hot gas from shocks is located adjacent to regions that are very cool because of adiabatic expansion. These highly inhomogeneous models energized only by acoustic waves without magnetic fields can account for the spatially and temporally averaged basal flux of the quiet Sun and suggest similar results for stars cooler than the Sun including red giant stars (Wedemeyer et al. 2017).

Further insight concerning the difference between active and nonactive stars is provided by the atmospheric heating calculations of Cuntz et al. (1999). They calculated the propagation and shock wave heating in the atmosphere of a K2 V star with a magnetic field strength of 2100 G covering a fraction of the photosphere area (filling factor) that decreases empirically with slower stellar rotation. Acoustic waves generated by convection propagate upward from the photosphere and then shock to heat the chromosphere (cf., Narain and Ulmschneider 1996). In the fraction of the photosphere with strong magnetic fields, magneto-acoustic waves in the form of longitudinal tube waves propagate upward and shock with much larger deposition of heat. With larger filling fractions, the flux tubes become crowded and therefore diverge with height less rapidly leading to stronger amplitude shocks and stronger heating that occurs lower in the chromosphere. As described in Sect. 3.2, there is empirical evidence that filling factors increase with activity from very low values of 0.001–0.01 for the relatively inactive Sun to values close to unity for active rapidly-rotating stars (e.g., Saar 1996; Cranmer and Saar 2011). Figure 2.3 shows

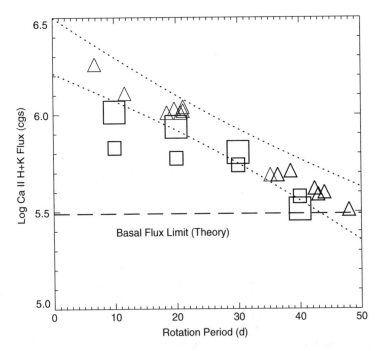

Fig. 2.3 Comparison of computed Ca II H and K emission for K2 V stars with observations. Triangles and the dotted 3σ range are observed Ca II fluxes (erg cm^{-2} s^{-1}) for K0 V to K3 V stars as a function of rotation period. The two-component chromosphere models are for acoustic and magneto-acoustic heating in flux tubes. The large squares are for uniformly distributed flux tubes and the small squares are for flux tubes in network regions. Figure from Cuntz et al. (1999). Reproduced by permission of the AAS

the resulting Ca II H and K line flux for two-component models with different magnetic filling factors determined by the rotation periods. The Ca II fluxes fit the observations very well and for the longest rotational periods, the filling factors approach zero. The model for $P_{rot} = 40$ days has nearly pure acoustic wave heating consistent with the observed basal (minimal) flux level for the Ca II emission. This calculation is consistent with our approach of identifying stellar activity as the result of magnetic heating processes. In the absence of magnetic heating, inactive stars will have chromospheres produced only by acoustic heating but with very low emission in Ca II and other activity indicators. The predicted heating by acoustic waves to 10^6 K temperatures is also very weak leading to very weak X-ray emission from acoustically heated coronae (Stepien and Ulmschneider 1989). However, the very old ($11.5^{+0.5}_{-1.5}$ Gyr), slowly rotating ($P_{rot} = 83.7 \pm 0.3$ days), metal poor M1.5 V Kapteyn's star has significant chromospheric and coronal emission (Guinan et al. 2016) indicating a low level of activity. The same conclusion is valid for stellar coronae as no main-sequence stars have X-ray surface fluxes below the quietest regions on the Sun (Schmitt and Liefke 2004).

While the conclusion that the basal fluxes of lower chromosphere emission lines of Ca II and Mg II can be explained by non-magnetic acoustic wave heating models (e.g., Buchholz et al. 1998; Schrijver 1995), the heating of the upper chromosphere presents a different picture. Judge et al. (2003) carefully analyzed solar internetwork regions to measure the lowest fluxes in the C II 133.5 nm multiplet formed in the upper chromosphere and the 133 nm ultraviolet continuum formed in the lower chromosphere. Comparison with radiative hydrodynamic simulations of acoustic wave heating shows that the basal continuum emission is consistent with the simulations, but that the basal C II emission is 4–7 times larger than predicted by the simulations. Inactive G stars such as τ Cet (G8 V) (Judge et al. 2004) have basal C II fluxes that are also higher than the simulations. This discrepancy led Judge et al. (2003) to conclude that for the Sun the upper chromosphere is heated magnetically, probably by weak internetwork fields not connected with the dynamo generated large scale magnetic fields. The formation region of the C II lines may also be heated by conduction down from the magnetically heated corona. They speculated that the heat source for the lower chromosphere of the Sun and nonactive G stars may have a small magnetic component in addition to the acoustic wave heating, but more realistic simulations are needed to definitively answer this question.

2.3 Effects of Activity on Exoplanet Measurements

Stellar activity can complicate the intercomparison of different portions of a star's spectrum observed at different times. This is an especially important effect for very young stars and M dwarfs which have high levels of activity. Since the flux in X-rays and the ultraviolet is very sensitive to activity, the comparison of high-energy radiation to the optical spectrum should be done with simultaneous data even for the less active stars.

Certain measurements of exoplanet properties can be biased by stellar activity. For example, Llama and Shkolnik (2015, 2016) considered the effects of variable surface brightness across a stellar surface on the light curve of an exoplanet as it crosses the disk of an active star. From a large number of simulations of an exoplanet traversing the Sun observed at different wavelengths, they find that the inferred planetary radius can differ significantly from its input value. For observations at X-ray, EUV, and FUV wavelengths (including Lyman-α), light curves become deeper when a planet occults bright active regions on the stellar surface leading to overestimates of the planet's radius. While the magnitude of this effect can be as large as 50% for the Sun, observed transit light curve depths at these wavelengths are often much larger than the effect of activity at the level seen in the Sun. For more active stars, the effect of activity can be larger than appears in the solar simulations. Simulations of light curves at optical wavelengths typically lead to accurate planetary radii as the effect on the light curve of traversing a starspot is small.

References

Buchholz, B., Ulmschneider, P., Cuntz, M.: Basal heating in main-sequence stars and giants: results from monochromatic acoustic wave models. Astrophys. J. **494**, 700 (1998)

Cardini, D.: Mg II chromospheric radiative loss rates in cool active and quiet stars. Astron. Astrophys. **430**, 303 (2005)

Carlsson, M., Stein, R.F.: Does a nonmagnetic solar chromosphere exist? Astrophys. J. Lett. **440**, L29 (1995)

Cranmer, S.R., Saar, S.H.: Testing a predictive theoretical model for the mass-loss rates of cool stars. Astrophys. J. **741**, 23 (2011)

Cuntz, M., Rammacher, W., Ulmschneider, P., Musielak, Z.E., Saar, S.H.: Two-component theoretical chromosphere models for K dwarfs of different magnetic activity: exploring the Ca II emission-stellar rotation relationship. Astrophys. J. **522**, 1053 (1999)

Guinan, E.F., Engle, S.G., Durbin, A.: Living with a red dwarf: rotation and X-ray and ultraviolet properties of the halo population Kapteyn's star. Astrophys. J. **821**, 81 (2016)

Hall, D.S., Henry, G.W.: The law of starspot lifetimes. International Amateur-Professional Photoelectric Photometry Communication, No. 55, p. 51 (1994)

Judge, P.G., Carlsson, M., Stein, R.F.: On the origin of the basal emission from stellar atmospheres: analysis of solar C II lines. Astrophys. J. **597**, 1158 (2003)

Judge, P.G., Saar, S.H., Carlsson, M., Ayres, T.R.: Comparison of the outer atmosphere of the "flat activity" star τ Ceti (G8 V) with the Sun (G2 V) and α Centauri A (G2 V). Astrophys. J. **609**, 392 (2004)

Llama, J., Shkolnik, E.L.: Transiting the Sun: the impact of stellar activity on X-Ray and ultraviolet transits. Astrophys. J. **802**, 41 (2015)

Llama, J., Shkolnik, E.L.: Transiting the Sun. II. The impact of stellar activity on Lyα transits. Astrophys. J. **817**, 81 (2016)

Narain, U., Ulmschneider, P.: Chromospheric and coronal heating mechanisms II. Space Sci. Rev. **75**, 453 (1996)

Oranje, B.J., Zwaan, C.: Magnetic structure in cool stars. VIII - The Mg II h and k surface fluxes in relation to the Mt. Wilson photometric Ca II H and K measurements. Astron. Astrophys. **147**, 265 (1985)

Pasachoff, J.M., Suer, T.-A.: The origin and diffusion of the H and K notation. J. Astron. Hist. Herit. **13**, 120 (2010)

Pasquini, L., de Medeiros, J.R., Girardi, L.: Ca II activity and rotation in F-K evolved stars. Astron. Astrophys. **361**, 1011 (2000)

Pérez Martìnez, M.I., Schröder, K.-P., Cuntz, M.: The basal chromospheric Mg II h+k flux of evolved stars: probing the energy dissipation of giant chromospheres. Mon. Not. R. Astron. Soc. **414**, 418 (2011)

Pérez Martínez, M.I., Schröder, K.-P., Hauschildt, P.: The non-active stellar chromosphere: Ca II basal flux. Mon. Not. R. Astron. Soc. **445**, 270 (2014)

Pevtsov, A.A., Fisher, G.H., Acton, L.W., Longcope, D.W., Johns-Krull, C.M., Kankelborg, C.C., Metcalf, T.R.: The relationship between X-Ray radiance and magnetic flux. Astrophys. J. **598**, 1387 (2003)

Saar, S.H.: Recent measurements of stellar magnetic fields. In: Strassmeier, K., Linsky, J. (eds.) IAU Symposium No. 176: Stellar Surface Structure, p. 237. Kluwer, Drodrecht (1996)

Schmitt, J.H.M.M., Liefke, C.: NEXXUS: a comprehensive *ROSAT* survey of coronal X-ray emission among nearby solar-like stars. Astron. Astrophys. **417**, 651 (2004)

Schrijver, C.J.: Magnetic structure in cool stars. XI - Relations between radiative fluxes measuring stellar activity, and evidence for two components in stellar chromospheres. Astron. Astrophys. **172**, 111 (1987)

Schrijver, C.J.: Basal heating in the atmospheres of cool stars. Astron. Astrophys. Rev. **6**, 181 (1995)

Schröder, K.-P., Mittag, M., Pérez Martínez, M.I., Cuntz, M., Schmitt, J.H.M.M.: Basal chromospheric flux and Maunder minimum-type stars: the quiet-Sun chromosphere as a universal phenomenon. Astron. Astrophys. **540**, A130 (2012)

Stepien, K., Ulmschneider, P.: X-ray emission from acoustically heated coronae. Astron. Astrophys. **216**, 139 (1989)

Strassmeier, K.G., Handler, G., Paunzen, E., Rauth, M.: Chromospheric activity in G and K giants and their rotation-activity relation. Astron. Astrophys. **281**, 855 (1994)

Wedemeyer, S., Freytag, B., Steffen, M., Ludwig, H.-G., Holweger, H.: Numerical simulation of the three-dimensional structure and dynamics of the non-magnetic solar chromosphere. Astron. Astrophys. **414**, 1121 (2004)

Wedemeyer, S., Kucinskas, A., Klevas, J., Ludwig, H.-G.: Three-dimensional hydrodynamical CO5BOLD model atmospheres of red giant stars. VI. First chromosphere model of a late-type giant. Astron. Astrophys. **606**, A26 (2017)

Chapter 3
Magnetic Fields–The Source of Stellar Activity

If the Sun did not have a magnetic field, it would be as uninteresting as most astronomers consider it to be.
(attributed to Robert Leighton)

Stellar magnetic field parameters depend on how they are measured.

Stellar activity controls exoplanet habitability.

These pithy quotes highlight the main reasons for discussing stellar magnetic fields in the context of exoplanet atmospheres and possible habitability. Magnetic fields are observed in most types of stars as shown in Fig. 3.1. Recent reviews of stellar magnetic fields by Donati and Landstreet (2009), Reiners (2012), Linsky and Scholler (2015), and Kochukhov et al. (2017) describe the presently known magnetic field properties of stars across the Hertzsprung-Russell (H-R) diagram. It is the conversion of magnetic energy to heat, nonthermal particles, and kinetic energy that is the power source for most, if not all, phenomena described as stellar activity (see Chap. 2). If the existence of an exoplanet's atmosphere and its benign chemical composition are prerequisites for habitability, then the host star's activity as manifested in its high-energy radiation and strong outflows energised by magnetic fields is essential for habitability. Even if an exoplanet retains an atmosphere, the high- energy radiation in stellar flares could deplete essential molecules including O_2 and O_3 from its atmosphere. Measurements of the magnetic properties of host stars are clearly important, but as we shall see, the intrinsic magnetic-field parameters for a star may differ greatly from the values typically obtained with conventional observing techniques.

© Springer Nature Switzerland AG 2019
J. Linsky, *Host Stars and their Effects on Exoplanet Atmospheres*,
Lecture Notes in Physics 955, https://doi.org/10.1007/978-3-030-11452-7_3

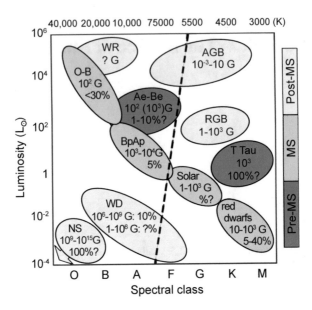

Fig. 3.1 A Hertzsprung-Russell diagram showing the range of magnetic field strengths for different star classes: WR (Wolf-Rayet stars), AGB (asymptotic giant branch stars), RGB (red giant branch stars), T Tau (young T Tauri stars), O-B (spectral types O and B), Ae-Be (Herbig Ae and Be emission line stars), BpAp (chemically peculiar B and A-type stars), solar (F and G-type stars), red dwarfs (K and M-type stars), WD (white dwarfs), and NS (neutron stars). For each star type there is a rough estimate of the percentage of stars with these magnetic fields. The dashed line separates stars with convective envelopes (to the right) from stars with radiative envelopes. Figure from Berdyugina (2009)

3.1   Magnetic Fields or Magnetic Fluxes

Table 3.1 summarizes the magnetic-field strengths observed in the Universe ranging over at least 21 orders of magnitude. Stellar magnetic fields lie in the mid-range of these field strengths. Spectro-polarimetric observations of the Sun as a star based, for example, on reflected sunlight from an asteroid, measure a magnetic flux equivalent to a homogeneous magnetic field with a strength of roughly 1 G.

Table 3.1 Magnetic field strengths in the Universe

Where	Field strength
Interstellar medium	3–$5\,\mu$G
Average Sun as a star	$\sim$1 G
Strong fields in the solar photosphere	$\sim$1.5 kG
Sunspots	3–4 kG
Strongest nondegenerate star (Babcock's star)	16 kG
Old neutron star surface	10^6 G
Magnetar crust	10^{14}–10^{15} G

On the other hand, magnetic field strengths measured with high-spatial resolution in solar active regions (also called plages) are roughly 1500 G, and in sunspots the fields are roughly 3000–4000 G. Integrated sunlight and high-spatial resolution observations imply very different magnetic-field properties of the Earth's host star. Since magnetic energy densities are proportional to the square of the magnetic field strength, the difference of 10^3 in the inferred-magnetic-field strength between the unresolved and resolved Sun corresponds to a factor of 10^6 difference in magnetic energy. This discrepancy must be understood before one can discuss the role of magnetic fields as the energy sources for stellar activity.

Figure 3.2 is a solar magnetogram obtained by the Michelson Doppler Imager (MDI) instrument on the Solar and Heliospheric Observatory (SOHO) that provides insight into the discrepancy between the different measurements of magnetic fields. This magnetogram of a solar active region shows magnetic fluxes measured by subtracting opposite circular-polarization signals in the Ni I 676.8 nm line. The solar magnetic field morphology is clearly filamentary even at the spatial resolution of this instrument (about 2 arcsec). Note that the largest flux measurements correspond to a uniform magnetic-field strength within the instrument's aperture of about 250 G. If, however, the magnetic-field structure is inhomogeneous on smaller spatial scales,

Solar magnetograms obtained with the Michelson Doppler Imager (MDI) instrument on SOHO

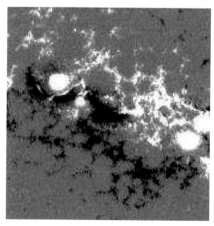

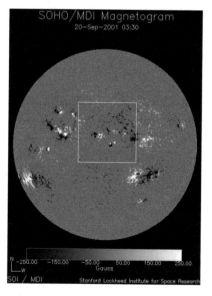

Fig. 3.2 *Right:* A net longitudinal magnetic flux density image of the Sun obtained with the MDI instrument on the Solar and Heliospheric Observatory (SOHO) spacecraft with a resolution of 2 arcsec. *Left:* Expanded field of the solar center region. White and black refer to opposite orientations of the net longitudinal magnetic-field flux density $< B_z >$. Figure courtesy of NSO/GSFC/SDO

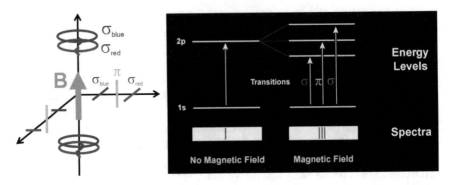

Fig. 3.3 (a) The splitting $\Delta\lambda = 0.004667g\lambda^2 B$ nm of a 2p level in the presence of a magnetic field, where B is in kGauss and g is the Landé-splitting factor. Since the splitting is proportional to the wavelength squared (λ^2), the best opportunity for actually measuring B is with infrared instrumentation. (b) Circular polarization (Stokes vector V) is observed when the magnetic field is aligned along the line of sight, and linear polarization (Stokes vectors Q and U) is observed when the magnetic field is observed perpendicular to the line of sight. Figure from Reiners (2012)

then the inferred-field strength would be much larger. How much larger is a critical question.

To answer this question, it is important to clarify how one measures magnetic fields. Figure 3.3 shows that magnetic fields parallel to the line of sight (longitudinal magnetic fields) Doppler-shift a spectral line into oppositely directed-circular polarization σ components. Magnetic fields perpendicular to the line of sight (transverse magnetic fields) Doppler-shift the oppositely directed-linearly polarized σ components, but do not shift the π component. The Zeeman splitting is $\Delta\lambda = 0.004667g\lambda^2 B$ nm, where g is the Landé-splitting factor typically between 0 and 3, B in the magnetic field strength, and λ is the wavelength in microns. The observing wavelength is critically important since Doppler shifts are proportional to λ^2. Most observations of solar and stellar magnetic fields are obtained at optical wavelengths where the Doppler shifts are small, producing only broadened lines in unpolarized light and weak line shifts in polarized light. On the other hand, unpolarized- infrared spectra clearly show the separation of the σ components and polarized-infrared spectra can measure magnetic field strengths rather than fluxes. Figure 3.4 shows what a simple absorption line, in this case the Na I D_2 (589 nm) line, looks like in unpolarized light (Stokes vector component I), linear polarization (Stokes Q and U), and circular polarization (Stokes V).

Since there are many examples in the literature of imprecise magnetic field terminology, I present in Table 3.2 an explicit, and I hope clear, set of definitions. With present technology, essentially all solar observations do not have sufficient spatial resolution to resolve the smallest scale structure present in the solar photo-sphere. Stellar observations are substantially more spatial resolution starved. Thus all magnetic observations are spatial averages, but there is a very large difference between the magnetic signal obtained from unpolarized spectra, for which there is

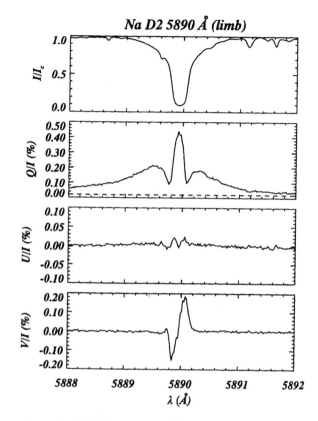

Fig. 3.4 Observations of the NaI D line at the solar limb. *Top panel:* unpolarized spectrum (Stokes I). *Second panel:* linear polarized spectrum (Stokes Q) normalized to the unpolarized spectrum. *Third panel:* linear-polarized spectrum (Stokes U) normalized to the unpolarized spectrum. *Bottom panel:* circularly polarized spectrum (Stokes V) normalized to the unpolarized spectrum. Figure from Trujillo Bueno and Manso Sainz (2002). Reproduced with permission of the Societa Italiana di Fisica

no cancellation of oppositely directed fields, and circularly and linearly polarized spectra, for which cancellation by oppositely directed fields can be very large.

For unpolarized spectra (Stokes I) of magnetically sensitive absorption lines, there is a difference between measurements of line broadening, when the separation of the magnetic components is small (typically in optical spectra) and measurements where the line is clearly split (typically in infrared spectra). In the former case, one measures the mean-unsigned magnetic-field strength $< B >$, which is the product of the magnetic field modulus, B_{MOD}, and the filling factor, f, if the field strength is uniform in part of the aperture and zero elsewhere. In the more realistic case of a distribution of magnetic-field strengths and corresponding filling factors, $< B >= \sum_i B_{MOD,i} * f_i$. If the spectral line is actually split, then one can separately measure $< B >$ and f or possibly the distribution functions of these quantities.

Table 3.2 Magnetic field terminology

Measurement type	Spatially unresolved observations	
	Filling factor included	Filling factor separate
Stokes I unsigned line-of-sight spectra (no cancellation)	Mean unsigned-magnetic-field strength (or flux density) (unpolarized Zeeman broadening) $\langle B \rangle = B_{MOD} f$	Unsigned magnetic-field modulus and filling factor (unpolarized Zeeman splitting) B_{MOD} and f
Stokes V or IVQU signed line-of-sight spectropolarimetry (with cancellation)	Net longitudinal magnetic-field strength (or flux density) (Zeeman broadening) $\langle B_z \rangle = B_{NET} f$	Net magnetic-field modulus and filling factor (Zeeman splitting) B_{NET} and f

Unpolarized spectra are broadened both by the σ components of the magnetic field along the line of sight and the σ components of the magnetic field perpendicular to the line of sight.

For spectra measured in circularly polarized light (Stokes V), cancellation of the oppositely polarized components provides a very different average of the magnetic field detected in the instrument's aperture. For observations when the absorption lines are spectrally unresolved, one measures the net longitudinal magnetic-field strength $< B_z >$ that is the product of B_{NET} and f or the product of the distributions of B_{NET} and f in the aperture, $< B_z >= \sum_i B_{NET,i} * f_i$. New instruments are beginning to obtain circularly polarized spectra in the infrared. Such data could provide separate measurements of B_{NET} and f or their distribution functions as seen through the instrumental aperture. Similar statements can be made for linearly polarized measurements (Stokes Q and U) now being obtained with new spectropolarimeters that measure all four Stokes vectors (e.g., Rosén et al. 2015).

Rabin (1992) has provided a clear demonstration of the different magnetic-field parameters that one obtains from spectrophotometry of unresolved and spectrally resolved spectral lines. Figure 3.5 shows magnetic fields measured by scanning the same solar active region with two different instruments at the same time and with the same spatial resolution (about 2 arcsec). The National Solar Observatory (NSO) magnetograph instrument measured the distribution of $< B_z >$ in an active region using the spectrally unresolved Fe I 638.8 nm line, while the Near-infrared magnetograph (NIR) observed the magnetic splitting of the Fe I 1.565 μm line ($g = 3.00$). The figure shows the distribution function of B_{NET} in the active region with filling factors generally less than 0.3. Note that typical values of $< B_z >$ lie in the range 0 to 250 G, similar to Fig. 3.2, whereas the peak of B_{NET} is in the range 1200–1500 G. The NIM data indicate that typical values of strong magnetic-field strengths in the solar photosphere are about 1500 G, which is close to equipartition between magnetic field pressure $B^2/8\pi$ and the local gas pressure.

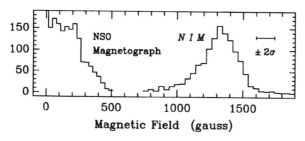

Fig. 3.5 Plotted on the right is the histogram (y axis) of the net magnetic-field modulus, B_{NET}, measured in a solar active region by the Near Infrared Magnetograph (NIM). The λ^2 splitting of the Zeeman σ components permits measurements of true magnetic-field strengths in the infrared. Plotted on the left is the histogram (also y axis) of the net-longitudinal magnetic-field strength (also called flux density), $\langle B_z \rangle$, in the same active region observed at the same time by the National Solar Observatory (NSO) magnetograph. The x-axis plots the corresponding magnetic-field strengths measured in the two different ways. The NIM was able to measure B_{NET} values from the complete splitting in Stokes V spectra of the Fe I 1.565 μm line with Landé $g = 3.00$, whereas the NSO magnetograph observing in the optical could not detect the splitting of spectral lines in Stokes V, resulting in field cancellation and spatial averaging of the inhomogeneous field. Figure from Rabin (1992). Reproduced by permission of the AAS

3.2 Magnetic Field Measurements from Unpolarized Spectra

Spectroscopic observations of magnetic fields in unpolarized light generally compare spectral lines with similar optical depths that are formed in the same temperature range but have very different Landé g factors. Figure 3.6 compares lines of Ti I near 2.2 μm that have g values of 1.58 to 2.08 with CO vibration-rotation lines that have g = 0. Valenti and Johns-Krull (2001) fitted the Ti I lines of the T Tauri star TW Hya with B_{MOD} components of 2, 4, 6, and 8 kG leading to $\sum_i B_{MOD,i} f = 2.6$ kG. Reiners (2012) includes a table of the unpolarized magnetic-field measurements up to that time, including $B_{MOD} f$ measurements as large as 4.5 kG for M dwarfs. Figure 3.7 shows that for G–M dwarfs, Stokes I measurements predict that B_{MOD} is close to its equipartition value for G–K dwarfs but not for premain sequence stars or sunspots. The filling factors for these data increase with rotational period with the empirical relation $f = 79 - 64 \log(P_{rot})$ percent (Valenti and Johns-Krull 2001). These simple relations of equipartition magnetic-field modulus and filling-factor dependence on rotational period and subsequent refinements by Cranmer and Saar (2011) are useful for predicting magnetic field properties in F–M dwarf stars.

An interesting property of stellar magnetic fields and activity indicators is the change from saturation levels observed in the most rapidly rotating active stars to a power law decrease with slower rotation and decreasing activity. The usual parameter for characterizing the evolution of stellar magnetic dynamos is the ratio of the turnover time of convective motions to the stellar rotational period called the Rossby number, $R_o = \tau/P_{rot}$. Figure 3.8 compares the observed dependence of the mean-unsigned magnetic-field strength $< B >$ with the Rossby number

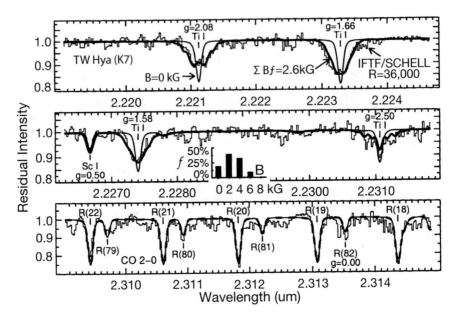

Fig. 3.6 Infrared spectrum of the T Tauri star TW Hya (thick line) compared to a synthetic spectrum with no magnetic fields. The Ti I lines have Landé g values of 1.58–2.50, and the CO lines have Landé g values of zero. The Ti I lines are fit with a distribution of magnetic field modulus $B_{MOD} = (0–8\,kG)$ and filling factors. Figure from Valenti and Johns-Krull (2001)

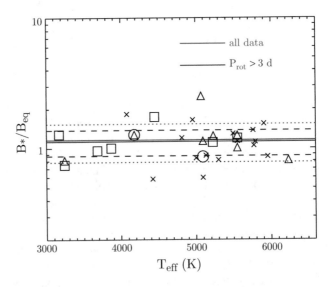

Fig. 3.7 Ratio of observed magnetic-field strengths to the equipartition magnetic field strengths in the photospheres of G0 V to M4.5 Ve stars. Symbols indicate data quality from crosses (highest) to triangles, squares, and circles (lowest). Dotted and dashed lines indicate regions within 1σ of the means (solid lines). Figure from Cranmer and Saar (2011). Reproduced by permission of the AAS

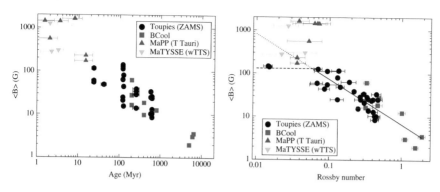

Fig. 3.8 *Left:* Average magnetic flux from Zeeman-Doppler Images of main sequence stars as a function of stellar age. *Right:* Same data but plotted as a function of Rossby number. Figures from Kochukhov et al. (2017)

and stellar age. Figures 3.9 and 3.10 show the dependence of f and Bf on R_0. These figures show saturation when $\log(R_o) < -0.1$ and a power-law decrease with increasing R_o, corresponding to the slower rotation periods that occur as stars spin down with increasing age. Saturation at small values of R_0 that is seen for many activity parameters (see Chap. 9) indicates that when stellar magnetic fields cover too large of the stellar surface area, there must be some feedback mechanism that suppresses further magnetic-field amplification.

What are the maximum magnetic field strengths that fully convective stars ($M \leq 0.35 M_\odot$ or later than M3.5 V) can have? New unpolarized magnetic field measurements by Shulyak et al. (2017) provide insight concerning this question. They found four late-M dwarfs with average magnetic field strengths $< B >$ greater than 4 kG with the extreme example of WX UMa, a rapidly-rotating M6 V star with $< B > \approx 7.0$ kG. They called attention to difference between fully convective stars that have simple dipolar magnetic fields and those stars with complex multipolar fields. Rapidly rotating stars with multipolar fields saturate at magnetic field strengths $< B > \approx 4$ kG, whereas stars with dipolar fields have field strengths that continue to increase with more rapid rotation rather than saturating at 4 kG. The underlying cause of this difference is an open question.

3.3 Magnetic Field Measurements from Spectropolarimetry and Magnetic Imaging

In the absence of spatial resolution, one can use a regular series of high-resolution spectra of a rotating star over at least one rotational period to resolve the large scale brightness structure across the star's surface. The fundamentals of Doppler imaging of a rotating star are shown in Fig. 3.11. Under nearly ideal conditions, Doppler

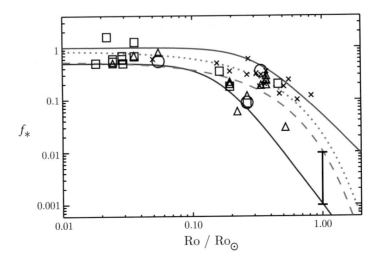

Fig. 3.9 Correlation of magnetic filling factors with Rossby number (compared to solar) for G0 V to M4.5 Ve stars. Symbols are the same as in Fig. 3.6. Solid lines indicate the lower and upper envelopes surrounding the data. The dotted and dashed lines are possible fitting relations. Figure from Cranmer and Saar (2011). Reproduced by permission of the AAS

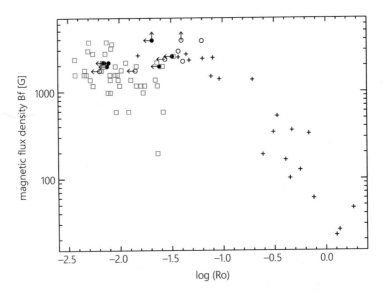

Fig. 3.10 Correlation of the mean-unsigned magnetic-field flux density, $< B >= B_{MOD} f$, with Rossby number R_0. Crosses are solar-type stars, and circles are M0–M6 stars. Red squares are M7–M9 stars. These data show a power-law rotation-activity relation for slowly rotating stars (large R_0) and saturation for rapidly rotating stars (small R_0). Figure from Reiners (2012)

Fig. 3.11 An illustration of Doppler imaging. The presence of a dark spot with a relatively faint continuum on the surface of a rotating star produces a spectrum with a partially filled-in absorption line at the wavelength corresponding to the radial velocity of the spot's longitude. Analysis of spectra obtained at many orbital phases as the spot rotates across the stellar surface can be reconstructed into a Doppler image showing the location of the spot on the stellar surface. Figure from Vogt and Penrod (1983)

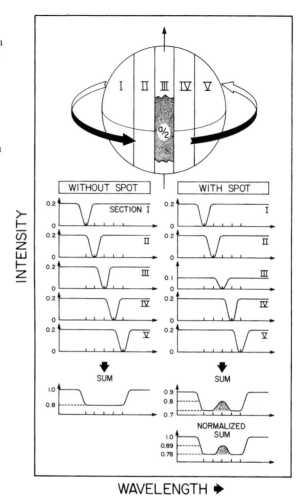

imaging can provide a reasonably valid recreation of the brightness distribution across the stellar surface. High-quality Doppler images require high signal/noise spectra, unblended spectral lines, approximate knowledge of the inclination angle (ι) of the stellar rotation axis with respect to the line of sight, observations covering at least one full rotation period with minimal data gaps, and a robust image reconstruction procedure. The number of spatial resolution elements in the rotational longitude direction, $N_{res} = v_{rot} \sin \iota / \text{FWHM}$, where v_{rot} is the equatorial rotation velocity and FWHM is the full width at half-maximum of the observed spectral line or lines. The inclination of the stellar rotational axis is important. When i is close to $0°$, (the line of sight is toward the rotational pole) all parts of the star have the same Doppler shift, and the resulting Doppler images have no spatial resolution. When i is close $90°$, there is no difference in Doppler shift for lines of sight in the northern and southern hemispheres at the same latitude. Doppler images show that starspots on rapidly rotating stars are much larger and often located near the rotational pole (unlike the Sun), that large spots or spot groups can vary on time scales of weeks or months, and that changes in the rotational periods of spots as they drift in latitude are useful indicators of stellar differential rotation (i.e., rotation rate as a function of latitude).

Extension of the Doppler imaging technique to circular polarized spectra, called Zeeman Doppler Images (ZDIs), can produce images of the large-scale stellar-vector magnetic field consisting of its radial, azimuthal, and meridional components. Figure 3.12 illustrates the different shapes of the Stokes V line profiles for two simple cases, a uniform radial field and a uniform azimuthal field across the stellar photosphere. For the case of a uniform radial field, the amplitude of the Stokes V signal of a simple absorption line decreases as the cosine of the angle from disk center, and the shape is the same toward both limbs. This behavior results from the circularly polarized signal being proportional to the projection of the radial magnetic-field vector along the line of sight. For the case of a uniform stellar

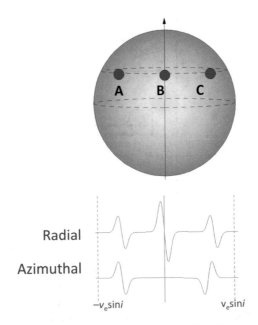

Fig. 3.12 A schematic representation of the circular-polarization Zeeman spectrum (V/I) of an absorption line formed in a magnetic field. If the magnetic field is radial relative to the star (and therefore longitudinal at disk center as viewed by an observer), the Zeeman pattern is strongest on the stellar meridian and retains the same symmetry, but weakens, as the radial field is observed towards the stellar limbs. If the magnetic field is azimuthal (E–W direction) relative to the star (and therefore transverse at disk center and longitudinal at the limbs), the Zeeman pattern is strongest at the limbs (but opposite symmetry) and zero at disk center. If the magnetic field is in the meridional direction (N–S direction, not shown in the figure), the Zeeman pattern is observed only toward the poles. Figure from Hussain (2004)

azimuthal field (aligned along lines of stellar latitude), the amplitude of the Stokes V signal is largest at the limbs and zero along the stellar meridian, and the shape is opposite at the two limbs. This behavior follows from the sign of the Stokes V signal depending on the direction of the magnetic field vector.

Donati and Landstreet (2009) have reviewed the ZDIs computed up to that time, but a large number have been synthesized since then, for example, by the BCool collaboration (Marsden et al. 2014) and other observations (cf., See et al. 2017). ZDIs are now available for main sequence stars with spectral types F to M, premain sequence stars including T Tauri stars, subgiants, and giant stars (see Kochukhov et al. 2017). Figure 3.13 illustrates that stars rotating much more rapidly than the Sun often have large-scale toroidal fields and that the magnetic field patterns change at least as often as yearly. Figure 3.14 is an example of a magnetic field reversal detected in ZDIs obtained only 1 year apart. Figure 3.15 provides a very rich overview of large scale magnetic patterns as functions of stellar mass and rotational period obtained from ZDI images.

While the ZDIs illustrate large-scale patterns in stellar magnetic fields, they have important limitations. They cannot measure small-scale magnetic-field strengths that are in the kilogauss range because of cancelation of oppositely directed circular polarization within the spatial resolution of the instrument. This cancellation effect is clearly shown by the inferred magnetic-field strengths of less than 100 G in Figs. 3.13 and 3.14 Irregular time sampling, data gaps, evolution of magnetic fields during an observing run, and possible cross-talk between the magnetic-field parameters can compromise the resulting ZDI. The presence of strong magnetic fields in starspots with dark continuum emission and plage regions with bright continuum emission can result in measured-average magnetic fluxes that are either too small or large. Including Stokes I data can improve the resulting ZDI (e.g., Carroll et al. 2012).

A clear demonstration of the limitations of the Zeeman-Doppler Imaging technique is shown in Fig. 3.16. In this simulation by Yadav et al. (2015), the left maps are the input high-resolution radial, meridional, and azimuthal magnetic-field strength, and the central columns are low-pass filters of the three original images. The images on the right are ZDIs obtained from the original images with the same spatial resolution as the low-pass filter images. While the ZDIs and low-pass images are remarkably similar and both faithfully reveal the coarse pattern of the original images, the inferred magnetic fields of the low-pass and ZDI images are much smaller than in the original magnetograms. The ZDI images for the three components of the magnetic field recover only 20% of the original magnetic flux, 80% is lost (cf. Reiners and Basri 2009). Figure 3.17 also illustrates the loss in inferred magnetic field strength and magnetic energy that results from measuring these magnetic field properties from ZDIs.

Until now most spectropolarimetric observing programs have measured only circular polarization (Stokes V) in the optical spectrum. New spectro-polarimeters coming on line will also measure linear polarization (Stokes Q and U) in addition to circular polarization (e.g., Rosén et al. 2015), which will lead to more accurate azimuthal and meridional fields. Kochukhov et al. (2017) describe the difference

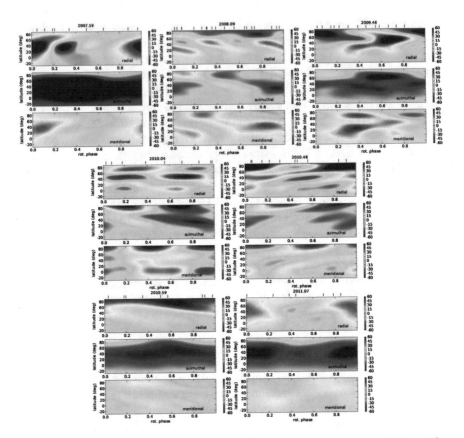

Fig. 3.13 Reconstructed net longitudinal magnetic-field strength $< B >$ maps of ξ Boo A for the radial, azimuthal and meridional components obtained at five times between 2007 and 2011 by Morgenthaler et al. (2012). The strong azimuthal component contains most of the magnetic energy when the star is most active (2007 and 2011). Reproduced with permission of ESO

in ZDIs constructed from both linear and circularly polarized data (Stokes IVQU) with ZDIs constructed from only circularly polarized data (Stokes IV) for the single-lined active binary system II Peg. They show that when the linearly polarized data are included, the inferred magnetic flux increases by a factor of 2.1–3.5, the inferred magnetic energy is larger, and the inferred magnetic structure is more complex. When infrared spectropolarimetry becomes available, the detection of magnetically split lines will lead to separate measurements of B_{NET} and f and almost certainly a better understanding of stellar magnetic fields and morphology.

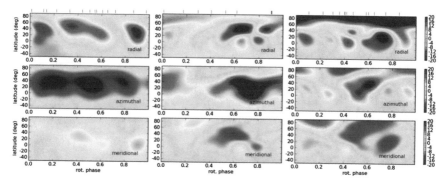

Fig. 3.14 Evidence for magnetic polarity reversal seen in the ZDIs of the G5 IV star HD 190771. The left, middle and right columns are for data obtained in 2007, 2008, and 2009. The top, middle and bottom rows are for the radial, azimuthal, and meridional components of the magnetic field. Figure from Petit et al. (2009). Reproduced with permission of ESO

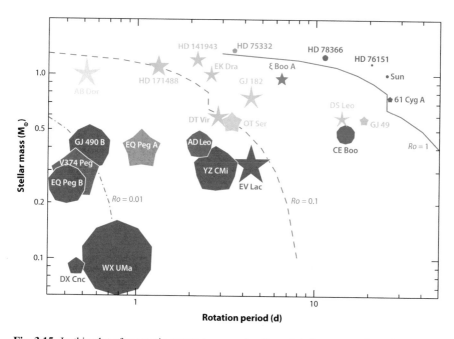

Fig. 3.15 In this plot of magnetic parameters as a function of stellar mass and rotation period, the symbol sizes indicate the net longitudinal magnetic energy $< B_z >^2$ in relative units. Note that $< B_z >^2$ generally increases with faster rotation and with decreasing stellar mass. The symbol colors indicate whether the large-scale field is poloidal (red), toroidal (blue), or a mixture (green). The importance of toroidal fields is clearly seen in the rapidly rotating solar-mass stars. Finally, the symbol shapes indicate whether the poloidal fields are axisymmetric (decagon) or nonaxisymmetric (star shape). The dashed lines are lines of constant Rossby number. Figure from Donati and Landstreet (2009)

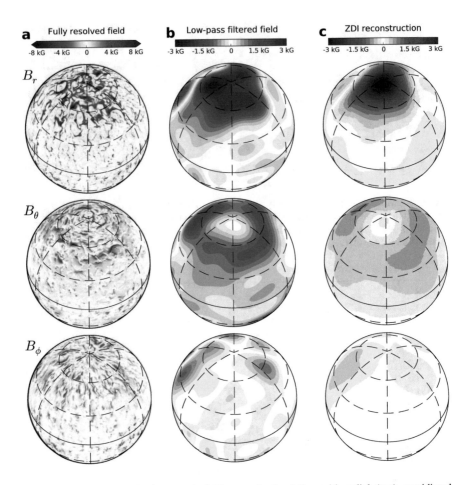

Fig. 3.16 (**a**) Fully resolved magnetic-field strength simulation with radial (top), meridional (middle), and azimuthal (bottom) components. (**b**) Low-pass filter maps of the input magnetic field. (**c**) Zeeman Doppler image reconstructions of the input magnetic field with the same spatial resolution as the low pass-filter images. Figure from Yadav et al. (2015). Reproduced by permission of the AAS

3.4 Combining Spectroscopic and Spectropolarimetric Data

Since Stokes I measurements of stellar magnetic-field strength lack spatial information and ZDI maps based on Stokes V measurements measure magnetic fluxes with large amounts of cancellation, the combination of these two techniques offers the promise of spatial mapping of magnetic field strengths. A major challenge is determining the area coverage of dark starspots, bright plages, and quiescent regions, all of which have different distributions of magnetic field strengths and filling factors.

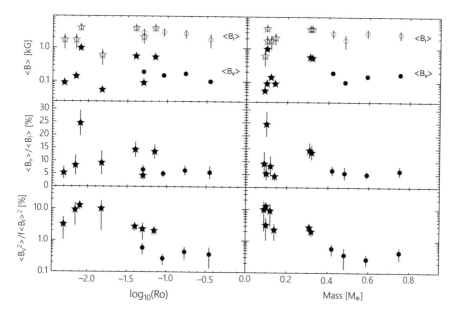

Fig. 3.17 *Top:* Comparison of mean-unsigned magnetic-field strengths ($< B_I >$) measured from unpolarized spectra (Stokes I) (open symbols) with net longitudinal magnetic-flux densities ($< B_V >$) measured from polarized spectra (Stokes V) (filled symbols). *Middle:* Percent of the unsigned magnetic-flux density ($< B_I >$) that is measured from the Stokes V data ($< B_V >$). *Bottom:* Percent of the total (unsigned) magnetic energy ($f < B_I >^2$) that is measured from Zeeman Doppler Images ($< B_V >^2$). Figure from Reiners (2012)

One forward step in computing more realistic representations of stellar magnetic fields and structures is to include continuum intensity maps, essentially unpolarized Doppler images, when computing ZDIs. In this way one can correct for the faint signal of very strong magnetic fields in the dark starspots (e.g., Carroll et al. 2012). Starspot magnetic fields and their fractional coverage of the stellar disk can also be studied by analyzing polarized spectra of molecules, for example MgH, TiO, CaH, and FeH (Afram and Berdyugina 2015).

Another approach is to analyze the Stokes I and Stokes QUV data obtained at the same time for a given star. From the high-resolution Stokes I spectrum, one can measure the unsigned magnetic-field modulus and filling factor and use these data to convert the magnetic fluxes in the ZDI to magnetic field strengths across the stellar surface. This process will involve subjective assignment of where the strong magnetic fields are located in the low spatial-resolution ZDI, but such models should be attempted. One further step could be to use the distribution of magnetic field strengths and filling factors to recalibrate the ZDI obtained at the same time. The distribution of magnetic field strengths and filling factors that Valenti and Johns-Krull (2001) obtained from spectra of the T Tauri star TW Hya is an example of the type of data needed for such analyses.

References

Afram, N., Berdyugina, S.V.: Molecules as magnetic probes of starspots. Astron. Astrophys. **576**, A34 (2015)

Berdyugina, S.V.: Stellar magnetic fields across the H-R diagram: observational evidence. In: IAU Symposium on Cosmic Magnetic Fields: From Planets, to Stars and Galaxies, vol. 259, p. 323 (2009)

Carroll, T.A., Strassmeier, K.G., Rice, J.B., Künstler, A.: The magnetic field topology of the weak-lined T Tauri star V410 Tauri. New strategies for Zeeman-Doppler imaging. Astron. Astrophys. **548**, A95 (2012)

Cranmer, S.R., Saar, S.H.: Testing a predictive theoretical model for the mass-loss rates of cool stars. Astrophys. J. **741**, 54 (2011)

Donati, J.-F., Landstreet, J.D.: Magnetic fields of nondegenerate stars. Ann. Rev. Astron. Astrophys. **47**, 333 (2009)

Hussain, G.A.J.: Stellar surface imaging: mapping brightness and magnetic fields. Astron. Nachr. **325**, 216 (2004)

Kochukhov, O., Petit, P., Strassmeier, K.G., Carroll, T.A., Fares, R., Folsom, C.P., Jeffers, S.V., Korhonen, H., Monnier, J.D., Morin, J., Rosén, L., Roettenbacher, R.M., Shulyak, D.: Surface magnetism of cool stars. Astron. Nachr. **338**, 428 (2017)

Linsky, J.L., Schöller, M.: Observations of strong magnetic fields in nondegenerate stars. Space Sci. Rev. **191**, 27 (2015)

Marsden, S.C., Petit, P., Jeffers, S.V. Morin, J., Fares, R., Reiners, A., do Nascimento, J.-D., Aurière, M., Bouvier, J., Carter, B.D., Catala, C., Dintrans, B., Donati, J.-F., Gastine, T., Jardine, M., Konstantinova-Antova, R., Lanoux, J., Lignières, F., Morgenthaler, A., Ramìrez-Vèlez, J.C.: A BCool magnetic snapshot survey of solar-type stars. Mon. Not. R. Astron. Soc. **444**, 3517 (2014)

Morgenthaler, A., Petit, P., Saar, S., Solanki, S.K., Morin, J., Marsden, S.C., Aurière, M., Dintrans, B., Fares, R., Gastine, T., Lanoux, J., Lignières, F., Paletou, F., Ramírez Vélez, J.C., Théado, S., Van Grootel, V.: Long-term magnetic field monitoring of the Sun-like star ξ Bootis A. Astron. Astrophys. **540**, A138 (2012)

Petit, P., Dintrans, B., Morgenthaler, A., Van Grootel, V., Morin, J., Lanoux, J., Aurière, M., Konstantinova-Antova, R.: A polarity reversal in the large-scale magnetic field of the rapidly rotating sun HD 190771. Astron. Astrophys. **508**, L9 (2009)

Rabin, D.: A true-field magnetogram in a solar plage region. Astrophys. J. Lett. **390**, 103 (1992)

Reiners, A.: Observations of cool-star magnetic fields. Living Rev. Sol. Phys. **9**, 1 (2012)

Reiners, A., Basri, G.: On the magnetic topology of partially and fully convective stars. Astron. Astrophys. **496**, 787 (2009)

Rosén, L., Kochukhov, O., Wade, G.A.: First Zeeman Doppler imaging of a cool star using all four stokes parameters. Astrophys. J. **805**, 169 (2015)

See, V., Jardine, M., Vidotto, A.A., Donati, J.-F., Boro Saikia, S., Fares, R., Folsom, C.P. Héebrard, É.M., Jeffers, S.V., Marsden, S.C., Morin, J., Petit, P., Waite, I.A., BCool Collaboration: Studying stellar spin-down with Zeeman-Doppler magnetograms. Mon. Not. R. Astron. Soc. **466**, 1542 (2017)

Shulyak, D., Reiners, A., Engeln, A., Malo, L., Yadav, R., Morin, J., Kochukhov, O.: Strong dipole magnetic fields in fast rotating fully convective stars. Nature Astron. **1**, 184 (2017)

Trujillo Bueno, J., Manso Sainz, R.: Remote sensing of chromospheric magnetic fields via the Hanle and Zeeman effects. Il Nuovo Cimento C **25**, 783 (2002)

Valenti, J.A., Johns-Krull, C.: Magnetic field measurements for cool stars. In: Mathys, G., Solanki, S.K., Wickramasinghe, D.T. (eds.) Magnetic Fields Across the Hertzsprung-Russell Diagram. ASP Conference Proceedings, vol. 248, p. 179. Astronomical Society of the Pacific, San Francisco (2001)

Vogt, S.S., Penrod, G.D.: Doppler imaging of spotted stars - application to the RS Canum Venaticorum star HR 1099. Publ. Astron. Soc. Pac. **95**, 565 (1983)

Yadav, R.K., Christensen, U.R., Morin, J., Gastine, T., Reiners, A., Poppenhaeger, K., Wolk, S.J.: Explaining the coexistence of large-scale and small-scale magnetic fields in fully convective stars. Astrophys. J. Lett. **813**, L31 (2015)

Chapter 4
Stellar Chromospheres: The Source of UV Emission

The lower layer of a star's atmosphere, its photosphere, has a thermal structure that decreases outward controlled by the balance of radiative and convection heat from below and the loss of radiation to space. With increasing height in a stellar atmosphere, magnetic heating processes become important in the energy balance, forcing the temperature to increase with height in a region called the chromosphere. Magnetic heating processes include the damping of different types of magnetic waves and the reconnection of magnetic fields. This is an broad topic with no clear convergence on the most likely heating mechanism is beyond the scope of this book. The interested reader can explore the topic starting with recent papers (e.g., Soler et al. 2017 and Leenaarts et al. 2018) and references therein. In this chapter, I describe the effects of magnetic heating to produce temperature-height distributions that are modeled in various ways and the emission of spectral lines and continua that are diagnostics of a chromosphere's thermal structure and provide tests for the models.

4.1 Chromospheric Radiation and Spectroscopic Diagnostics

Emission from stellar chromospheres is observed in spectral lines and continua formed at temperatures between 3000 and about 20,000 K. Most of the well-studied spectral features are resonance line transitions between the ground state and the first-excited state of an atom or ion. In the optical spectrum, the emission cores of the Ca II H and K lines (see Fig. 4.1) are very useful diagnostics of chromosphere properties. The H and K lines in G-type stars show weak emission features in the centers of broad photospheric absorption lines, but, as shown in Fig. 4.1, the core emission is stronger in stars that are young and rapidly rotating. In M stars (see Fig. 4.2) the absorption wings are much fainter and the emission core is far brighter, especially for more active M stars. Emission in the Ca II H and K lines

© Springer Nature Switzerland AG 2019
J. Linsky, *Host Stars and their Effects on Exoplanet Atmospheres*,
Lecture Notes in Physics 955, https://doi.org/10.1007/978-3-030-11452-7_4

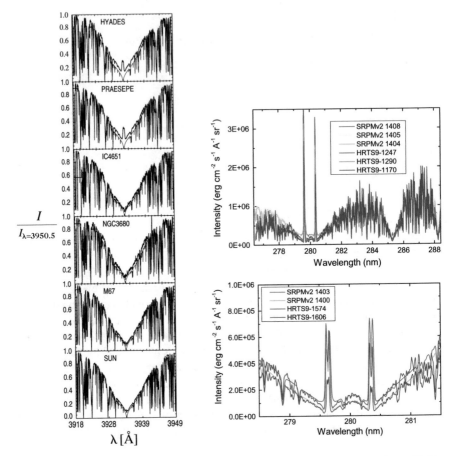

Fig. 4.1 *Left:* Spectra of the Ca II K line for solar-mass stars in clusters with ages from 600 Myr (Hyades and Praesepe) to M67 (4.8 Gyr) and the Sun. The chromospheric emission component in at the center of the broad photospheric absorption line. Thin lines are the simulated radiative equilibrium solar spectrum. Figure from Pace and Pasquini (2004). *Upper Right:* Mg II h and k lines observed (HRTS) and computed (SRPM) for dim cell interior and bright plages on the Sun. *Lower Right:* Mg II h and k lines observed and modelled for very active regions on the Sun. Data from Fontenla et al. (2011). Reproduced with permission of ESO

cores ratioed to the nearby continuum (the S-index) or to the stellar bolometric luminosity (the R'_{HK} parameter) is commonly used as an indicator of stellar activity (see Sect. 13.1). Other resonance lines in the optical region including the Na I D_1 588.995 nm and D_2 589.592 nm and Ca I 657.278 nm lines also can show core emission features, especially in the coolest stars. Table 4.1 lists the wavelengths and formation temperatures of these and other chromospheric lines, and Linsky (2017) provides more detail concerning these and other chromospheric spectral diagnostics and how to analyze these spectral features.

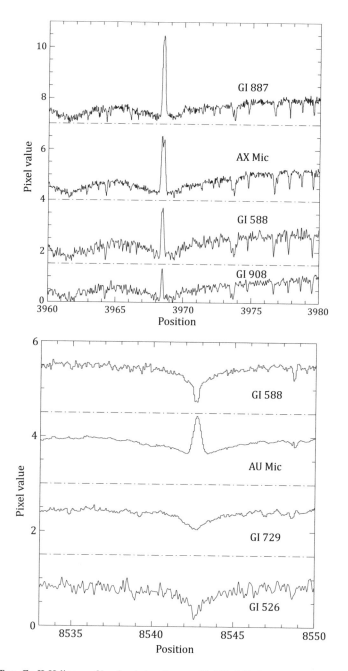

Fig. 4.2 Top: Ca II H line profiles for 4 dwarf stars: Gl 887 (M2V), AX Mic (M0Ve), Gl 588 (M2.5V), and Gl 908 (M2V) showing bright core emission produced in their chromospheres and faint absorption wings produced in their photospheres. Bottom: the 854 nm Ca II infrared triplet line for 4 dwarf stars: Gl 588 (M2.5V), AU Mic (M1Ve), Gl 729 (M3.5Ve), and Gl 526 (M3V) showing that active M dwarfs can have bright chromospheric emission cores. Figure from Houdebine (2009). Reproduced from MNRAS by permission of the Oxford University Press

Table 4.1 Emission lines formed in stellar chromospheres and transition regions

Wavelength (nm)	Ion	Formation log T^a	Surface flux[b] (10^3 ergs cm^{-2} s^{-1})		
			α Cen A (G2 V)	ϵ Eri (K2 V)	AU Mic (dM1e)
30.392	He II (Lyα)	~4.9	(4.7)	(33)	(340)
58.4334	He I	~4.65			
97.702	C III	4.72	5.20	13.5	79.6
102.572	H I (Lyβ)	~4.5			
103.193	O VI	5.43	2.21	14.5	109
103.634	C II	4.30	0.188	0.502	3.64
103.702	C II	4.30	0.283	1.04	6.76
103.762	O VI	5.43	1.06	7.67	55.1
117.6mult	C III	4.72	2.48	10.9	66.0
120.6510	Si III	4.45	4.39	12.1	36.9
121.567	H I (Lyα)	~4.5	234	1632	5336
121.8344	O V]	5.31	0.619	2.28	9.52
123.8821	N V	5.22	0.666	3.15	23.5
124.2804	N V	5.22	0.329	1.54	11.0
130.2169	O I		1.87		9.98
130.4858	O I		1.99		25.4
130.6029	O I		2.06		25.9
133.4532	C II	4.30	3.10	6.30	35.8
133.5708	C II	4.30	4.28	16.2	74.8
135.1657	Cl I		0.286	0.663	3.80
135.5598	O I		0.437		2.70
137.1292	O V	5.31	0.050	0.352	1.82
139.3755	Si IV	4.78	2.04	7.71	25.6
140.1156	O IV]	5.15	0.217	0.606	2.08
140.2770	Si IV	4.78	1.09	4.05	15.2
154.8195	C IV	4.98	4.18	18.7	111
155.0770	C IV	4.98	2.32	9.45	59.0
156.1mult	C I		1.84		20.2
164.3mult	He II (Hα)	~4.9	0.587		137
166.6153	O III]	4.84	0.294	0.405	0.676
167.0787	Al II		1.68		9.67
189.2030	Si III]	4.45		4.19	
190.87	C III]	4.72			
232.5mult	C II]	4.30			
234.0mult	Si II]	4.00		28.7	
279.5523	Mg II k	~3.9	614	1022	1457
280.2697	Mg II h	~3.9	307	777	728
393.3663	Ca II K	~3.8	440	950	
396.8468	Ca II H	~3.8		720	
587.57	He I (D3)				

(continued)

Table 4.1 (continued)

Wavelength (nm)	Ion	Formation log T^a	Surface flux[b] (10^3 ergs cm^{-2} s^{-1}) α Cen A (G2 V)	ϵ Eri (K2 V)	AU Mic (dM1e)
588.995	Na I (D1)				
589.592	Na I (D2)				
656.280	H I (Hα)	~3.9			
849.802	Ca II (IR triplet)	~3.8			
854.209	Ca II (IR triplet)	~3.8		3100	
866.214	Ca II (IR triplet)	~3.8			
1083.0mult	He I				

[a]Formation temperatures are from Sim and Jordan (2005). For the optically lines of H I, He I, He II, and Mg II, the formation temperatures refer to the line cores
[b]Surface flux to observed flux ratios: 2.32×10^{15} (α Cen A), 3.35×10^{16} (ϵ Eri), and 5.20×10^{17} (AU Mic)

The ultraviolet (UV) spectral range (120–300 nm) contains many important emission lines formed in stellar chromospheres including the HI Lyman-α line and the resonance multiplets of O I, C II, and Si II. These UV resonance lines are formed in the chromosphere because the energy differences between the ground and first excited states of neutral atoms and low ionization stages are typically 5–10 eV and can be collisionally excited by thermal electrons in chromospheres where the temperatures are 5000–20,000 K. Figure 4.3 (Pagano et al. 2004) shows the FUV (117–160 nm) spectrum of the solar-like star α Cen A and Fig. 4.4 (Pagano et al. 2000) shows the FUV spectrum of the M0 V star AU Mic. Despite the large difference in effective temperature of the two stars, their FUV spectra look very similar and consist of the same chromospheric lines and higher temperature lines formed in their transition regions (see below).

In the NUV (170–300 nm), the Mg II h and k lines are analogous to the Ca II H and K lines but are formed at slightly higher temperatures with more easily observed emission because of the fainter line wings and continuum at these shorter wavelengths (see Fig. 4.1). The resonance lines of He I (58.4334 nm) and He II (30.392 nm) are observed in solar spectra but are not usually detected in stellar spectra because of interstellar absorption. These lines are typically formed by collisional excitation, but ambipolar diffusion (AD) and recombination following photoexcitation by EUV (10–91.2 nm) radiation can be important. As a result, these lines are not useful indicators of the local temperature and density in stellar chromospheres.

There are a number of transitions between excited states, often called subordinate transitions, that can be analyzed to determine chromospheric temperatures and densities, but they require multilevel radiative-transfer calculations to be properly fit. The most commonly studied line is the Hα transition between levels 2 and 3 of hydrogen. This line is seen in absorption in inactive stars or in emission in active stars (see Fig. 4.5). Since the Hα line is observed in contrast to the adjacent continuum, which is faint in M dwarfs, the Hα line appears in emission for most

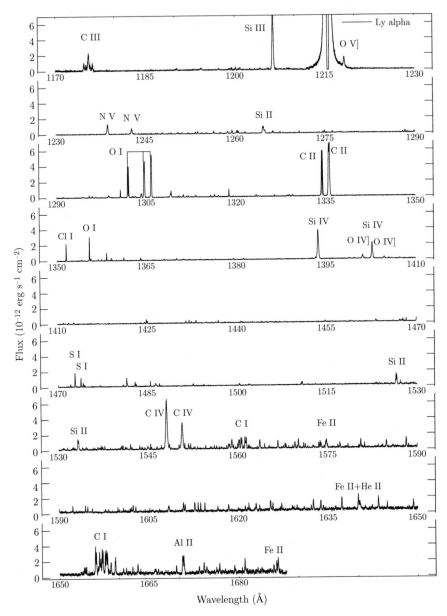

Fig. 4.3 High-resolution (2.6 km s^{-1}) spectrum of α Cen A (G2 V) obtained with the STIS instrument on HST by Pagano et al. (2004). Included in this spectrum are chromospheric lines of 17 atoms and ions, transition region lines of 17 ions, the 124.19 nm coronal line of Fe XII and molecular lines of CO and H_2. Reproduced with permission of ESO

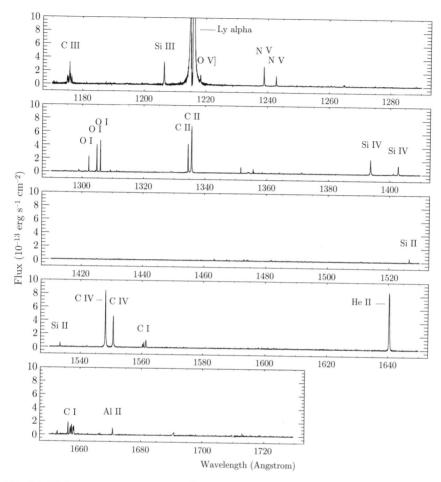

Fig. 4.4 Moderate-resolution (6.55 km s^{-1}) spectrum of AU Mic (M0 V) obtained with the STIS instrument on HST by Pagano et al. (2000). Included in this spectrum are chromospheric lines of 12 atoms and ions, transition region lines of 10 ions, the 135.408 nm coronal line of Fe XXI, and molecular lines of CO and H$_2$. Reproduced by permission of the AAS

M stars, except for the least active stars. Houdebine (2009) shows examples of hydrogen Balmer and Paschen series lines in M dwarf stars. The infrared triplet lines of Ca II (849.802 nm, 854.209 nm, and 866.214 nm) are transitions between the upper states of the H and K lines and the metastable 3^2D state. These lines are generally in absorption, but their line cores can fill-in and go into emission in the most active stars and during flares. The upper states of the 1083 nm subordinate line of He I and the 164 nm Hα line of He II are populated mostly by recombination following coronal X-ray and EUV photoionization, as seen in the Sun (Andretta and Jones 1997; Centeno et al. 2008; Golding et al. 2014) and active stars, but also by electron collisions, which are more important in inactive and metal-poor stars (Takeda and Takada-Hidai 2011). Emission line ratios comparing permitted

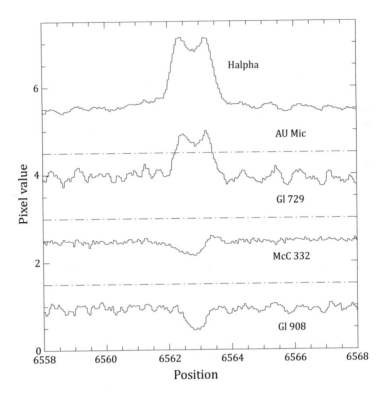

Fig. 4.5 Hα line profiles for 4 dwarf stars: AU Mic (M1Ve), Gl 729 (M3.5Ve), McC 332 (dK7+dM3.5), and Gl 908 (M2V) showing that Hα is an absorption line for inactive M stars and an emission line for active M stars. Figure from Houdebine (2009). Reproduced from MNRAS by permission of the Oxford University Press

to forbidden transitions, such as those involving lines of O IV] near 140 nm, are diagnostic of electron densities (e.g., Linsky et al. 1995).

At wavelengths less than 150 nm for G stars or 250 nm for M dwarfs, continuum emission formed in the chromosphere is brighter than the continuum radiation formed in the photosphere. The cold photospheric temperatures of M dwarfs produce very weak continuum emission at wavelengths below 250 nm. Chromospheric continuum emission formed by recombination to the ground and lower-excited states of C I, Si I, Mg I, and Fe I is observed in the 120–170 nm spectral region (Linsky et al. 2012b; Loyd et al. 2016). These continua are fit in solar models, but except for ε Eri, they are usually too faint to be fit with stellar models. The hydrogen Lyman continuum (60–91.2 nm) is observed in solar spectra, but is completely absorbed in stellar spectra. The sub-mm to cm wavelength free-free continuum is potentially very useful for modeling solar and stellar chromospheres, because the emission is thermal and formed in local thermal equilibrium (LTE) as described in Sect. 4.3.

The 91–118 nm wavelength region contains a wealth of chromospheric emission lines including the hydrogen Lyman series beginning with Lyman-β (102.572 nm)

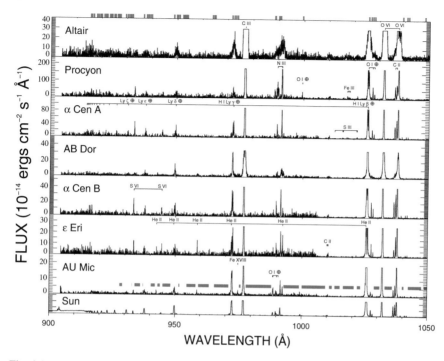

Fig. 4.6 Moderate-resolution 98–118 nm spectra of seven dwarfs stars obtained with FUSE by Redfield et al. (2002). Included are the chromospheric lines of the HI Lyman series and the C II multiplet, transition region lines of C III and O VI, and the coronal lines of Fe XVIII (97.486 nm) and Fe XIX (111.806 nm). Reproduced by permission of the AAS

and lines of C II (103.634 nm and 103.712 nm) (see Fig. 4.6). Transition region lines of C III (97.7026 and 117.5 nm multiplet) and OVI (103.195 nm and 103.763 nm) are also present in this spectral region, as are coronal lines of Fe XVIII (97.5 nm) and Fe XIX (111.8 nm). Between 1999 and 2007, the Lyman Far Ultraviolet Explorer (FUSE) satellite obtained spectra of A7 IV to M0 V dwarf stars (Redfield et al. 2002) and F0 II to M2 Iab luminous stars (Dupree et al. 2005). The ratio of the C III 117 nm to 97.3 nm lines is a useful tool for determining the electron density in the transition region. The O VI lines of luminous stars are asymmetric probably due to outflowing plasma in their winds (Dupree et al. 2005). Osten et al. (2006) used FUSE fluxes of the C III, O VI, and Fe XVIII lines in their emission-measure model of M4 V flare star EV Lac.

4.2 Stellar Atmosphere Regions

Beginning in the 1920s, astronomers began constructing physical models of stellar atmospheres to describe how the flow of energy from deep layers in a star is converted to observable continuum and spectral-line radiation. An important impetus for these studies was the goal of determining chemical abundances relative to hydrogen. These classical models assumed that a stellar atmosphere can be described by a grid of plane-parallel layers in which all physical parameters depend only on height, pressure, or mass column density above some reference level. These models also assumed steady-state conditions (all parameters are time independent), hydrostatic equilibrium (no bulk flows or magnetic pressure support), and no heating sources in addition to radiative and convective heat transport from below. The latter assumption rules out heating by magnetic shock waves or magnetic-field recombination events. I refer to these two processes as nonradiative heating. The early models assumed that the populations of all excited levels and the ionization states of all atoms depend only on the local temperature, an approximation known as local thermodynamic equilibrium (LTE). Following the pioneering work of Thomas and Athay (1961), Hummer (1962), and Avrett and Hummer (1965) the LTE approximation was replaced by statistical equilibrium equations (non-LTE) that describe how each atomic level is populated by line and continuum radiation, cascades from higher levels, and electron collisions.

Deep in a stellar atmosphere, classical models provide a sensible description of atmospheric properties, but at higher levels they become inadequate approximations. Departures from the classical atmosphere assumptions are the root causes of the complex phenomena that comprise stellar activity (see Chap. 2). It is, therefore, useful to introduce specific terms that characterize the different regions of a stellar atmosphere according to whether the classical assumptions are valid. I describe these atmospheric regions in terms of plane-parallel layers, but if the Sun is a useful guide, the geometry is three dimensional and is likely controlled by the magnetic field structure. These thermal regimes are controlled by the local energy balance and may be structured along magnetic loop geometries. While it is convenient to study these regions in isolation, they are not just boundary conditions to adjacent regions because radiation, energy flows, and magnetic fields couple these regions together (Judge 2010).

Photosphere   This term refers to that portion of a stellar atmosphere where the visible continuum is formed. In the photosphere, possible energy input from magnetic or acoustic shocks and magnetic field heating is small compared to the upward radiative/convective energy flow and does not significantly alter the energy balance. The resulting negative temperature gradient with height ($dT/dh < 0$) that drives the outward radiative/convective flux produces an absorption line spectrum, because the more optically thick line cores are formed higher in the photosphere where the local temperature is cooler and excitation by line photons is weaker. The solar absorption line spectrum is called the Fraunhofer spectrum. Magneto-convection in stellar photospheres produces inhomogeneous velocity and brightness patterns across a stellar surface. The resulting time variations in spectral line shapes are a major source of noise for exoplanet searches based on precision radial velocity measurements (e.g., Cegla et al. 2018).

Lower Chromosphere   Above the photosphere, the density and optical depth in the continuum decrease with height. Following the solar example, I identify the base of the lower chromosphere as beginning where the continuum at optical wavelengths becomes optically thin at the stellar limb, and spectral lines including Hα and the Ca II H and K lines appear in emission against the faint continuum. This type of spectrum was first seen during solar eclipses and was called the flash spectrum. In the lower chromosphere, magnetic and other nonradiative heating is present but is not sufficiently strong to overpower the temperature decrease with height (dT/dh < 0). In the lower chromosphere, hydrogen is mostly neutral and the metals are neutral or singly ionized depending on the local temperature. The broad absorption wings of the Ca II and Mg II resonance lines are formed in the lower chromospheres of G and K stars. Since for M stars the temperatures in the lower chromosphere are too cool for significant abundances of singly ionized metals, the Ca II and Mg II resonance lines have narrower wings than for the G and K stars. The temperature reaches a minimum at the height where the sum of radiative'convective heat flow from below and nonradiative heating balances radiative losses producing a neutral temperature gradient (dT/dh = 0). In the Sun, cooling by CO depresses the minimum temperature below that predicted by radiative equilibrium without molecular opacities. The thermal structure of the lower chromosphere can also be studied at millimeter and submillimeter wavelengths by ALMA, the Atacama Large Millimeter/submillimeter Array.

Upper Chromosphere Above the temperature minimum, nonradiative heating dominates over radiative cooling as the density further decreases with height, producing a positive temperature gradient (dT/dh > 0). The emission cores of the Ca II, Mg II, Hα, and Lyman-α emission lines are formed here. Also, the hydrogen Lyman continuum in the EUV, ultraviolet continua of C I, Mg I, Al I, Si I, and Fe I, and free-free continuum emission at submm to cm wavelengths are produced here. The major sources of radiative cooling in the upper chromosphere are the many emission lines of Fe II, Ca II, Mg II, and H I.

Transition Region As temperatures increase further with height, neutral hydrogen and singly ionized Fe, Ca, and Mg are themselves ionized, but the more highly ionized species are less efficient cooling agents. In the absence of efficient cooling, nonradiative heating and thermal conduction from the hot corona produce a region of steep temperature gradient called the transition region. Emission lines of C II-IV, Si III-IV, N V, and O V are formed in this region at temperatures of 20,000–300,000 K, as shown in Figs. 4.2 and 4.3). Ambipolar diffusion (AD) of neutral H and neutral and singly ionized He into hotter regions can enhance the strength of their emission lines (Fontenla et al. 1990, 1993). Transition region emission lines show increasing redshifts as stars rotate faster (e.g., Linsky et al. 2012a).

Corona Eventually temperatures become sufficiently large that highly ionized metals, in particular Fe and Ni, and thermal bremsstrahlung can balance the local nonradiative heating. This energy balance produces the geometrically extended multimillion-degree coronae seen in the Sun and cooler stars. The X-ray spectrum of the solar corona includes emission lines of Fe VII to Fe XVI formed at temperatures of 1–2 million degrees, while active stars and stellar flares show emission lines of Fe XV to Fe XXVI formed at temperatures of 10 million degrees and even hotter (See Chap. 5). The X-ray emission is primarily from closed magnetic-field regions where the magnetic field-lines are open. Expansion of the stellar wind against gravity where the magnetic-field line are open is a cooling source.

Chromospheres as indicated by emission in the cores of the Ca II, Mg II, and Hα lines are present in all dwarf and subgiant stars from spectral type A7 V, for example Altair (HD 187642), throughout the F, G, K, and M spectral classes and into the early L-type brown dwarfs. Giant and supergiant stars as hot as the A9 II star Canopus (HD 45348) have chromospheres, but the strength of the chromospheric emission becomes weaker for giant stars cooler than spectral type K2 III (Linsky and Haisch 1979). The strength of the chromospheric emission depends on many factors including spectral type, gravity, rotation rate, and age. See Chap. 9 for a discussion of these factors and their correlations.

4.3 Semiempirical Models of the Solar Atmosphere

When heating and cooling are in balance and time independent, the thermal structure of a stellar chromosphere is stable, but shock waves and rapid heating during flares can strongly perturb this simple scenario (see Chap. 9). Even for model atmospheres without additional heating sources, it is important to compute the emergent spectrum and net cooling rates in non-LTE by solving the multilevel statistical equilibrium equations for the populations of the ground and excited levels of all important atoms and ions including line absorption and scattering processes. This is a challenging endeavor as there are a large number of atoms and ions to consider and some important species such as Fe I and Fe II have a very large number of transitions to include in the calculation, and many lines are optically thick. Vernazza et al. (VAL) (1973, 1976, 1981) pioneered this approach with solar models computed with their PANDORA code. An additional complication is that for optically thick lines, including the Ca II H and K, Mg II h and k, and H I Lyman lines, scattering is more important than absorption. Solutions of the radiative transfer equations must include coherent scattering in the ion's rest frame and Doppler redistribution in the observer's frame. The development of this partial frequency redistribution (PRD) formalism was pioneered in papers by Milkey and Mihalas (1973a,b, 1974) and Milkey et al. (1975). Most models computed after this time included PRD radiative transfer.

Semiempirical models have been computed to fit stellar spectral line and continuum observations without consideration of energy balance, since there is no agreed prescription for the nonradiative heating rates. The objective of these semiempirical models is to find a temperature-height distribution that is consistent with ionization and excitation equilibria, including radiative transfer in spectral lines, and that fits the observed spectral line and continuum emission of the Sun or star. In the optical and near infrared spectrum, the important chromospheric spectral lines to fit are the Hα and other Balmer lines of H I, the Ca II H, K, and infrared triplet lines, and the He I 1083.0 nm multiple. The UV spectral region (120–300 nm) contains a very rich spectrum of chromospheric emission lines including HI Lyman-α and higher lines in the Lyman series, and emission lines of O I, Si I, Si II, Mg II, and Fe II. The 120–200 nm region contains free-bound continuum emission from Si I, Mg I, and Fe I, and the 60–91.2 nm region includes bright emission in the Lyman continuum. Linsky (2017) has reviewed the development of solar and stellar models based on this semiempirical approach.

The pioneering set of solar models computed by Vernazza et al. (1973, 1976, 1981) fit all of these spectral diagnostics for the quiet Sun (their model C), for active regions of the Sun (solar plages), for very quiet regions in solar granulation cells, and for sunspots. Figure 4.7 shows the temperature-height structure of Model C in their

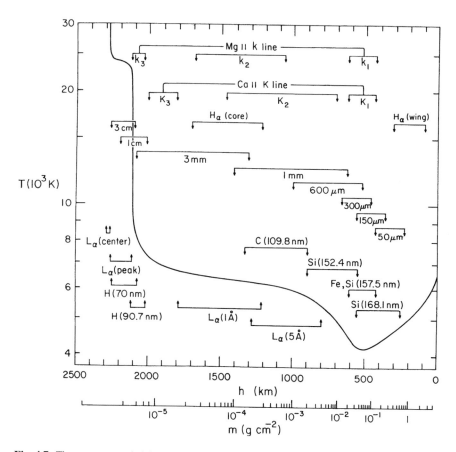

Fig. 4.7 The temperature-height structure of average Sun Model C computed by Vernazza et al. (1981). Horizontal lines indicate the range of heights in this chromosphere where important spectral lines are formed. Also shown are the range of heights where the free-bound continuum edges of C I, Si I, and Fe I are formed and the locations where submm and mm radio-continuum emission is formed. Reproduced by permission of the AAS

1981 paper including the locations where important spectral features are formed. In addition to the cores and wings of the H I, Ca II, and Mg II lines, the figure also shows where the bound-free edges of C I, Si I, and Fe I in the UV spectrum are formed. The figure also shows where radio emission at different wavelengths is produced. Since this free-bound emission is thermal, it provides an important test of the chromospheric thermal structure derived by fitting the optical and UV emission. In particular, the shortest wavelengths observable by ALMA can test the thermal structure of the temperature minimum region without the complexity of non-LTE spectral line formation (Loukitcheva et al. 2015; Wedemeyer et al. 2016). Early ALMA observations of the solar-like stars α Cen A (G2 V) and α Cen B (K1 V) demonstrate the feasibility of this approach (Liseau et al. 2015).

More recently, Fontenla et al. (2009, 2011, 2014, 2015) have computed a grid of semiempirical solar models using a non-LTE model atmosphere code that includes 53 elements and ions and an extensive set of spectral line and continuum opacities. These models computed using the Solar-Stellar Radiation Physical Modeling (SSRPM) tools include ambipolar diffusion of neutral hydrogen into higher temperature regions that can explain the bright emission of the He I (58.4 nm) and He II (30.3 nm) resonance lines and the formation of the Lyman-α line without introducing a 20,000 K temperature plateau in the VAL models. An alternative way of understanding the anomalously bright HeI and HeII resonance lines is that they are collisionally excited by a nonthermal distribution of electrons (Smith and Jordan 2002; Smith 2003). The introduction of new opacities of the molecules NH, CH, and OH in Fontenla et al. (2015) appears to have resolved the missing opacity problem of overly high emission in the NUV emission computed by earlier models.

The Fontenla models describe different regions on the Sun identified by the strength of Ca II K line emission from intergranular lanes with the weakest K line emission to plage regions with very bright K line emission. Since the K line is a good indicator of the chromospheric heating rate, this grid of seven models represents a sequence of increasing nonradiative heating. As shown in Fig. 4.8, the models all have the same general shape, but they differ in three ways with increasing nonradiative heating rates: (1) the temperature minimum moves inward to larger column densities, (2) temperatures at the base of the upper chromosphere are hotter, and (3) the steep temperature rise above 8000 K moves inward to larger column densities. The latter change means that emission lines formed in the upper chromosphere, transition region, and corona are formed at higher densities and are, therefore, much brighter as the emission from optically thin lines is proportional to density squared.

An important question for understanding stellar activity is whether the range in emission line and continuum flux seen in different regions on the Sun is representative of the observed range in emission of solar-mass stars with a wide variety of ages, rotational periods, and other activity characteristics. By selecting a sample of stars with nearly the same mass, radius, and chemical composition, one can isolate the effect of nonradiative heating as the driver of stellar activity. In particular, is the emission of solar plages, regions on the solar disk with the strongest magnetic fields

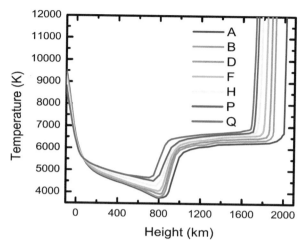

Fig. 4.8 Semempirical chromospheric models for different regions on the Sun computed by Fontenla et al. (2011). The models are for regions with a range of Ca II K-line emission from very weak emission near the center of granular cells (Model A) to very strong emission in active regions called plages (Model Q). Note the very similar shapes of these temperature-height distributions

except for sunspots representative of the bright chromospheric and coronal emission of very active young solar-mass stars? Ribas et al. (2005) have compiled the X-ray to UV emission of 5 solar-mass stars and the Sun that span a range of ages from β Hyi (slightly older than the Sun), and to β Com, κ Cet, π^1 UMa, and EK Dra with ages of 1.5 Gyr to 0.1 Gyr. For the youngest star EK Dra, the 92–120 nm emission is 15 times brighter than the quiet Sun and the 0.1–2 nm X-ray emission is about 1000 times brighter. For the plage model computed by Fontenla et al. (2015), the corresponding flux increases are TBD and TBD.

Linsky et al. (2012b) compared the 120–150 nm continuum flux of six solar-mass stars with the Sun (see Fig. 4.9). The stars have rotational periods from 25.4 days (α Cen A) to 2.6 days (EK Dra) and 1.47 days (HII314). HII314 is a member of the Pleiades cluster with an age of about 140 Myr. The continuum flux of α Cen A is very similar to that of the quiet Sun, and the continuum fluxes of the two fastest rotating and youngest solar-mass stars are similar to that of the solar plage model of Fontenla et al. (2015). This result supports the hypothesis that solar plages with their strong magnetic fields have chromospheric emission and, therefore, nonradiative heating rates that are similar to young stars with the same mass, radius, and chemical composition as the Sun.

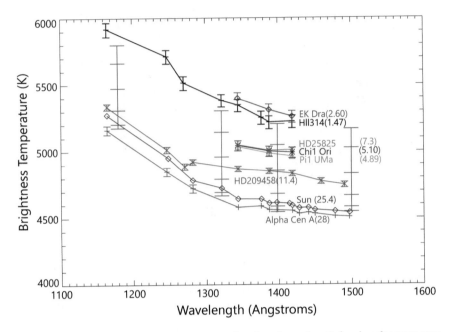

Fig. 4.9 Continuum brightness temperature as a function of wavelength for six solar-mass stars and the Sun with a range of indicated rotational periods (days) from 28 days for α Cen A to 1.47 days for the Pleiades star HII314. Horizontal lines with bars show the brightness temperatures at each wavelength for solar models (Fontenla et al. 2011) with faintest to brightest Ca II K-line emission. Figure from Linsky et al. (2012b). Reproduced by permission of the AAS

4.4   Semiempirical Models of Stellar Atmospheres and Their Spectroscopic Diagnostics

The first semiempirical stellar chromosphere models computed in the 1970s, constructed to fit the available ground-based Ca II H and K lines, could only describe the stellar thermal structure at temperatures below about 8000 K. They provide a rough estimate of the pressure in the transition region and corona. UV Spectrometers on several rocket and balloon experiments and the *Copernicus* satellite followed by the *International Ultraviolet Explorer (IUE)* satellite observed the Lyman-α and Mg II k lines and the FUV emission lines of C II, C IV, Si III-IV, and N V. Based on these data, the next generation of models could determine thermal structures that included the upper chromosphere and transition region. The review article by Linsky (2017) includes a comprehensive list of models for F–M dwarfs and luminous stars. Several bright stars have been modeled by different authors using different data sets, including Procyon (F5 IV-V), α Cen A (G2 V), ξ Boo A (G8 V), ϵ Eri (K2 V), and AU Mic (M1 V). There are a number of model grids that cover specific ranges of spectral type or activity, including G2–5 V stars (Vieytes et al. 2005), K0–2 V stars (Vieytes et al. 2009), and M dwarfs (Doyle et al. 1994; Houdebine and Stempels

1997; Walkowicz et al. 2008; Schmidt et al. 2015). The M-dwarf models computed by Fuhrmeister et al. (2005) and Peacock et al. (2019) are based on the PHOENIX code upgraded in the latter paper to include partial redistribution formation of strong emission lines like Lyman-α. There are photospheric models for L and T dwarfs (e.g., Tremblin et al. 2016), but these models do not have chromospheres or coronae. The emission-measure analysis technique pioneered by Gabriel and Jordan (1975) and Brown and Jordan (1981) is often used to model transition regions and coronae by including X-ray and FUV data.

The UV spectrographs on the *Hubble Space Telescope (HST)* beginning with the Goddard High Resolution Spectrograph (GHRS 1990–1997) and continuing with the Space Telescope Imaging Spectrograph (STIS 1997–present) and the Cosmic Origins Spectrograph (COS 2009–present) have provided very high quality data upon which to develop the next generation of semiempirical models. The StarCAT (Ayres 2010) and ASTRAL[1] catalogs include the highest quality HST spectra. STIS obtains UV spectra with resolution as high as $3 \, km \, s^{-1}$ that can measure line widths and Doppler shifts in addition to the line flux measurements obtained with the previous instruments. For studies of the hydrogen Lyman-α line, STIS is the instrument of choice because its narrow slit minimizes geocoronal emission, and its high spectral resolution permits a clean separation of interstellar and astrosphere absorption components in the observed line profile (see Chap. 6). With this information, there are several techniques for reconstructing the intrinsic stellar Lyman-α emission line as described in Sect. 6.1. France et al. (2013) found that the intrinsic Lyman-α flux can be comparable to the rest of the UV emission from an M dwarf star.

With its far higher throughput and very low background, COS can measure much fainter emission lines than GHRS and STIS and about 100 times fainter than *IUE* (France et al. 2016). One of the many accomplishments of the COS instrument has been to observe the FUV spectra of K and M dwarf stars. For example, the MUSCLES[2] Treasury Survey (France et al. 2016) observed the complete UV spectra of 4 K-dwarf and 7 M-dwarf stars, which have known exoplanets, together with simultaneous or near-simultaneous observations at X-ray, optical, and infrared wavelengths. Near-simultaneous observations are essential for M dwarfs as these stars are highly variable.

With the excellent data set for GJ 832 (M1.5 V) from COS and STIS, Fontenla et al. (2016) computed a semiempirical model of the chromosphere, transition region, and corona using a modification of their SSRPM[3] radiative transfer code with the addition of 20 diatomic molecules and about 2,000,000 molecular transitions. Inclusion of these molecular lines, in particular TiO, is essential for fitting the continuum adjacent to the Ca II H and K lines and the Hα line. Their model for GJ 832 (Fig. 4.10) has a much cooler photosphere and lower chromosphere than the

---

[1] casa.colorado.edu/$\sim$ayres/ASTRAL/.

[2] Measurements of the ultraviolet spectral characteristics of low-mass exoplanetary systems.

[3] Solar-stellar radiation physical modeling tools.

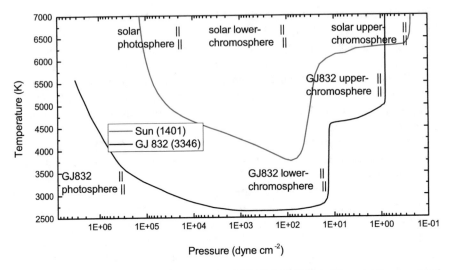

Fig. 4.10 Comparison of semiempirical models of GJ 832 (M1.5 V) and the quiet Sun computed by Fontenla et al. (2016). The figure shows the locations in the two stars of the photosphere, lower chromosphere, and upper chromosphere. Reproduced by permission of the AAS

quiet Sun model. Since Ca and Mg are almost entirely neutral in these layers, the wings of the Ca II and Mg II lines are much narrower than for the much warmer Sun. Temperatures at the base of the upper chromosphere are also cooler than the quiet Sun, but the steep temperature rise in the upper chromosphere and transition region forces the emission lines to be formed at higher densities. The effect is to produce bright emission cores in the optically thick Ca II and Mg II lines and especially in the optically thin higher temperature emission lines. For these lines, emission is proportional to electron density squared because the collision excitation rate and column density are both proportional to the electron density.

4.5   Energy Balance in Stellar Chromospheres

In a pioneering set of calculations, Anderson and Athay (1989) computed solar models with a range of nonradiative heating rates per atom. As shown in Fig. 4.11, their radiative equilibrium model without additional non-radiative heating has temperatures that decrease outwards to very low values driven by efficient cooling by CO below 4400 K. As the assumed non-radiative heating rate increases, the models show a temperature rise in the upper chromosphere with increasing temperature and deeper location for the temperature rise. With sufficient heating, their models approximate the thermal structure of Model C computed by Vernazza et al. (1981). As shown in Fig. 4.12, the major sources of radiative cooling in the

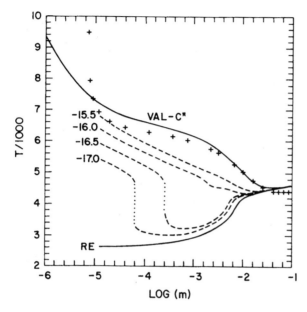

Fig. 4.11 Comparison of a solar radiative equilibrium model (lower solid line marked RE) with chromospheric models with increasing amounts of non-radiative heating per atom (dashed lines). These models are computed by Anderson and Athay (1989). The upper solid line is the Vernazza et al. (1981) semiempirical Model C, and pluses are from model computed by Maltby et al. (1986). Reproduced by permission of the AAS

upper chromosphere are lines of Fe II, Mg II, and Ca II (cf. Linsky 2001). Above temperatures of 8000 K, hydrogen lines and continua become important cooling sources.

4.6   Composite, Multidimensional and Time-Dependent Models

Time-independent, single component, plane-parallel geometry models are clearly approximations to the reality of dynamic inhomogeneous stellar atmospheres if the Sun is a relevant role model. One approach to constructing more realistic models is to describe a stellar atmosphere as consisting of a number of static plane-parallel components. For example, Fontenla et al. (2009) showed that integrated sunlight at the minimum of the solar magnetic cycle can be fit by the sum of models for cell interiors (87%), average network (10%), and bright network and plage (3%). At cycle maximum, the contribution of network and plage models is larger. During the Maunder minimum when the essentially no sunspots were observed and the Sun was very inactive, Bolduc et al. (2014) reconstructed a solar spectrum with no plage component.

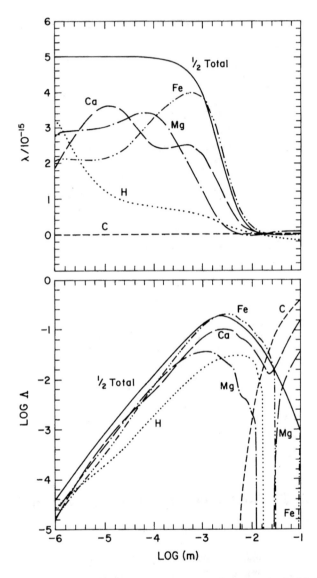

Fig. 4.12 Net cooling rates (positive numbers) of five abundance species in Model C*. Upper panel gives rates per atom (in log units), and the lower panel gives rates in log (erg cm^{-3} s^{-1}) in the models computed by Anderson and Athay (1989). Reproduced by permission of the AAS

Constructing composite models for stars is more difficult as there are no spectra of isolated stellar plages and no direct measurements of the fraction of the stellar surface covered by plages. To get around these problems, Houdebine (2009, 2010a,b) assumed that for M dwarfs the ratio of Ca II K equivalent widths of plage and quiet regions is the same as for the Sun, or they assumed arbitrarily that the plage coverage factor is 30% (Houdebine and Doyle 1994). Sim and Jordan (2003, 2005) assumed a plage coverage factor of 20% for their composite models of ϵ Eri (K2 V). Cuntz et al. (1999) tried a different approach for computing composite models of K dwarfs. They used an empirical relation between the magnetic coverage factor in the photosphere and stellar rotation rate and then extrapolated this coverage factor upwards in the atmosphere assuming horizontal magnetic and gas pressure balance. They computed the thermal structure of the plage and quiet regions by propagating magneto-hydrodynamic (MHD) waves upward in the different magnetic geometries. For stars with large magnetic filling factors in the photosphere, the magnetic cross-sectional area increases slowly with height in plage regions leading to shock-wave heating occurring lower in the atmosphere at higher densities, thereby producing enhanced Ca II and Mg II emission.

Although the physical processes and nonradiative heating rates in stellar chromospheres and coronae are still uncertain after many years of study, many investigators have computed quasi steady-state models heated by upwardly propagating MHD shocks or highly time-dependent flare models in which shocks or electron beams propagate downward to heat the chromosphere and even the photosphere. Examples of heating by upwardly propagating MHD shocks include models by Ulmschneider (2001), Rammacher and Ulmschneider (2003), Carlsson and Stein (2002), and their previous papers. Carlsson and Stein argued that the observed solar Ca II K line emission can be explained by the high temperatures in shocks and the high temperature sensitivity of the Ca II emission. They argued that emission from shocks can explain the time-averaged Ca II K line emission from the Sun without requiring a temperature rise with height in a one-component chromosphere model, but this conclusion is controversial. The three-dimensional radiative hydrodynamics simulations of Wedemeyer et al. (2004) produce shock waves that heat regions of the chromosphere that can produce the observed emission lines and where there have been no recent shocks regions that are cool enough to explain the observed emission from molecules like CO.

Modeling solar and stellar flares is particularly challenging because flaring plasmas: (1) heat rapidly but cool more slowly leading to departures from ionization and excitation very different from LTE conditions, (2) nonthermal electron energy distributions can enhance collisional ionization and excitation rates by orders of magnitude larger than for thermal Maxwell-Boltzmann distributions, and (3) inhomogeneous atmospheric properties occur on spatial scales too small to be resolved on the Sun. Beam-heating models by Abbett and Hawley (1999) and by Allred et al. (2005, 2006, 2015) show very bright chromospheric emission lines that are broad and typically redshifted, indicating rapidly downflowing material. In addition to high-intensity X-ray and EUV emission, these models include beam penetration into the photosphere that produce white light emission (Kowalski et al. (2015) that is observed.

4.7 Does the Sun Have a Twin?

While the proximity of the Sun benefits high-spatial resolution and high signal/noise spectroscopy at all wavelengths without interstellar absorption, it is often difficult to obtain observations of the Sun as a point source for direct comparison with stellar observations. Measurements of solar irradiation requires either time-consuming integration over the solar surface, observations of the Sun reflected by an asteroid or moon, or observations by telescopes specifically designed to study integrated sunlight (e.g., Pevtsov et al. 2014). Each of these techniques may introduce measurement errors that are different from stellar observations. Several authors have instead searched for reasonably bright stars with mass, radius, effective temperature, gravity, chemical composition, age, rotation rate, and activity indicators very close to solar values, in other words a solar twin. Porto de Mello and da Silva (1997) identified 18 Sco (HR 6060) as the best candidate for this designation. This star is bright ($m_V = 5.5$) and nearby (14 pc). Observations of 18 Sco can, therefore, be used as a proxy for integrated sunlight. Meléndez and Ramírez (2007) found that HIP 56948 and HIP 73815 are likely even better candidates for solar twins, although both stars are much fainter than 18 Sco. HD 101364, HD 197027, and 51 Peg are also solar twin candidates on the basis of their chromospheric activity and age (Mittag et al. 2016). There are STIS and COS spectra of 18 Sco and *GALEX* observations of most of the solar twins in the *HST* MAST data archive. For a listing of more recently identified solar twins see Galarza et al. (2016) and references therein.

References

Abbett, W.P., Hawley, S.L.: Dynamic models of optical emission in impulsive solar flares. Astrophys. J. **521**, 906 (1999)

Allred, J.C., Hawley, S.L., Abbett, W.P., Carlsson, M.: Radiative hydrodynamic models of the optical and ultraviolet emission from solar flares. Astrophys. J. **630**, 573 (2005)

Allred, J.C., Hawley, S.L., Abbett, W.P., Carlsson, M.: Radiative hydrodynamic models of the optical and ultraviolet emission from M dwarf flares. Astrophys. J. **644**, 484 (2006)

Allred, J.C., Kowalski, A.F., Carlsson, M.: A unified computational model for solar and stellar flares. Astrophys. J. **809**, 104 (2015)

Anderson, L.S., Athay, R.G.: Model solar chromosphere with prescribed heating. Astrophys. J. **346**, 1010 (1989)

Andretta, V., Jones, H.P.: On the role of the solar corona and transition region in the excitation of the spectrum of neutral helium. Astrophys. J. **489**, 375 (1997)

Avrett, E.H., Hummer, D.G.: Non-coherent scattering, II: line formation with a frequency independent source function. Mon. Not. R. Astron. Soc. **130**, 295 (1965)

Ayres, T.R.: StarCAT: a catalog of space telescope imaging spectrograph ultraviolet echelle spectra of stars. Astrophys. J. Suppl. **187**, 149 (2010)

Bolduc, C., Charbonneau, P., Barnabé, R., Bourqui, M.S.: A reconstruction of ultraviolet spectral irradiance during the Maunder Minimum. Sol. Phys. **289**, 2891 (2014)

Brown, A., Jordan, C.: The chromosphere and corona of Procyon (α CMi, F5 IV-V). Mon. Not. R. Astron. Soc. **196**, 757 (1981)

Carlsson, M., Stein, R.F.: Dynamic hydrogen ionization. Astrophys. J. **572**, 626 (2002)

Cegla, H.M., Watson, C.A., Shelyag, S., Chaplin, W.J., Davies, G.R., Mathioudakis, M., Palumbo, M.L., III, Saar, S.H., Haywood, R.D.: Stellar surface magneto-convection as a source of astrophysical noise II. Center-to-limb parameterisation of absorption line profiles and comparison to observations. Astrophys. J. **866**, 55 (2018)

Centeno, R., Trujillo Bueno, J., Uitenbroek, H., Collados, M.: The Influence of coronal EUV irradiance on the emission in the He I 10830 Å and D3 Multiplets. Astrophys. J. **677**, 742 (2008)

Cuntz, M., Rammacher, W., Ulmschneider, P., Musielak, Z.E., Saar, S.H.: Two-Component theoretical chromosphere models for K Dwarfs of different magnetic activity: exploring the Ca II emission-stellar rotation relationship. Astrophys. J. **522**, 1053 (1999)

Doyle, J.G., Houdebine, E.R., Mathioudakis, M., Panagi, P.M.: Lower chromospheric activity in low activity M dwarfs. Astron. Astrophys. **285**, 233 (1994)

Dupree, A.K., Lobel, A., Young, P.R., Ake, T.B., Linsky, J.L., Redfield, S.: The far-ultraviolet spectroscopic survey of luminous cool stars. Astrophys. J. **622**, 629 (2005)

Fontenla, J.M., Avrett, E.H., Loeser, R.: Energy balance in the solar transition region. I - hydrostatic thermal models with ambipolar diffusion. Astrophys. J. **355**, 700 (1990)

Fontenla, J.M., Avrett, E.H., Loeser, R.: Energy balance in the solar transition region. III - helium emission in hydrostatic, constant-abundance models with diffusion. Astrophys. J. **406**, 319 (1993)

Fontenla, J.M., Curdt, W., Haberreiter, M., Harder, J., Tian, H.: Semiempirical models of the solar atmosphere. III. Set of non-LTE models for far-Ultraviolet/extreme-ultraviolet irradiance computation. Astrophys. J. **707**, 482 (2009)

Fontenla, J.M., Harder, J., Livingston, W., Snow, M., Woods, T.: High-resolution solar spectral irradiance from extreme ultraviolet to far infrared. J. Geophys. Res. **116**, D20108 (2011)

Fontenla, J.M., Landi, E., Snow, M., Woods, T.: Far- and extreme-UV solar spectral irradiance and radiance from simplified atmospheric physical models. Solar. Phys. **289**, 515 (2014)

Fontenla, J.M., Stancil, P.C., Landi, E.: Solar spectral irradiance, solar activity, and the near-ultraviolet. Astrophys. J. **809**, 157 (2015)

Fontenla, J., Linsky, J.L., Witbrod, J., France, K., Buccino, A., Mauas, P., Vieytes, M., Walkowicz,L.: Semi-empirical modeling of the photosphere, chromosphere, transition region, and corona of the M-dwarfs host star GJ 832. Astrophys. J. **830**, 154 (2016)

France, K., Froning, C.S., Linsky, J.L., Roberge, A., Stocke, J.T., Tian, F., Bushinsky, R., Désert, J.-M., Mauas, P., Vietes, M., Walkowicz, L.: The ultraviolet radiation environment around M dwarf exoplanet host stars. Astrophys. J. **763**, 149 (2013)

France, K., Loyd, R.O.P., Youngblood, A., Brown, A., Schneider, P.C., Hawley, S.L., Froning, C.S., Linsky, J.L., Roberge, A., et al.: The MUSCLES Treasury Survey I: motivation and overview. Astrophys. J. **820**, 89 (2016)

Fuhrmeister, B., Schmitt, J.H.M.M., Hauschildt, P.H.: PHOENIX model chromospheres of mid- to late-type M dwarfs. Astron. Astrophys. **439**, 1137 (2005)

Gabriel, A.H., Jordan, C.: Analysis of EUV observations of regions of the quiet and active corona at the time of the 1970 March 7 eclipse. Mon. Not. R. Astron. Soc. **173**, 397 (1975)

Galarza, J.Y., Meléndez, J., Cohen, J.G.: Serendipitous discovery of the faint solar twin Inti 1. Astron. Astrophys. **589**, A65 (2016)

Golding, T.P., Carlsson, M., Leenaarts, J.: Detailed and simplified nonequilibrium helium ionization in the solar atmosphere. Astrophys. J. **784**, 30 (2014)

Houdebine, E.R.: Observation and modelling of main-sequence star chromospheres - XII. Two-component model chromospheres for five active dM1e stars. Mon. Not. R. Astron. Soc. **397**, 2133 (2009)

Houdebine, E.R.: Observation and modelling of main-sequence star chromospheres. IX. Two-component model chromospheres for nine M1 dwarfs. Astron. Astrophys. **509**, A65 (2010a)

Houdebine, E.R.: Observation and modelling of main-sequence star chromospheres - X. Radiative budgets on Gl 867A and AU Mic (dM1e), and a two-component model chromosphere for Gl 205 (dM1). Mon. Not. R. Astron. Soc. **403**, 2157 (2010b)

Houdebine, E.R., Doyle, J.G.: Observation and modelling of main sequence star chromospheres. 2: Modelling of the AU MIC (dM2.5e) hydrogen spectrum. Astron. Astrophys. **289**, 185 (1994)

Houdebine, E.R., Stempels, H.C.: Observation and modelling of main sequence stellar chromospheres. VI. Hα and Ca II line observations of M1 dwarfs and comparison with models. Astron. Astrophys. **326**, 1143 (1997)

Hummer, D.G.: Non-coherent scattering: I. The redistribution function with Doppler broadening. Mon. Not. R. Astron. Soc. **125**, 21 (1962)

Judge, P.G.: The chromosphere: gateway to the corona? ...Or the purgatory of solar physics? Memorie della Societa Astronomica Italiana **81**, 543 (2010)

Kowalski, A.,F., Hawley, S.L., Carlsson, M., Allred, J.C., Uitenbroek, H., Osten, R.A., Holman, G.: New insights into white-light flare emission from radiative-hydrodynamic modeling of a chromospheric condensation. Solar Phys. **290**, 3487 (2015)

Leenaarts, J. de la Cruz Rodríguez, J., Danilovic, S., Scharmer, G. Carlsson, M.: Chromospheric heating during flux emergence in the solar atmosphere. Astron. Astrophys. **612**, 28 (2018)

Linsky, J.L.: In: Ulmschneider, P., Priest, E.R., Rosner, R. (ed.) Mechanisms of Chromospheric and Coronal Heating, p. 166. Springer, Berlin (2001)

Linsky, J.L.: Stellar model chromospheres and spectroscopic diagnostics. Ann. Rev. Astron. Astrophys. **55**, 159 (2017)

Linsky, J.L., Haisch, B.M.: Outer atmospheres of cool stars. I - the sharp division into solar-type and non-solar-type stars. Astrophys. J. Lett. **229**, 27 (1979)

Linsky, J.L., Wood, B.E., Judge, P., Brown, A., Andrulis, C., Ayres, T.R.: The transition regions of Capella. Astrophys. J. **442**, 381 (1995)

Linsky, J.L., Bushinsky, R., Ayres, T., France, K.: Ultraviolet spectroscopy of rapidly rotating solar-mass stars. Emission-line redshifts as a test of the solar-stellar connection. Astrophys. J. **754**, 69 (2012a)

Linsky, J.L., Bushinsky, R., Ayres, T., Fontenla, J., France, K.: Far-ultraviolet continuum emission: applying this diagnostic to the chromospheres of solar-mass stars. Astrophys. J. **745**, 25 (2012b)

Liseau, R., Vlemmings, W., Bayo, A., Bertone, E., Black, J.H., del Burgo, C., Chavez, M., Danchi, W., De la Luz, V., et al.: ALMA observations of α Centauri. First detection of main-sequence stars at 3 mm wavelength. Atron. Astrophys. **573**, L4 (2015)

Loukitcheva, M., Solanki, S.K., Carlsson, M., White, S.M.: Millimeter radiation from a 3D model of the solar atmosphere. I. Diagnosing chromospheric thermal structure. Astron. Astrophys. **575**, A15 (2015)

Loyd, R.O.P., France, K., Youngblood, A., Schneider, C., Brown, A., Hu, R., Linsky, J., Froning, C.S., Redfield, S., Rugheimer, S., Tian, F.: The MUSCLES Treasury Survey III: X-ray to infrared spectra of 11 M and K stars hosting planets. Astrophys. J. **824**, 102 (2016)

Maltby, P., Avrett, E.H., Carlsson, M., Kjeldseth-Moe, O., Kurucz, R.L., Loeser, R.: A new sunspot umbral model and its variation with the solar cycle. Astrophys. J. **306**, 284 (1986)

Meléndez, J., Ramírez, I.: HIP 56948: a solar twin with a low lithium abundance. Astrophys. J. **669**, 89 (2007)

Milkey, R.W., Mihalas, D.: Resonance-line transfer with partial redistribution: a preliminary study of Lyman α in the solar chromosphere. Astrophys. J. **185**, 709 (1973a)

Milkey, R.W., Mihalas, D.: Calculation of the solar chromospheric Lα profile allowing for partial redistribution effects. Solar Phys. **32**, 361 (1973b)

Milkey, R.W., Mihalas, D.: Resonance-line transfer with partial redistribution. II - the solar Mg II lines. Astrophys. J. **192**, 769 (1974)

Milkey, R.W., Ayres, T.R., Shine, R.A.: Resonance line transfer with partial redistribution. III Mg II resonance lines in solar-type stars. Astrophys. J. **197**, 143 (1975)

Mittag, M., Schröder, K.-P., Hempelmann, A., González-Pérez, J.N., Schmitt, J.H.M.M.: Chromospheric activity and evolutionary age of the Sun and four solar twins. Astron. Astrophys. **591**, 89 (2016)

Osten, R.A., Hawley, S.L., Allred, J., Johns-Krull, C.M., Brown, A., Harper, G.M.: From radio to X-ray: the quiescent atmosphere of the dMe flare star EV Lacertae. Astrophys. J. **647**, 1349 (2006)

Pace, G., Pasquini, L.: The age-activity-rotation relationship in solar-type stars. Astron. Astrophys. **426**, 1021 (2004)

Pagano, I., Linsky, J.L., Carkner, L., Robinson, R.D., Woodgate, B., Timothy, G.: HST/STIS echelle spectra of the dM1e star AU Microscopii outside of flares. Astrophys. J. **532**, 497 (2000)

Pagano, I., Linsky, J.L., Valenti, J., Duncan, D.K.: HST/STIS high resolution echelle spectra of α Centauri A (G2 V). Astron. Astrophys. **415**, 331 (2004)

Peacock, S., Barman, T., Shkolnik, E.L., Hauschildt, P.H., Baron, E.: Predicting the extreme ultra-violet environment of exoplanets around low-mass stars: the TRAPPIST-1 system. Astrophys. J. **871**, 235 (2019)

Pevtsov, A.A., Bertello, L., Marble, A.R.: The Sun-as-a-star solar spectrum. AN **335**, 21 (2014)

Porto de Mello, G.F., da Silva, L.: HR 6060: the closest ever solar twin? Astrophys. J. Lett. **482**, L89 (1997)

Rammacher, W., Ulmschneider, P.: Time-dependent ionization in dynamic solar and stellar atmospheres. I. Methods. Astrophys. J. **589**, 988 (2003)

Redfield, S., Linsky, J.L., Ake, T.B., Dupree, A.K., Robinson, R.D., Wood, B.E., Young, P.R.: A far-ultraviolet spectroscopic explorer survey of late-type dwarf stars. Astrophys. J. **581**, 626 (2002)

Ribas, I., Guinan, E.F., Güdel, M., Audard, M.: Evolution of the solar activity over time and effects on planetary atmospheres. I. High-energy irradiances (1-1700 Å). Astrophys. J. **622**, 680 (2005)

Schmidt, S.J., Hawley, S.L., West, A.A., Bochanski, J.J., Davenport, J.R.A., Ge, J., Schneider, D.P.:BOSS ultracool dwarfs. I. Colors and magnetic activity of M and L dwarfs. Astron. J. **149**, 158 (2015)

Sim, S.A., Jordan, C.: On the filling factor of emitting material in the upper atmosphere of ε Eri (K2 V). Mon. Not. R. Astron. Soc. **346**, 846 (2003)

Sim, S.A., Jordan, C.: Modelling the chromosphere and transition region of ε Eri (K2 V). Mon. Not. R. Astron. Soc. **361**, 1102 (2005)

Smith, G.R.: Enhancement of the helium resonance lines in the solar atmosphere by suprathermal electron excitation - II. Non-Maxwellian electron distributions. Mon. Not. R. Astron. Soc. **341**, 143 (2003)

Smith, G.R., Jordan, C.: Enhancement of the helium resonance lines in the solar atmosphere by suprathermal electron excitation - I. Non-thermal transport of helium ions. Mon. Not. R. Astron. Soc. **337**, 666 (2002)

Soler, R., Terradas, J., Oliver, R., Ballester, J.L.: Propagation of torsional Alfvén Waves from the photosphere to the corona: reflection, transmission, and heating in expanding flux tubes. Astrophys. J. **840**, 20 (2017)

Takeda, Y. Takada-Hidai, M.: Chromospheres in metal-poor stars evidenced from the He I 10830Å line. Publ. Astron. Soc. Jpn. **63**, 547 (2011)

Thomas, R.N., Athay, R.G.: Physics of the Solar Chromosphere. Interscience Monographs and Texts in Physics and Astronomy. Interscience Publication, New York (1961)

Tremblin, P., Amundsen, D.S., Chabrier, G., Baraffe, I., Drummond, B., Hinkley, S., Mourier, P., Venot, O.: Cloudless atmospheres for L/T dwarfs and extrasolar giant planets. Astrophys. J. **817**, 19 (2016)

Ulmschneider, P.: In: Ulmschneider, P., Priest, E.R., Rosner, R. (ed.) Mechanisms of Chromospheric and Coronal Heating, p. 328. Springer, Berlin (2001)

Vernazza, J.E., Avrett, E.H., Loeser, R.: Structure of the solar chromosphere. Basic computations and summary of the results. Astrophys. J. **184**, 605 (1973)

Vernazza, J.E., Avrett, E.H., Loeser, R.: Structure of the solar chromosphere. II - the underlying photosphere and temperature-minimum region. Astrophys. J. Suppl. **30**, 1 (1976)

Vernazza, J.E., Avrett, E.H., Loeser, R.: Structure of the solar chromosphere. III - models of the EUV brightness components of the quiet-sun. Astrophys. J. Suppl. **45**, 635 (1981)

Vieytes, M., Mauas, P., Cincunegui, C.: Chromospheric models of solar analogues with different activity levels. Astron. Astrophys. **441**, 701 (2005)

Vieytes, M.C., Mauas, P.J.D. Díaz, R.F.: Chromospheric changes in K stars with activity. Mon. Not. R. Astron. Soc. **398**, 1495 (2009)

Walkowicz, L.M., Johns-Krull, C.M., Hawley, S.L.: Characterizing the near-UV environment of M dwarfs. Astrophys. J. **677**, 593 (2008)

Wedemeyer, S., Freytag, B., Steffen, M., Ludwig, H.-G., Holweger, H.: Numerical simulation of the three-dimensional structure and dynamics of the non-magnetic solar chromosphere. Astron. Astrophys. **414**, 1121 (2004)

Wedemeyer, S., Bastian, T., Brajsa, R., Hudson, H., Fleishman, G., Loukitcheva, M., Fleck, B., Kontar, E.P., De Pontieu, B., et al.: Solar science with the Atacama Large Millimeter/Submillimeter Array—a new view of our sun. Space Sci. Rev. **200**, 1 (2016)

Chapter 5
Stellar Coronae: The Source of X-ray Emission

The temperature distribution in a star decreases monotonically from the nuclear burning core to the top of its photosphere as heat leaks out into cold space. However, the hottest region of a star can be its outermost layer, the corona with temperatures of 10^6 K to 10^7 K or even hotter. The heating mechanism responsible for hot coronae has been debated since the existence of hot coronal gas was first identified in the early 1940s. There are two general schools of thought concerning the heating mechanism: damping of compressional or Alfvén waves, and conversion of magnetic energy to heat and accelerated particles by magnetic reconnection events during flares. Eventually one of these two mechanisms or perhaps a combination of the two will be shown to explain phenomena seen in solar and stellar coronae, but in the meantime we concentrate here on the radiation emitted by the hot gas in stellar coronae, which has critically important effects on exoplanet atmospheres and habitability.

5.1 X-ray Observations Across the Hertzsprung-Russell Diagram

Hot gas in stellar coronae emits at X-ray wavelengths, mostly at wavelengths shortward of 10 nm. Broad-band fluxes and moderate resolution X-ray spectra obtained from satellites are the prime diagnostics for identifying the thermal and density structures in coronae, but coronal emission is also observed at other wavelengths. For example, coronal emission lines are an important component of EUV emission as described in Sect. 6.2.4. Later in this chapter, we will describe coronal emission at UV, optical, and radio wavelengths. First, we explore X-ray emission from the Sun and stars. Two excellent reviews of stellar X-ray astronomy are Güdel (2004) and Güdel and Nazé (2009).

© Springer Nature Switzerland AG 2019
J. Linsky, *Host Stars and their Effects on Exoplanet Atmospheres*,
Lecture Notes in Physics 955, https://doi.org/10.1007/978-3-030-11452-7_5

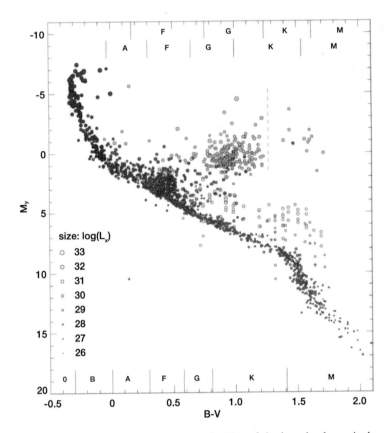

Fig. 5.1 Detected X-ray emitting stars plotted as a function of absolute visual magnitude vs. B-V color like a classical Hertzsprung-Russell diagram. Symbol sizes indicate X-ray luminosity and symbol colors represent different source catalogs. Figure from Güdel (2004)

The Hertzsprung-Russell diagram for X-ray detected stars in Fig. 5.1 shows that X-rays are detected from nearly all star types from a variety of emission processes:

O and early-B stars: There is no evidence that these stars have coronae similar to the Sun, but they do have very strong radiation driven winds that contain X-ray emitting shocks. The X-ray emission from these shocks can be enhanced in the youngest stars with magnetic fields. The X-ray emission from supergiants hotter than B1 I is very luminous with $\log L_x > 32$ (ergs s^{-1}) and $\log(L_x/L_{bol}) \approx -7.2$ (e.g., Cassinelli et al. 1981).

Non-magnetic B2–A5 stars: These stars are not detected as X-ray sources presumably because they have very thin convective zones in which the $\alpha\Omega$ dynamo generation of magnetic fields does not occur. X-ray upper limits are very low, for example, $\log L_x/L_{bol} < -8.8$ for the A0 V star HR 7329A observed with *Chandra* (Stelzer et al. 2006) and $\log L_x/L_{bol} < -10.1$ for the bright A0 V star Vega (Pease et al. 2006). In a survey of 74 A-type stars observed in deep X-ray images by *ROSAT*, Simon et al. (1995) found five presumably single A-type stars (B–V = 0.0–0.20) that are X-ray sources. However, the occasional detection of X-rays from B2–A5 stars can be explained by emission from a previously unknown cooler binary companion.

Magnetic chemically peculiar Bp and Ap stars: These are stars with kiloGauss magnetic fields and pronounced chemical peculiarities such as weak or strong He and strong Si, Cr, Sr, and rare Earth elements. They exhibit strong gyrosynchrotron radio emission, have strong magnetic fields and $\log L_x/L_{bol} \approx -6$. X-rays are detected from the He-weak, Si-strong stars, and He-strong stars with $\log L_x/L_{bol}$ in the range -7 to -6, but not from the cooler SrCrEu stars (Drake et al. 1994). θ^1 Ori C (O7 V), a periodic X-ray source with $\log L_x/L_{bol} = -6.75$, may be a high mass analog of the chemically peculiar stars (Gagné et al. 1997).

Spectral type A7–early F stars: As T_{eff} decreases from about 8500 K, stars develop subphotospheric convective zones that with rapid rotation generate magnetic fields by dynamo processes. The hottest star with measured X-ray emission is Altair (A7 V) with $\log L_x/L_{bol} = -7.4$ and a relatively cool corona (Robrade and Schmitt 2009). At the equator of this rapidly rotating star, $T_{eff} \approx 7600$ K. Another rapidly rotating star α Cep (A8 V) is also a detected X-ray source.

Sun and Sun-like stars: The Solar corona is a relatively weak X-ray source with $\log L_x/L_{bol} \approx -7$ at sunspot minimum and $\log L_x/L_{bol} \approx -6$ at sunspot maximum. Over the 11 year magnetic cycle, the Sun's mean X-ray luminosity is $\log L_x = 27.3$. By comparison, young solar-type stars have $\log L_x/L_{bol} \approx -3$, for example the rapidly rotating EK Dra, a 50 Myr old G1.5 V star with $\log L_x = 30.2$ (Ayres 2015).

mid-F to late M stars: With decreasing T_{eff} stellar convection zones deepen until dwarf stars become fully convective at $T_{eff} \approx 3200$ K near spectral type M4 V. In the warmer stars, the $\alpha\Omega$ dynamo is thought to generate magnetic fields near the tachocline that separates the base of the convective zone from the top of the radiative core. These stars have coronae and X-ray emission, which is especially strong for stars that are young and rapidly rotating. Stars cooler than about spectral type M4 V likely have a different dynamo mechanism, perhaps a distributed Ω^2 dynamo, as they do not have radiative cores and thus an interface between a radiative core and a convective envelope. Stelzer et al. (2013) detected X-ray emission from 28 M dwarf stars within 10 pc of the Sun with spectral types between M0.5 V ($T_{eff} = 3725$ K) and M6 V ($T_{eff} = 2860$ K) in the *ROSAT* all-sky survey and *XMM-Newton* archive. The X-ray luminosities of these M dwarfs decrease rapidly with age, $\log L_x \propto t^{-1.1\pm0.02}$, and there is no obvious change in the X-ray properties of M dwarfs near spectral type M4 V where the stars become fully convective. Loyd et al. (2016) obtained X-ray observations of 7 M dwarf and 4 K dwarf planet hosting stars, yielding stellar spectral energy distributions which they used to estimate molecular photodissociation rates in the exoplanetary atmospheres.

Brown dwarfs: Stars with masses less than about $0.07 M_\odot$ corresponding to $T_{eff} \leq 2800$ K and spectral type M7 V cannot support hydrogen burning in their cores and are referred to as brown dwarfs. The highest mass brown dwarfs, $M = (0.05 - 0.07) M_\odot$, especially young brown dwarfs in the TW Hya association (e.g., Kastner et al. 2016) and other young populations, have been detected by *Chandra* and *XMM-Newton*. Examples include HR 7329 B (M7.5 V) and Gl 569 Bab (M8.5 V and M9 V) (Stelzer et al. 2006) and DENIS-PJ1048 (M9 V) (Stelzer et al. 2012). Searches for X-ray emission from cooler brown dwarfs have usually been unsuccessful (e.g., Berger et al. 2010), but the Kelu-1 binary, consisting of an L2 star ($T_{\rm eff} \approx 2000$ K) and an L3 star ($T_{\rm eff} \approx 1800$ K) was detected by Audard et al. (2007). The X-ray luminosities of brown dwarfs are usually low, for example $\log L_x = 25.1$ for DENNIS-PJ1048 and $\log L_x = 25.46^{+0.17}_{-0.26}$ for Kelu-1, but an exception is the rapidly rotating young binary NLTT 33370 (M7 Ve) with $\log L_x = 27.7$ outside of flares and $\log L_x = 29.0$ during a flare (Williams et al. 2015). Although the X-ray properties of brown dwarfs depend on both $T_{\rm eff}$ and age, Stelzer et al. (2006) and Berger et al. (2010) found a clear decrease in both L_x and $L_x/L_{\rm bol}$ for brown dwarfs cooler than spectral type M9.

Giants and supergiants: Ayres et al. (1981) called attention to the absence of soft X-ray emission in giant and supergiant stars cooler than about spectral type K1 III. This result, based on observations with the *Einstein X-ray Observatory*, is similar to the absence or weak C IV emission, and thus minimal amounts of 10^5 K plasma, previously found by Linsky and Haisch (1979) in giant stars cooler than the same spectral type. A volume-limited survey of X-ray emission from luminosity class III giants located within 25 pc showed that essentially all luminosity class III giants with B–V< 1.2 corresponding to about spectral type K3 III are X-ray sources and nearly all cooler giants were not detected (Hünsch et al. 1996). Subsequent observations have supported this result. For example, a *Chandra* High Resolution Camera observation of Arcturus (K2 III) provided a possible detection at the very low level of $\log L_x/L_{\rm bol} = -10.2$ but an upper limit of $\log L_x/L_{\rm bol} = -10.1$ for Aldebaran (K5 III) (Ayres et al. 2003a). There are a few other cool giant stars with detected X-ray emission, although this emission may come from unknown M dwarf binary companions. Whether giants and supergiants cooler than about spectral type K2 III have no 10^6 K plasma or very little such plasma in their outer atmospheres or perhaps have hot plasma that is buried below cooler absorbing material (Ayres et al. 2003a) is an open question.

Close binary stars: Tides can force spin-orbit synchronous rotation of binaries on relatively short timescales. Stars with orbital periods less than about 20 days are usually synchronous rotators. This means that subgiant and giant stars of spectral types G and K in short period binary systems rotate much faster than predicted from their increase in radius with evolution and angular momentum loss from their winds. The class of RS CVn binaries, typically consisting of a G dwarf and G or K subgiant, exhibit very strong X-ray emission. HR 1099 (G5 IV+K1 IV) and UX Ari (G5 V+K0 IV) are typical examples. In the *ROSAT* all-sky survey of X-ray emission, Dempsey et al. (1993) detected X-ray emission from 112 of 136 known RS CVn systems and noted that the X-ray surface fluxes decrease with orbital period and thus rotational period, $F_x \propto P_{\rm rot}^{-1}$. BY Dra binary systems, consisting of M dwarfs in short period orbits, were detected in the *ROSAT* all-sky survey by Dempsey et al. (1997) as strong X-ray emitters with $F_x \propto P_{rot}^{-0.6}$.

5.2 Ionization Equilibria in Stellar Coronal Plasmas

Plasmas in stellar coronae are usually assumed to be optically thin and in collisional ionization equilibrium (CIE), meaning that the ionization equilibria of all species are time independent and determined by a balance between collisional ionization and radiative and dielectronic recombination. The latter process involves two steps, the first is electronic recombination of an ion (X^{i+1}) to a doubly excited state of the next lower ion (X^i) followed by the emission of a line photon to a singly excited state of X^i, vis

$$X^{i+1} + e \rightarrow X^i \text{(doubly excited)} \rightarrow X^i \text{(singly excited)} + h\nu. \tag{5.1}$$

Dielectronic recombination rates often exceed radiative recombination rates by large factors, for example Fe XV $\rightarrow$ Fe XIV. Figure 5.2 shows the CIE fractional abundances of iron between Fe I (neutral) to Fe XXVII (fully ionized) calculated by Bryans et al. (2009). The figure shows that near 10^7 K the ionization stages of Fe XVII through Fe XXIII are all abundant and that Fe XVII (Ne-like) and Fe XXV

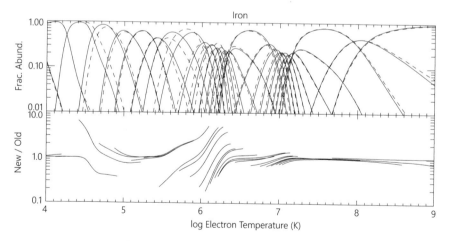

Fig. 5.2 Fractional abundance of iron ions computed in collisional ionization equilibrium by Bryans et al. (2009). Ionization stages are from Fe I (lower left) to fully ionized Fe XXVII (upper right). Solid lines are new calculations and dashed lines are from an earlier calculation. The lower plot is a comparison of the new to old calculations. Figure from Bryans et al. (2009). Reproduced by permission of the AAS

(He-like) are each important over a wide temperature range. For a CIE plasma, He-like oxygen (O VII) has peak abundance at $\log T_e = 6.3$ and H-like oxygen (O VIII) has peak abundance at $\log T_e = 6.5$. When CIE is valid, the electron temperature (T_e) and ionization temperatures (T_{ion}) are equal. Del Zanna and Mason (2018) provide additional examples of coronal ionization equilibria.

Transient ionization states can occur in stellar coronae and winds. During flares and the passage of shocks, electrons are heated faster than the ionization equilibria can respond to the additional heat leading to a time dependent ionizing plasma in which $T_e > T_{ion}$. The reverse situation is a recombining plasma that cools faster than the ionization equilibria can respond to the change in temperature. In this case $T_e < T_{ion}$. Finally, at very low plasma densities with very low collisional ionization and recombination rates as occur in stellar winds, the populations in ionization states are frozen to their values in the higher density corona. In such frozen-in plasmas, local values of T_e and T_{ion} are unrelated to each other.

5.3 X-ray Spectroscopy and Spectral Inversion

This is a very rich topic that has been developed to analyze first solar spectra and then stellar spectra. This section provides an overview of some of the important diagnostic techniques, but for a comprehensive review of these techniques with applications to solar spectra see Del Zanna and Mason (2018). They provide examples of solar X-ray and EUV spectra, describe ionization and excitation equilibria both for thermal and non-Maxwellian plasmas, and review spectral diagnostic techniques for measuring temperatures, densities, and abundances in coronal plasmas.

X-ray spectra consist of continua and emission lines that are spontaneous de-excitations following either electron collisional excitation from an ion's ground state to the excited state or recombination and collisional cascade to this excited state. Since excited states of highly ionized ions are typically 1 keV or more above the ground state, million degree electrons in coronae are required to excite these states and the emission occurs at keV energies in the X-ray spectrum. Unlike optical spectra of neutral and singly ionized atoms, line emission from highly excited ions often occurs in transitions that are not allowed by the usual electronic dipole rules, for example the allowed resonant line $1s2p\,^1P_1 \rightarrow 1s^2\,^1S_0$ transition of He-like O VII. Intercombination transitions involve a change in electron spin and orbital angular momentum, for example the $1s2p\,^3P_1 \rightarrow 1s^2\,^1S_0$ transition of O VII. Forbidden transitions often involve a change of electronic spin without a change in orbital angular momentum, for example the $1s2s\,^3S_1 \rightarrow 1s^2\,^1S_0$ transition of O VII as shown in Fig. 5.3.

Figure 5.4 shows examples of X-ray spectra obtained with *XMM-Newton* of the RS CVn system HR 1099 (G5 IV+K1 IV) with a hot coronal plasma, Capella (G8 III+G1 III) with a somewhat cooler plasma, and Procyon (F5 IV-V) with the coolest coronal plasma. As shown in the figure, the brightest lines are usually

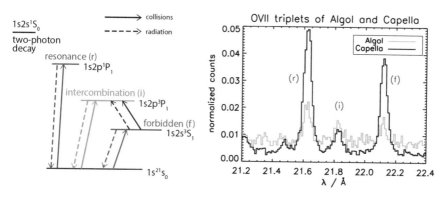

Fig. 5.3 *Left:* Energy level diagram of He-like O VII showing resonance (r), intercombination (i) and forbidden (f) line transitions. *Right:* Spectra of the O VII lines for Algol (B8 V+K2 IV) and Capella (G8 III+G1 III). Figure from Güdel (2004)

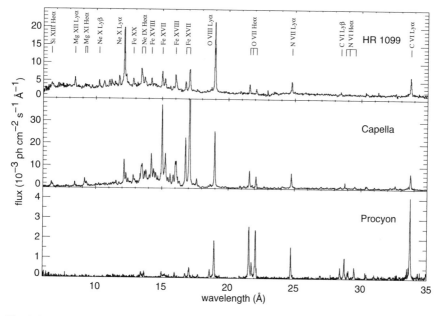

Fig. 5.4 X-ray spectra of HR 1099, Capella, and Procyon obtained with *XMM-Newton*. The hot corona of HR 1099 shows strong continuum and line emission at short wavelengths, whereas Procyon with the coolest corona shows only line emission mostly from lower stages of ionization at longer wavelengths. Figure from Güdel (2004)

Lyman-α transitions of H-like C VI, N VII, O VIII, and Ne X or He-like transitions of O VII, Ne IV, and Mg XI.

If the Sun is an appropriate role model, plasmas in stellar coronae are not isothermal but rather exhibit a range of temperatures with plasma confined in closed magnetic structures being hotter than plasma in open field regions. Since observations of stellar coronae generally have no spatial resolution, a common way of writing the distribution of plasma with temperature is using the differential emission measure, $Q(T) = n_e n_H dV/dlnT$, where V is the volume and T is the temperature. $Q(T)$ is proportional to density squared because the emission in a line is proportional to the plasma density where the ionization stage is abundant and to the collisional excitation rate, which is proportional to the electron density. $Q(T)$ is determined as a best fit to the observed fluxes in emission lines F_j formed over a wide range of temperatures,

$$F_j = (1/4\pi d^2) \int AG_j(T)Q(T)dlnT, \tag{5.2}$$

where A is a constant that includes the atomic abundance, $G_j(T)$ is the product of the temperature dependent ionization fraction for the ion and the collisional excitation rate for the line, and d is the stellar distance. These calculations typically assume collisional ionization equilibrium, the electron velocity distribution is Maxwellian, and the abundances are uniform in the corona.

The procedure for computing $Q(T)$ from emission line flux is outlined in Sect. 6.2.4, but a more complete description can be found in Güdel (2004) and Güdel and Nazé (2009). The method assumes that coronal emission lines are optically thin, CIE is valid, and the atomic data are accurate. A serious problem is that the inversion of observed spectra into physical parameters is degenerate with a range of $Q(T)$ shapes that can fit the same spectral data (Craig and Brown 1976; Telleschi et al. 2005).

Nevertheless, much can be learned from differential emission measure analyses. As shown in Fig. 5.5, the differential emission measure distributions have very different shapes for G-type stars of different ages and evolution. $Q(T)$ increases with temperature until reaching a peak temperature of at least $\log T = 7.5$ for a young weak-lined T Tauri star (HD 283572) and the zero age main sequence star EK Dra (G1.5 V), whereas the peak temperature for the 5 Gyr old Sun (G2 V) is only $\log T = 6.2$ (Scelsi et al. 2005). The moderate age binary ξ Boo (G8 V+K4 V), for which the dominant X-ray emitter is the G8 V star (Wood and Linsky 2010), shows a peak temperature of $\log T = 6.6$, and $Q(T)$ for the rapidly rotating G0 III giant star 31 Com is similar to the young stars. Telleschi et al. (2005) have identified the following statistical relations among the X-ray luminosity (L_x), stellar rotational period ($P_{\rm rot}$), and the emission measure weighted mean temperature ($\bar{T}$ in megaKelvins) for solar-type dwarf stars, vis

$$L_x \approx 1.61 \times 10^{26} \bar{T}^{4.05 \pm 0.25}\ erg\ s^{-1} \tag{5.3}$$

$$\bar{T} \approx 12.2 P_{\rm rot}^{-0.50 \pm 0.08}\ MK \tag{5.4}$$

$$L_x = 4.04 \times 10^{30} P_{\rm rot}^{-2.03 \pm 0.35}\ erg\ s^{-1} \tag{5.5}$$

$$L_R \approx 1.69 \times 10^9 \bar{T}^{5.29 \pm 0.74}\ erg\ s^{-1}\ Hz^{-1}. \tag{5.6}$$

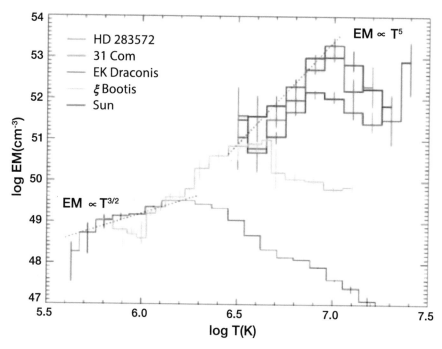

Fig. 5.5 Differential emission measure distributions ($Q(T)$) for four solar-type stars and the Sun. The $Q(T)$ for the youngest star (HD 283572) is peaked at $\log T = 7.5$ or perhaps higher, whereas the oldest star (the Sun) shows a peak in $Q(T)$ at $\log T = 6.2$. Figure from Güdel and Nazé (2009)

Johnstone and Güdel (2015) found a similar relation to Eq. (5.3) using stellar X-ray surface fluxes F_x (erg cm^{-2} s^{-1}), vis

$$\bar{T} = 0.11 F_x^{0.26},\qquad(5.7)$$

and Wood et al. (2018) called attention to the rapid increase in F_x with increasing coronal temperature.

It is important to compare the properties of the solar corona with that of stars of different ages, rotation, T_{eff} and gravity. However, stellar X-ray observatories cannot observe the Sun, and solar X-ray observatories cannot observe stars and have difficulty integrating the solar X-ray emission over the solar disk to simulate the Sun as a star. The most important problem, however, is that different X-ray observatories observe X-ray spectra with different biases, because they have different effective areas as a function of wavelength. For example, *ROSAT* is only sensitive to energies below 2 keV, *Chandra* is sensitive to higher energy photons than *XMM-Newton*, and solar observatories, e.g., *ASCA* and *Yohkoh*, are sensitive to higher energy photons than *Chandra*. Thus the X-ray luminosity and inferred parameters of multi-thermal models of coronae all depend on the spectral bandpass of the observatory.

In a survey of planet hosting stars within 30 pc of the Sun, Poppenhaeger et al. (2010) intercompared L_x values inferred from *XMM-Newton* observations of 36 stars with L_x values inferred from *ROSAT* observations of 34 stars. They used estimates of the coronal temperature to transform the *XMM-Newton* data to the *ROSAT* L_x scale, taking into account that *XMM-Newton* is much less sensitive than *ROSAT* to the low energy X-ray emission from cool coronae. Another approach to comparing coronal parameters measured for the Sun and stars by different spacecraft in a selfconsistent manner was developed by Peres et al. (2000). Their approach is to infer $Q(T)$ for each pixel in *Yohkoh* satellite images of the Sun, then fold each $Q(T)$ through the spectral response of a stellar X-ray observatory, in this case *ROSAT*, and finally to integrate over the solar surface. The resulting L_x and derived temperatures and mean emission measures for multi-thermal components of stellar coronae and the solar corona observed as a star should then be properly intercomparable.

The He-like ions play an important role in understanding coronal plasmas because the very different collisional de-excitation rates for the r, i, and f lines lead to an important electron density diagnostic. In statistical equilibrium, the number of collisional and radiative excitations from the ground state (l) to the upper state (u) must be balanced by the number of collisional and radiative de-excitations, viz

$$n_l(n_e C_{lu} + B_{lu} J) = n_u(A_{ul} + n_e C_{ul}), \qquad (5.8)$$

where n_i, n_u, and n_e are the densities of the lower and upper levels of the ion and the electron density, C_{lu} and C_{ul} are the collisional excitation and de-excitation rates, A_{ul} is the radiative de-excitation rate, and $B_{lu} J$ is the radiative excitation rate, which is usually small and can be ignored. When the electron density is low, $A_{ul} > n_e C_{ul}$ and $n_u/n_l \propto n_e$. Since $A_r \gg A_i \gg A_f$, there are density ranges where n_u/n_l is density sensitive. This ratio is a useful density indicator in the range $\log n_e = 9.5 - 11.5$ for O VII, $\log n_e = 11.0 - 13.0$ for Ne IX, and $\log n_e = 12.0 - 14.0$ for Mg XI. If one can measure the electron density from the He-like triplet line ratios or other line ratios, then the volume of the emitting plasma in a stellar corona at a temperature T is proportional to $Q(T) n_e^3$.

Knowledge of $Q(T)$ and the electron density at a given temperature from one He-like ion, or better n_e at several temperatures from several He-like ions, allows one to infer a spatially homogeneous coronal model. The very inhomogeneous solar corona with its bright loops suggests that the basic geometry of a stellar corona is hot plasma confined in magnetic loops. Rosner et al. (1978) developed a loop model assuming uniform heating and hydrostatic equilibrium, which is valid when the vertical loop semi-length (L) is small compared to the pressure scale height. In this case, the temperature at the top of the loop, $T = 1400(pL)^{1/3}$, where p is the gas pressure and the volumetric heating rate is $\epsilon = 9.8 \times 10^4 p^{7/6} L^{-5/6}$. For large loops with heights exceeding the pressure scale height, Serio et al. (1981) developed modified scaling laws. If the solar corona is a useful guide, stellar coronae probably consist of many loops of different sizes with lower temperature and density plasma between the loops. For this geometry, the X-ray emission from a stellar corona is likely dominated by emission from the loops. Coronal image reconstructions of the eclipsing binary YY Gem (dM1e+dM1e) obtained from *XMM-Newton* observations at many orbital phases (Güdel et al. 2001; Güdel 2004) show extended X-ray emitting structures between the two stars consistent with emission primarily from extended loops.

Correlations of coronal X-ray properties with stellar parameters (e.g., mass, age, and rotation period) are described in Sect. 9.1, and correlations with other activity indicators are discussed in Sect. 9.2.

5.4 Coronal Emission Lines at Wavelengths Outside of the X-ray Region

While the X-ray region of the spectrum is rich in coronal emission lines from a variety of ionization states formed over a wide range of temperatures, the low spectral resolution of existing spectrographs (*XMM-Newton* and *Chandra*), less than $R = \lambda/\Delta\lambda = 1000$ corresponding to $300\,\mathrm{km\,s^{-1}}$, makes it difficult to measure spectral line widths and Doppler shifts. However, the wavelength stability of *Chandra's* High Energy Transmission Grating Spectrometer allowed Ayres et al. (2001) to detect the $50\,\mathrm{km\,s^{-1}}$ orbital radial velocity variation of the K1 IV star in the HR 1099 binary system. Observations of coronal emission lines at longer wavelengths where spectrometers have much higher spectral resolution, therefore, can provide important diagnostics of coronal plasmas. Solar spectra (e.g., Feldman et al. 2000) provide a useful starting point for identifying coronal emission lines. For example, emission lines of Fe X (637.45 nm) and Fe XIV (530.28 nm) are often observed during solar flares at optical wavelengths and the Fe II emission lines (124.22 nm and 134.96 nm) are observed in the FUV.

The high resolution spectrographs on *HST* are the prime instruments to search for stellar coronal emission lines. The early GHRS spectrum of the dM0e flare star AU Mic obtained by Maran et al. (1994) detected the Fe XXI 135.408 nm line for the first time in a star other than the Sun. After deblending from the adjacent C I 135.429 nm line, the Fe XXI line showed no bulk motion or profile assymmetry. The Fe XXI line width was consistent with thermal broadening in a $\log T = 7.1$ plasma and an upper limit of $38\,\mathrm{km\,s^{-1}}$ for nonthermal broadening. In the subsequent high signal/noise observation of AU Mic with STIS, Pagano et al. (2000) measured FWHM $= 116 \pm 3\,\mathrm{km\,s^{-1}}$ for the Fe XXI line, which is slightly larger than the expected $\approx 90\,\mathrm{km\,s^{-1}}$ thermal width indicating that turbulent broadening, if present, would be subsonic. The measured radial velocity relative to the stellar photosphere is $-2.2 \pm 2.5\,\mathrm{km\,s^{-1}}$ including only measurement errors, but the uncertainty in the laboratory wavelength of the Fe XXI line is about $11\,\mathrm{km\,s^{-1}}$. As a result, the emitting coronal plasma could be in motion relative to the photosphere.

Ayres et al. (2003b) used STIS to survey 29 F–M dwarf, giant, and super-giant stars for ultraviolet coronal emission lines, detecting Fe XII (124.22 nm and 134.96 nm) emission from 10 dwarf stars (F7 V to M5.5 V) and Fe XXI (135.408 nm) emission from 17 dwarf and 7 giant stars including Capella A (G8 III), Capella B (G1 III), ι Cap (G8 III), and β Cet (K0 III). Both Fe XII and Fe XXI were

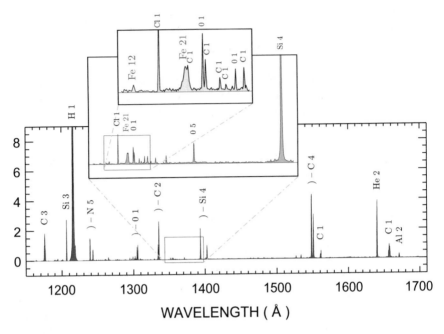

Fig. 5.6 Fe XII and Fe XXI coronal emission lines observed in a STIS spectrum of AD Leo (M4 V) by Ayres et al. (2003b). Reproduced by permission of the AAS

detected in seven stars (see Fig. 5.6), all dwarfs. Ayres et al. (2003b) also searched for but did not find emission from other coronal emission lines previously observed in solar spectra (e.g., Feldman et al. 2000). Except for the very rapidly rotating star 31 Com (G0 III), the Fe XII and Fe XXI emission lines are not significantly displaced from the photospheric rest velocity and have line widths similar to only thermal broadening at the expected line formation temperature. These results are consistent with emission in closed magnetic field loops as in the solar corona rather than stellar wind shocks. The Fe XII and Fe XXI line fluxes are correlated with the X-ray emission with power law indices near 0.5 and 1.0, respectively. In an extensive study of the young solar-mass star EK Dra (G1.5 V) with both STIS and COS, Ayres (2015) found that the central wavelength of the Fe XXI line is close to the photospheric radial velocity given the uncertain laboratory wavelength, but the line's FWHM of 120 km s^{-1} is larger than the expected thermal width of about 90 km s^{-1}, suggesting either enhanced rotational broadening from an extended corotating corona or turbulent broadening.

Redfield et al. (2003) searched for coronal emission lines from 26 A4–M0 dwarf, giant, and supergiant stars in the 91–118 nm spectral range of the *FUSE* satellite. They identified both Fe XVIII (97.485 nm) and Fe XIX (111.808 nm) in five dwarf and five giant stars. The detection list largely overlaps the list of detected *HST* stars. They searched for but did not detect other coronal emission observed in solar spectra. As seen in the Ayres et al. (2003b) *HST* survey, the velocity

Table 5.1 Examples of coronal emission lines at longer wavelengths

Wavelength (nm)	Ion	Formation (log T)	Wavelength (nm)	Ion	Formation (log T)
11.72	Fe XXII	7.1	62.4	Mg X	6.0
12.87	Fe XXI	7.0	97.485	Fe XVIII	6.9
13.29	Fe XXIII	7.3	111.808	Fe XIX	7.0
13.57	Fe XXII	7.1	124.22	Fe XII	6.2
17.107	Fe IX	5.9	134.96	Fe XII	6.2
17.453	Fe X	6.0	135.408	Fe XXI	7.1
19.19	Fe XXIV	7.6	338.80	Fe XIII	6.2
25.50	Fe XXIV	7.6	530.28	Fe XIV	6.3
28.415	Fe XV	6.3	637.45	Fe X	6.0
33.541	Fe XVI	6.4	1074.68	Fe XIII	6.2
36.080	Fe XVI	6.4	1079.79	Fe XIII	6.2
49.9	Si XII	6.1			

centroids of the coronal emission lines are consistent with the photospheric radial velocities, implying formation in closed magnetic field structures in the corona with no bulk flows. The line widths are generally consistent with thermal broadening and perhaps some turbulent broadening, but the rapidly rotating stars 31 Com, AB Dor, and Capella have broader profiles suggesting line formation in co-rotating coronal regions extending out to 0.4–1.3 stellar radii. The flux in the Fe XVIII line is correlated with the soft X-ray flux from the target stars observed by *ROSAT*.

Extreme ultraviolet stellar spectra observed with the *EUVE* satellite (Craig et al. 1997; Monsignori Fossi et al. 1996) contain many coronal emission lines including strong lines of Fe IX, Fe X, Fe XV, Fe XVI, and even Fe XXIV as listed in Table 5.1. For example, Fuhrmeister et al. (2007) detected the Fe XIII (338.80 nm) line during a flare and outside of large flares on CN Leo (M5.5 V). Spectropolarimetry of the near-infrared lines of Fe XIII (1074.68 and 1079.79 nm) will be used to measure coronal magnetic fields with the new *DKIST* solar telescope.

The wavelengths listed in Table 5.1 are from various sources including solar flare spectra observed by Feldman et al. (2000) and the EUV spectrum of AU Mic (M0 V) (Monsignori Fossi et al. 1996). The line formation temperatures are the peak ionization temperatures in the collisional ionization equilibrium calculations of Bryans et al. (2009).

5.5 Coronal Emission at Radio Wavelengths

Radio emission from stellar coronae is often detected at cm wavelengths by the *Very Large Array (VLA)* and other radio telescopes. Güdel (2002) has summarized the possible emission processes at radio wavelengths including bremsstrahlung and gyroresonance emission from thermal electrons, gyrosynchrotron emission from a hot thermal plasma and from a power-law distribution of relativistic electrons, and various coherent emission processes. Gyrosynchrotron emission from a power-law distribution of relativistic electrons spiraling in coronal magnetic fields is usually identified as the dominant emission mechanism from stellar coronae on the basis of the spectral shape of the continuum emission at cm wavelengths. The typical 10^9 K brightness temperature characterizes the strong radio emission available from even a relatively small number of high energy electrons. This emission is a function of the coronal magnetic field strength and the power-law distribution of electron energies. Gyrosynchrotron radio emission is detected from a wide variety of active stars including young stars, G–M dwarfs, brown dwarfs, RS CVn and other short period binaries, and even some OB stars and Wolf-Rayet stars.

Many authors have called attention to the correlation of radio and X-ray emission for different classes of stars. Figure 5.7 shows that the correlation of X-ray luminosity with radio luminosity, typically at 5–8 GHz, is linear over ten orders

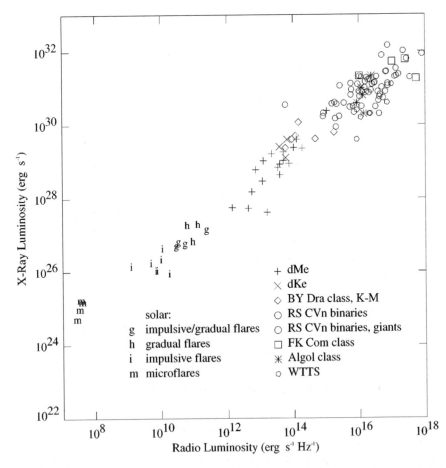

Fig. 5.7 Comparison of X-ray and radio luminosities for a variety of solar flares (letter symbols) and stars. Figure from Güdel (2002)

of magnitude in radio luminosity from solar flares to very active binaries and young stars. This correlation can be expressed as $L_x \propto L_R^{1.02\pm0.17}$ and $L_x/L_R \approx 10^{15\pm1}$ (Güdel and Benz 1993; Benz and Güdel 1994).

What makes these correlations very interesting is that the empirical correlations indicate that coronal heating and the acceleration of relativistic electrons must occur in parallel and presumably by the same process (Güdel and Benz 1993). Since the correlations are independent of stellar age, rotation rate, binarity, and the strength of activity indicators, the heating and acceleration process must be basically the same for all of these star types and likely involve flaring in magnetic structures on a wide range of temporal and spatial scales.

Although the Güdel-Benz relations accurately describe the relation of L_x to L_R over ten orders of magnitude, there are interesting exceptions. For example, quiescent radio emission from the Sun, which is produced by either bremsstrahlung or gyroresonant emission from thermal electrons, does not fit the Güdel-Benz relations presumably because of the absence of relativistic electrons in the quiet solar corona. On the other hand, stars cooler than M7 do not fit these relations because their quiescent X-ray emission rapidly declines in the cooler stars while their radio emission is remains consistent with that of the warmer stars as shown in Fig. 5.8 (Berger et al. 2010; Stelzer et al. 2012) and often includes a beamed

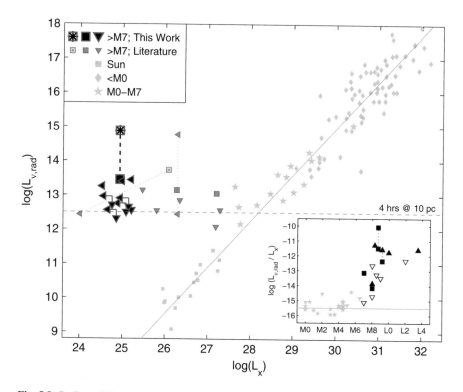

Fig. 5.8 Radio and X-ray luminosities for stars cooler than spectral type M7 compared to warmer stars (lighter symbols and dashed line). Figure from Berger et al. (2010). Reproduced by permission of the AAS

coherent component (Hallinan et al. 2008). This implies that for these brown dwarfs the process that accelerates electrons to relativistic energies does not efficiently heat the corona (Berger et al. 2010).

References

Audard, M., Osten, R.A., Brown, A., Briggs, K.R., Güdel, M., Hodges-Kluck, E., Gizis, J.E.: A Chandra X-ray detection of the L dwarf binary Kelu-1. Simultaneous Chandra and Very Large Array observations. Astron. Astrophys. **471**, L63 (2007)

Ayres, T.R.: The flare-ona of EK Draconis. Astron. J. **150**, 7 (2015)

Ayres, T.R., Linsky, J.L., Vaiana, G.S., Golub, L., Rosner, R.: The cool half of the H-R diagram in soft X-rays. Astrophys. J. **250**, 293 (1981)

Ayres, T.R., Brown, A., Osten, R.A., Huenemoerder, D.P., Drake, J.J., Brickhouse, N.S., Linsky, J.L.: Chandra, EUVE, HST, and VLA multiwavelength campaign on HR 1099: instrumental capabilities, data reduction, and initial results. Astrophys. J. **549**, 554 (2001)

Ayres, T.R., Brown, A., Harper, G.M.: Buried alive in the coronal graveyard. Astrophys. J. **598**, 610 (2003a)

Ayres, T.R., Brown, A., Harper, G.M., Osten, R.A., Linsky, J.L., Wood, B.E., Redfield, S.: Space telescope imaging spectrograph survey of far-ultraviolet coronal forbidden lines in late-type stars. Astrophys. J. **583**, 963 (2003b)

Benz, A.O., Güdel, M.: X-ray/microwave ratio of flares and coronae. Astron. Astrophys. **285**, 621 (1994)

Berger, E., Basri, G., Fleming, T.A., Giampapa, M.S., Gizis, J.E., Liebert, J., Martin, E., Phan-Bao, N., Rutledge, R.E.: Simultaneous multi-wavelength observations of magnetic activity in ultracool dwarfs. III. X-ray, radio, and Hα activity trends in M and L dwarfs. Astrophys. J. **709**, 332 (2010)

Bryans, P., Landi, E., Savin, D.W.: A new approach to analyzing solar coronal spectra and updated collisional ionization equilibrium calculations. II. Updated ionization rate coefficients. Astrophys. J. **691**, 1540 (2009)

Cassinelli, J.P., Waldron, W.L., Sanders, W.T., Harnden Jr., F.R., Rosner, R., Vaiana, G.S.: X-ray emission from Of stars and OB supergiants. Astrophys. J. **250**, 677 (1981)

Craig, I.J.D., Brown, J.C.: Fundamental limitations of X-ray spectra as diagnostics of plasma temperature structure. Astron. Astrophys. **49**, 239 (1976)

Craig, N., Abbott, M., Finley, D., Jessop, H., Howell, S.B., Mathioudakis, M., Sommers, J., Vallerga, J.V., Malina, R.F.: The extreme ultraviolet explorer stellar spectral atlas. Astrophys. J. Suppl. Ser. **113**, 131 (1997)

Del Zanna, G., Mason, H.E.: Solar UV and X-ray spectral diagnostics. Living Rev. Sol. Phys. **15**, 5 (2018)

Dempsey, R.C., Linsky, J.L., Fleming, T.A., Schmitt, J.H.M.M.: The ROSAT all-sky survey of active binary coronae. I – Quiescent fluxes for the RS Canum Venaticorum systems. Astrophys. J. Suppl. Ser. **86**, 599 (1993)

Dempsey, R.C., Linsky, J.L., Fleming, T.A., Schmitt, J.H.M.M.: The ROSAT all-sky survey of active binary coronae. III – Quiescent coronal properties for the BY Draconis-type binaries. Astrophys. J. **478**, 358 (1997)

Drake, S.A., Linsky, J.L., Schmitt, J.H.M.M., Rosso, C.: X-ray emission from chemically peculiar stars. Astrophys. J. **420**, 387 (1994)

Feldman, U., Curdt, W., Landi, E., Wilhelm, K.: Identification of spectral lines in the 500–1600 Å wavelength range of highly ionized Ne, Na, Mg, Ar, K, Ca, Ti, Cr, Mn, Fe, Co, and Ni emitted by flares ($T_e \geq 3 \times 10^6$ K) and their potential use in plasma diagnostics. Astrophys. J. **544**, 508 (2000)

Fuhrmeister, B., Liefke, C., Schmitt, J.H.M.M.: Simultaneous XMM-Newton and VLT/UVES observations of the flare star CN Leonis. Astron. Astrophys. **468**, 221 (2007)

Gagné, M., Caillault, J.-P., Stauffer, J.R., Linsky, J.L.: Periodic X-ray emission from the O7 V star θ^1 Orionis C. Astrophys. J. Lett. **478**, L87 (1997)

Güdel, M.: Stellar radio astronomy: probing stellar atmospheres from protostars to giants. Ann. Rev. Astron. Astrophys. **40**, 217 (2002)

Güdel, M.: X-ray astronomy of stellar coronae. Astron. Astrophys. Rev. **12**, 71 (2004)

Güdel, M., Benz, A.O.: X-ray/microwave relation of different types of active stars. Astrophys. J. Lett. **405**, L63 (1993)

Güdel, M., Nazé, Y.: X-ray spectroscopy of stars. Astron. Astrophys. Rev. **17**, 309 (2009)

Güdel, M., Audard, M., Magee, H., Franciosini, E., Grosso, N., Cordova, F.A., Pallavicini, R., Mewe, R.: The XMM-Newton view of stellar coronae: coronal structure in the Castor X-ray triplet. Astron. Astrophys. **365**, 344 (2001)

Hallinan, G., Antonova, A., Doyle, J.G., Bourke, S., Lane, C., Golden, A.: Confirmation of the electron cyclotron maser instability as the dominant source of radio emission from very low mass stars and brown dwarfs. Astrophys. J. **684**, 644 (2008)

Hünsch, M., Schmitt, J.H.M.M., Schroeder, K.-P., Reimers, D.: ROSAT X-ray observations of a complete, volume-limited sample of late-type giants. Astron. Astrophys. **310**, 801 (1996)

Johnstone, C.P., Güdel, M.: The coronal temperatures of low-mass main-sequence stars. Astron. Astrophys. **578**, 129 (2015)

Kastner, J.H., Principe, D.A., Punzi, K., Stelzer, B., Gorti, U., Pascucci, I., Argiroffi, C.: M stars in the TW Hya Association: stellar X-rays and disk dissipation. Astron. J. **152**, 3 (2016)

Linsky, J.L., Haisch, B.M.: Outer atmospheres of cool stars. I – The sharp division into solar-type and non-solar-type stars. Astrophys. J. Lett. **229**, 27 (1979)

Loyd, R.O.P., France, K., Youngblood, A., Schneider, C., Brown, A., Hu, R., Linsky, J., Froning, C.S., Redfield, S., Rugheimer, S., Tian, F.: The MUSCLES Treasury Survey III: X-ray to infrared spectra of 11 M and K stars hosting planets. Astrophys. J. **824**, 102 (2016)

Maran, S.P., Robinson, R.D., Shore, S.N., Brosius, J.W., Carpenter, K.G., Woodgate, B.E., Linsky, J.L., Brown, A., Byrne, P.B., et al.: Observing stellar coronae with the Goddard High Resolution Spectrograph. 1: The dMe star AU microscopii. Astrophys. J. **421**, 800 (1994)

Monsignori Fossi, B.C., Landini, M., Del Zanna, G., Bowyer, S.: A time-resolved extreme-ultraviolet spectroscopic study of the quiescent and flaring corona of the flare star AU Microscopii. Astrophys. J. **466**, 427 (1996)

Pagano, I., Linsky, J.L., Carkner, L., Robinson, R.D., Woodgate, B., Timothy, G.: HST/STIS echelle spectra of the dM1e star AU Microscopii outside of flares. Astrophys. J. **532**, 497 (2000)

Pease, D.O., Drake, J.J., Kashyap, V.L.: The darkest bright star: Chandra X-ray observations of Vega. Astrophys. J. **636**, 436 (2006)

Peres, G., Orlando, S., Reale, F., Rosner, R., Hudson, H.: The Sun as an X-ray star. II. Using the Yohkoh/Soft X-ray telescope-derived solar emission measure versus temperature to interpret stellar X-ray observations. Astrophys. J. **528**, 537 (2000)

Poppenhaeger, K., Robrade, J., Schmitt, J.H.M.M.: Coronal properties of planet-bearing stars. Astron. Astrophys. **515**, 98 (2010)

Redfield, S., Ayres, T.R., Linsky, J.L., Ake, T.B., Dupree, A.K., Robinson, R.D., Young, P.R.: A Far Ultraviolet Explorer survey of coronal forbidden lines in late-type stars. Astrophys. J. **585**, 993 (2003)

Robrade, J., Schmitt, J.H.M.M.: Altair – the "hottest" magnetically active star in X-rays. Astron. Astrophys. **497**, 511 (2009)

Rosner, R., Tucker, W.H., Vaiana, G.S.: Dynamics of the quiescent solar corona. Astrophys. J. **220**, 643 (1978)

Scelsi, L., Maggio, A., Peres, G., Pallavicini, R.: Coronal properties of G-type stars in different evolutionary phases. Astron. Astrophys. **432**, 671 (2005)

Serio, S., Peres, G., Vaiana, G.S., Golub, L., Rosner, R.: Closed coronal structures. II – Generalized hydrostatic model. Astrophys. J. **243**, 288 (1981)

Simon, T., Drake, S.A., Kim, P.D.: The X-ray emission of A-type stars. Publ. Astron. Soc. Pac. **107**, 1034 (1995)

Stelzer, B., Micela, G., Flaccomio, E., Neuhäuser, R., Jayawardhana, R.: X-ray emission of brown dwarfs: towards constraining the dependence on age, luminosity, and temperature. Astron. Astrophys. **448**, 293 (2006)

Stelzer, B., Alcalá, J., Biazzo, K., Ercolano, B., Crespo-Chacón, I., López-Santiago, J., Martínez-Arnáiz, R., Schmitt, J.H.M.M., Rigliaco, E., Leone, F., Cupani, G.: The ultracool dwarf DENIS-P J104814.7-395606. Chromospheres and coronae at the low-mass end of the main-sequence. Astron. Astrophys. **537**, A94 (2012)

Stelzer, B., Marino, A., Micela, G., López-Santiago, J., Liefke, C.: The UV and X-ray activity of the M dwarfs within 10 pc of the Sun. Mon. Not. R. Astron. Soc. **431**, 2063 (2013)

Telleschi, A., Güdel, M., Briggs, K., Audard, M., Ness, J.-U., Skinner, S.L.: Coronal evolution of the Sun in time: high-resolution X-ray spectroscopy of solar analogs with different ages. Astrophys. J. **622**, 653 (2005)

Williams, P.K.G., Berger, E., Irwin, J., Berta-Thompson, Z.K., Charbonneau, D.: Simultaneous multiwavelength observations of magnetic activity in ultracool dwarfs. IV. The active, young binary NLTT 33370 AB (= 2MASS J13142039+1320011). Astrophys. J. **799**, 192 (2015)

Wood, B.E., Linsky, J.L.: Resolving the ξ Boo binarity with Chandra, and revealing the spectral type dependence of the coronal "FIP effect". Astrophys. J. **717**, 1279 (2010)

Wood, B.E., Laming, J., Warren, H.P., Poppenhaeger, K.: A Chandra/LETGS survey of main-sequence stars. Astrophys. J. **862**, 66 (2018)

Chapter 6
Reconstructing the Missing Stellar Emission

Unfortunately, the two of the most important features in stellar spectra, the Lyman-α line and the EUV, are the most difficult to observe. The Lyman-α emission line dominates the UV spectra of stars cooler than the Sun. For the Sun, the intrinsic Lyman-α line flux is about equal to the rest of the FUV flux, whereas for an M star like GJ 876 (M4 V) the Lyman-α line flux is more than twice as large as the rest of the FUV and nearly as large as the entire FUV+NUV flux excluding this line (France et al. 2012). However, for stars other than the Sun, the Lyman-α line is heavily obscured by interstellar absorption. The EUV spectrum is critically important for driving hydrodynamic mass loss from exoplanets, but the long wavelength portions of stellar EUV spectra are completely absorbed by interstellar hydrogen. Since these important spectral features are needed to understand exoplanet mass loss and photochemistry, this chapter summarizes the various methods that are being used to reconstruct these features from other portions of a star's spectrum and from stellar models.

6.1 Lyman-α Emission Line

Figure 6.1 shows that the solar Lyman-α line is a bright emission feature with broad wings and a self-reversal near line center produced by the low excitation of hydrogen near the top of the chromosphere. This figure shows Lyman-α irradiance spectra obtained with the Solar Ultraviolet Measurement of Emitted Radiation (SUMER) instrument on the *Solar and Heliospheric Observatory (SOHO)* spacecraft. *SOHO* is located in a halo orbit around the Sun-Earth L1 point far from the Earth, but there is still a significant amount of Lyman-α airglow due to hydrogen surrounding the Earth. Lyman-α observations obtained from spacecraft in low Earth orbit, however, show a weak emission feature near line center produced by solar Lyman-α photons scattered by hydrogen atoms surrounding the Earth.

© Springer Nature Switzerland AG 2019
J. Linsky, *Host Stars and their Effects on Exoplanet Atmospheres*,
Lecture Notes in Physics 955, https://doi.org/10.1007/978-3-030-11452-7_6

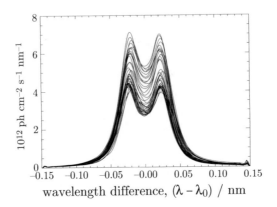

Fig. 6.1 Observations by the SUMER instrument on the *SOHO* spacecraft of the solar irradiance Lyman-α line at different times during solar cycle 23. The brightest and faintest emission occurs at maximum (year 2001) and minimum (years 1997 and 2009) of this solar magnetic cycle. The spectral resolution is 0.0086 nm. Figure from Lemaire et al. (2015). Reproduced with permission of ESO

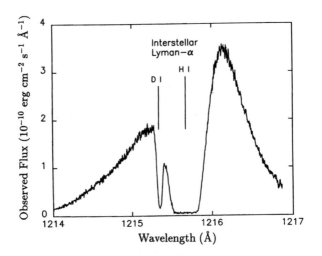

Fig. 6.2 The GHRS high-resolution spectrum of Capella (G5 III + G0 III) showing the stellar Lyman-α emission line and interstellar absorption by hydrogen and deuterium Lyman-α. Figure from Linsky et al. (1993). Reproduced by permission of the AAS

Figure 6.2 shows the Lyman-α line of the binary star Capella observed by the Goddard High Resolution Spectrograph (GHRS) instrument on *HST*. Interstellar hydrogen completely absorbs the stellar emission in a 0.04 nm wavelength region near line center with an optical depth of roughly 3×10^5 near line center. This large optical depth is for a star located only 12.9 pc away with a hydrogen column density of $\log N(\text{H I})=18.2$ (Redfield and Linsky 2008). Absorption by the Lyman-α line of deuterium in the interstellar medium is at -0.033 nm (-81.4 km/s) from

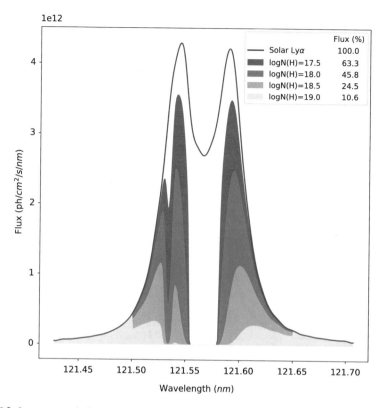

Fig. 6.3 Lyman-α emission line profiles observed in the quiet Sun and with increasing amounts of interstellar hydrogen absorption. The observed flux through the interstellar medium is included in the box. Figure courtesy of Dennis Tilipman

the hydrogen line. The abundance of deuterium relative to hydrogen within about 100 pc of the Sun is D/H=$(1.56 \pm 0.4) \times 10^{-5}$ (Wood et al. 2004).

The signature of interstellar absorption in the stellar Lyman-α emission line depends on the column density of interstellar hydrogen $N(\text{H I})$ along the line of sight to the star and the velocity difference between the interstellar hydrogen and the star. Figure 6.3 shows the effects of increasing $N(\text{H I})$ on the solar Lyman-α profile (Lemaire et al. 2015) when there is no velocity difference. With increasing $N(\text{H I})$, the observed fraction of the initial Lyman-α flux decreases and the width of the interstellar hydrogen absorption core increases until the interstellar deuterium line is obscured when $\log N(\text{H I}) \approx 18.7$. Figure 6.4 shows the effect of velocity differences of $\pm 30 \, \text{km s}^{-1}$ when $\log N(\text{H I})=18.0$. The effect is to produce an asymmetric Lyman-α emission line with a brighter blue peak when the interstellar flow is at positive velocities relative to the star and a brighter red peak when the flow is at negative velocities. Typically the velocity difference is less than $30 \, \text{km s}^{-1}$, but there are cases with much larger velocity differences.

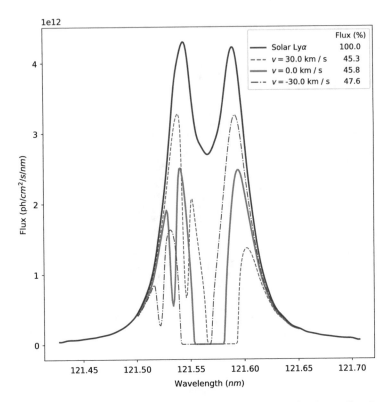

Fig. 6.4 Lyman-α emission line profiles observed in the quiet Sun, after interstellar absorption with log N(H I)=18.0 and interstellar flow velocities of -30, 0, and $+30\,\mathrm{km\,s^{-1}}$ relative to the star. Figure courtesy of Dennis Tilipman

Most of the intrinsic Lyman-α line can be observed in the few stars with very high radial velocities relative to the interstellar medium. One example shown in Fig. 6.5 is Kapteyn's star (sdM1) with a radial velocity of $+245\,\mathrm{km\,s^{-1}}$ (Guinan et al. 2016). Since most of the stellar Lyman-α emission is Doppler shifted away from the interstellar absorption and the geocoronal emission, Youngblood et al. (2016) was able to measure the line profile in the low-resolution spectrum.

Observations of spectroscopic binary stars with large orbital radial velocity variations provides another way of studying the core emission of the stellar Lyman-α line. Figure 6.6 shows Lyman-α spectra of the spectroscopic binary HR 1099 obtained at nearly opposite orbital phases (0.24 and 0.85) by Piskunov et al. (1997). Since the star with the brightest Lyman-α emission is the K0 IV star whose radial velocity changes from $+47\,\mathrm{km\,s^{-1}}$ at orbital phase 0.24 to about $-70\,\mathrm{km\,s^{-1}}$ at phase 0.85, the asymmetry of the combined emission changes with phase. The interstellar absorption and faint geocoronal emission feature near line center are the same in both spectra. The radial velocity of the G0 IV star at orbital phase 0.85 is not sufficient to reveal the line core emission, but the spectrum at this phase reveals the inner wing emission.

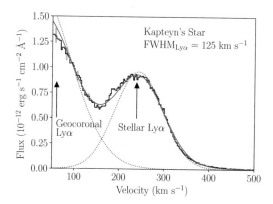

Fig. 6.5 Lyman-α spectrum of the high-radial velocity star called Kapteyn's star. Figure from Youngblood et al. (2016). Reproduced by permission of the AAS

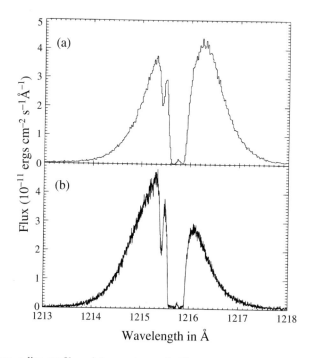

Fig. 6.6 Lyman-α line profiles of the spectroscopic binary system HR 1099 (K0 IV + G5 5) observed with the GHRS instrument on *HST*. The top profile is for orbital phase 0.24 and the bottom profile is for orbital phase 0.85. The maximum radial velocity separation of the two stars is about 113 km s⁻¹. Figure from Piskunov et al. (1997). Reproduced by permission of the AAS

6.1.1 Using Interstellar Medium Absorption Data

In a series of papers, Wood and colleagues have reconstructed the intrinsic Lyman-α absorption from observed line profiles and measurements of interstellar absorption. High-resolution spectra of the Fe II and Mg II spectral lines reveal the velocity structure and column densities in each velocity component along the line of sight to the star. The deuterium Lyman-α line differs from the hydrogen line only in the D/H abundance and the line thermal widths. Since for nearby stars the deuterium line is either optically thin or with optical depth about unity near line center, the interstellar optical depth as a function of wavelength can be computed from the deuterium and metal lines. This information permits reconstruction of the hydrogen Lyman-α close to line center, provided the deuterium absorption line is not obscured by very broad interstellar hydrogen absorption which occurs when $\log N(\mathrm{H\ I}) >$ 18.7. For the line center region, Wood and collaborators typically assumed the same shape as the Mg II resonance lines that are also very optically thick and formed in the chromosphere. Solar observations show that the Lyman-α and Mg II lines have similar shapes. Figure 6.7 shows examples of these reconstructions in Wood et al. (2005).

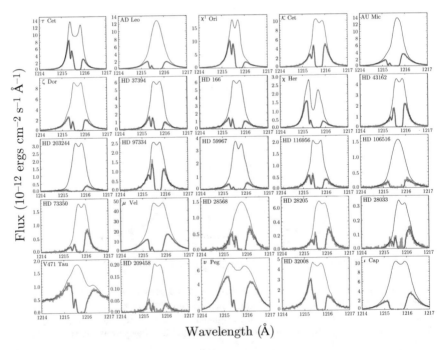

Fig. 6.7 Comparison of observed Lyman-α line profiles (thick lines) from nearby stars with reconstructed line profiles (thin lines). Figure from Wood et al. (2005). Reproduced by permission of the AAS

These examples show that for nearby stars the flux in the reconstructed line profiles can be from two to ten times larger than the observed flux. Also, the fraction of the intrinsic line flux that is observed through the interstellar medium increases with the radial velocity difference between the star and the interstellar medium flow speed. When there are high-resolution spectra available that show the deuterium interstellar absorption line and metal lines, this reconstruction technique should result in high quality Lyman-α line profiles and fluxes with an uncertainty of about 20%.

6.1.2 Solving for the Stellar Lyman-α Profile and Interstellar Parameters Simultaneously

For many stars of interest there are no available high-resolution spectra containing the deuterium Lyman-α line and/or the Fe II and Mg II lines. Even for *HST*, many interesting exoplanet host stars are too faint to obtain high-resolution spectra. Instead, there may be only be lower resolution spectra with modest S/N in the Lyman-α emission feature but very noisy data where the deuterium absorption should be present. In order to analyze such spectra, France et al. (2012) developed a second technique to solve for the stellar Lyman-α profile and the interstellar absorption simultaneously. The assumptions underlying the technique are that the stellar emission line can be approximated by one or two Gaussians and that the D/H ratio is fixed at the cannonical value. They then solve for the stellar Gaussian parameters and the interstellar N(H I) and velocity that minimize the differences between the intrinsic profile absorbed by the interstellar medium and the observed profile. Figure 6.8 shows an example of this reconstruction technique for the star GJ 876 (M4 V) observed with the COS instrument on HST. Note the excellent

Fig. 6.8 Comparison of the observed Lyman-α line profile of the star GJ 876 (black histogram) with the reconstructed line profile (dashed blue) and the reconstructed profile corrected for interstellar absorption (red line). Figure from France et al. (2012). Reproduced by permission of the AAS

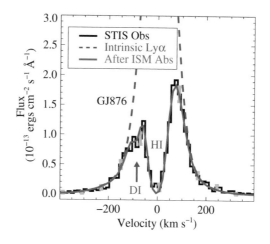

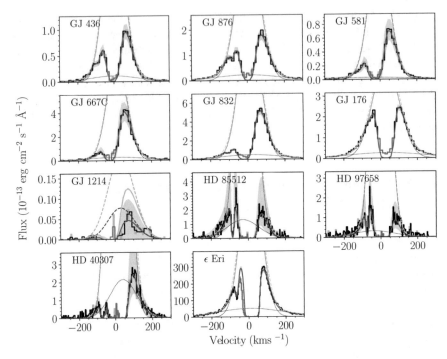

Fig. 6.9 Lyman-α line spectra for the 11 stars observed in the MUSCLES Treasury Survey. Observed spectra are black with green error bars, reconstructed line profiles are dash-dot blue, and the reconstructed line profiles folded through the interstellar medium are red. The gray-shaded areas represent uncertainties in the fits. Figure from Youngblood et al. (2016). Reproduced by permission of the AAS

agreement between the absorbed and observed profiles and the much larger flux in the reconstructed profile.

Youngblood et al. (2016) then modified this technique to reconstruct the Lyman-α line profiles of the 11 exoplanet host stars observed with STIS during the MUSCLES Treasury Survey. Their modification was to assume a single Voigt profile shape rather than one or two Gaussians for the intrinsic stellar emission line. As Fig. 6.9 shows, they found good agreement with the observed Lyman-α line profiles despite the lower S/N and spectral resolution compared to the brighter stars that Wood et al. (2005) had analyzed using interstellar absorption lines. Youngblood et al. (2016) also assessed possible systematic errors including the effects of multiple interstellar velocity components, which their technique does not include, and inadequate spectral resolution.

The good agreement between the observed and absorbed reconstructed line profiles for the well exposed stars in Fig. 6.9 suggests that the reconstructed fluxes are accurate to 30% or perhaps better. To test this accuracy estimate, there are two stars for which the two reconstruction techniques can be intercompared. From the analysis of high-resolution spectra of ϵ Eri (K2 V) obtained with the GHRS instrument on *HST*, Dring et al. (1997) used the first technique to measure $\log N(\text{H I})$ = 17.88 ± 0.07 and a Lyman-α flux of 4.88×10^{-11} ergs cm^{-2} s^{-1} erg cm^{-2} s^{-1} (cf. Wood et al. 2005). From the high-resolution STIS spectrum of ϵ Eri, Youngblood et al. (2016) used the second technique to obtained $\log N(\text{H I}) = 17.93 \pm 0.02$ and a Lyman-α flux of $(6.1^{+0.3}_{-0.2}) \times 10^{-11}$ erg cm^{-2} s^{-1}. While the interstellar column densities agree, the Lyman-α flux is about 25% larger in the Youngblood et al. (2016) analysis. The two techniques provide excellent agreement for the bright M0 V star AU Mic. Using the first technique, Wood et al. (2005) found $\log N(\text{H I}) = 18.356 \pm 0.002$ and Lyman-α flux of 1.03×10^{-11} erg cm^{-2} s^{-1}. Using the second technique, Youngblood et al. (2016) found $\log N(\text{H I}) = 18.35 \pm 0.001$ and Lyman-α flux = $1.07 \pm 0.04 \times 10^{-11}$ erg cm^{-2} s^{-1}. These two test cases suggest that the both reconstruction technique yield results that are probably accurate to better than 25% for well-exposed spectra, but both techniques have larger uncertainties when analyzing noisier data.

6.1.3 Inverting Fluorescent Spectra

In the FUV spectra of classical T Tauri stars (CTTSs), M dwarfs, and cool supergiants, there are many examples of fluorescent emission lines pumped by the bright Lyman-α emission line. Molecular hydrogen emission is a prime example of Lyman-α pumped emission, for example in both planet-hosting and non-planet-time M dwarfs where the location of the fluorescence may be in starspots or the lower chromosphere (Kruczek et al. 2017). CO in T Tauri stars (France et al. 2011) and S I, Fe II, Cr II and other species in cool supergiants (e.g., Carpenter et al. 2014) are also pumped by Lyman-α. Other emission lines including Lyman-β and C IV can also initiate fluorescent emission processes.

The energy level diagram of H_2 in Fig. 6.10 (Herczeg et al. 2002) shows the transitions from the vibration-rotation levels in the ground electronic state that overlap in wavelength with the Lyman-α line initiating fluorescent emission in many transitions from the pumped upper states that are observed in CTTS spectra. Figure 6.11 (Yang et al. 2011) shows the observed Lyman-α lines of two CCTSs with the wavelengths of the H_2 pumping transitions indicated. In many cases there is obvious absorption at the wavelengths of the pumping transitions. With a model for the location of the pumped gas, whether uniformly distributed around the star or in a disk, one can use the observed fluorescent emission line fluxes to predict the Lyman-α emission at the wavelengths of the pumping transitions. This technique is useful when a portion of the stellar Lyman-α line is observed to test the validity of the assumed geometry of the fluorescing gas (Herczeg et al. 2004). Schindhelm et al. (2012) applied this technique to T Tauri stars with the result that the Lyman-α flux dominates the far-UV spectra of these stars.

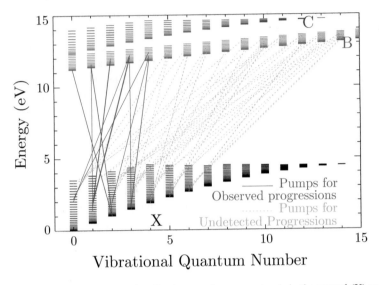

Fig. 6.10 The horizontal lines are the vibration-rotation energy levels in the ground (X) and two excited electronic states (B and C) of H_2. The solid black lines are transitions at wavelengths coincident with and pumped by the stellar Lyman-α line. Fluorescent H_2 emission lines originate in the upper states of the pumping transitions. Figure from Herczeg et al. (2002). Reproduced by permission of the AAS

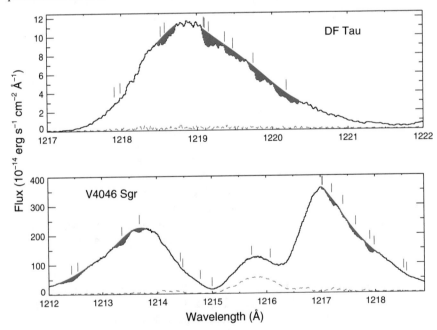

Fig. 6.11 Lyman-α line profiles observed from the CTTSs DF Tau and V4046 Sgr. Vertical red lines identify the wavelengths of transitions that pump H_2 and the blue depressions in the stellar line profile are produced by absorption in the pumping transitions. The dashed red lines indicate uncertainty levels and geocoronal emission for V4046 Sgr. Figure from Yang et al. (2011). Reproduced by permission of the AAS

6.2 Extreme Ultraviolet Emission

6.2.1 The Solar EUV Spectrum and the Effects of Interstellar Absorption on Stellar EUV Spectra

Since 2008, the EUV Variability Experiment (EVE) instrument onboard the *Solar Dynamics Observatory (SDO)* has been obtaining full-disk irradiance spectra of the Sun viewed as an unresolved distant star. EVE spectra include the 1–106 nm region with 0.1 nm resolution at high temporal cadence as described by Woods et al. (2012). EVE also observes the total Lyman-α flux. The solar EUV spectrum has also been observed by previous instruments including the EIS instrument on *HINODE* and the CDS instrument on the *Solar and Heliospheric Observatory (SOHO)*.

Figure 6.12 shows a typical quiet Sun irradiance spectrum (see Linsky et al. 2014). The EUV portion of the spectrum includes the H I recombination continuum often called the Lyman continuum visible from the photoionization threshold at 91.2 nm down to roughly 60 nm and the He I recombination continuum extending from 50.4 nm down to about 45 nm. The brightest emission line in the EUV spectral region is the resonance line of He II (30.39 nm) formed in the lower transition region. Other chromospheric or transition region emission lines include He I (58.4 nm), O III (72 and 84 nm), O IV (79 nm), O V (630 nm), and lines of Ne VII (46.5 nm), and Ne VIII (76–78 nm). Higher temperature coronal lines include Mg X (60.99 and 62.23 nm), Si XII (49.93 and 52.11 nm), and many emission lines at shorter wavelengths (See Table 6.1). During a solar flare all of the EUV emission lines and continua become much brighter than during quiescent times. In the solar flare spectrum observed by Milligan et al. (2014) with the EVE instrument (see Fig. 6.13), the spectral slope of the Lyman continuum has a color temperature of

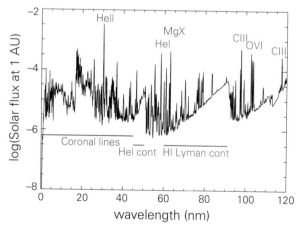

Fig. 6.12 The quiet Sun EUV and X-ray spectrum with important emission lines and continua indicated. Figure from Linsky et al. (2014). Reproduced by permission of the AAS

Table 6.1 AU Mic emission line fluxes ratioed to He II 30.39 nm

Ion	Wavelength (nm)	Flux ratio
Fe XXI + Fe XIX	9.12	0.13
Fe XVIII + Fe XX	9.39	0.27
Fe XXI	9.82	0.16
Fe XIX	10.16	0.12
Fe XXI	10.24	0.21
Fe XVIII	10.40	0.12
Fe XIX	10.84	0.21
Fe XIX	11.00	0.06
Fe XXII	11.44	0.07
Fe XXII	11.72	0.37
Fe XX + Fe XXI	11.87	0.16
Fe XIX	12.02	0.12
Fe XX	12.18	0.18
Fe XXI	12.87	0.31
Fe XXIII + Fe XX	13.29	1.16
Fe XXII	13.57	0.31
Fe XXI	14.22	0.07
Fe IX	17.11	0.24
Fe XXIV	19.19	0.97
Fe XXIV	25.50	0.34
Fe XV	28.44	0.24
He II	30.39	1.00
Fe XVI	33.50	0.18

14,500 K indicating that the continuum is formed near the top of the chromosphere. In the quiet Sun spectrum, the corresponding color temperature is 13,230 K (Linsky et al. 2014).

The complete solar EUV spectrum is readily observed because there is minimal absorption by atomic hydrogen in the outer corona and interplanetary medium along the line of sight from the Sun and to the Earth. It is not possible to observe the entire EUV spectra of even the nearest stars because interstellar hydrogen, helium and metals completely absorb the long wavelength portion and partially absorb the shorter wavelength portion. Analysis of high-resolution ultraviolet spectra of nearby stars obtained with STIS allowed Redfield and Linsky (2008) to measure the hydrogen column density $N(H I)$ in the lines of sight to nearby stars. The smallest value of $N(H I)$ obtained in their survey was log $N(H I)$=17.2 toward Sirius (2.6 pc). For the line of sight to the nearest stellar system, α Cen (1.3 pc), log $N(H I)$ = 17.6, and for the line of sight to ϵ Eri (3.3 pc) log $N(H I)$ = 18.0. More distant stars generally have log $N(H I)$ values similar to or larger than for ϵ Eri.

Absorption by interstellar hydrogen, helium, and metals severely limit our ability to observe EUV emission even from nearby host stars. Figure 6.14 shows that optical depth unity due to hydrogen in the interstellar medium toward the host star Proxima Cen occurs near 50 nm and for ϵ Eri it occurs near 35 nm. For Trappist-

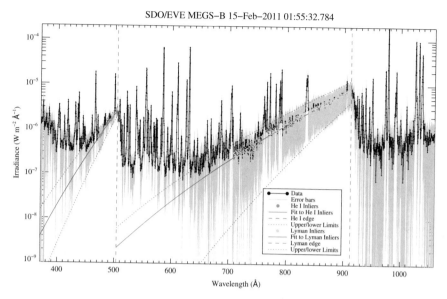

Fig. 6.13 Solar EUV spectrum during a flare showing the spectral slopes of the Lyman recombination continuum (blue) and the He I recombination continuum (red). Figure from Milligan et al. (2014). Reproduced by permission of the AAS

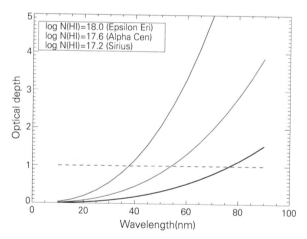

Fig. 6.14 Optical depth as a function of wavelength for hydrogen column densities to nearby stars. The blue line is for the line of sight to ϵ Eri (3.3 pc), red line for α Cen (1.3 pc), and the black line for Sirius (2.6 pc). The dashed line is optical depth unity

1 located at 12.1 pc, $\log N(H\ I) = 18.3 \pm 0.2$, obtained by fitting the Lyman-α line (Bourrier et al. 2017), corresponds to optical depth unity near 30 nm. Rapidly increasing interstellar optical depths at longer wavelengths make detections of stellar EUV spectra more difficult at longer wavelengths. Figure 6.15 shows the

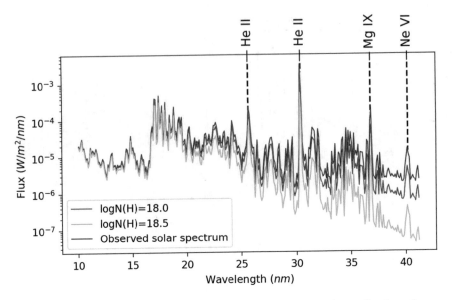

Fig. 6.15 The observed solar EUV spectrum and the effect of increasing interstellar absorption on the spectrum. Figure courtesy of Dennis Tilipman

effect of increasing interstellar absorption on observed stellar EUV spectra using the solar EUV spectrum as the background spectrum.

The *Extreme Ultraviolet Explorer (EUVE)* satellite obtained stellar spectra in the 7–76 nm range with a modest resolution of $\lambda/\Delta\lambda \approx 200$ (Bowyer and Malina 1991). The EUV stellar spectral atlas (Craig et al. 1997) contains *EUVE* spectra of many nearby stars extending in some cases to 36 nm, but most stars observed by *EUVE* had no detected EUV emission lines, Nearby stars typically show detections of the He II resonance line (30.4 nm) and two lines of Fe XVI (33.5 and 36.0 nm) before interstellar absorption prevents observations of longer wavelength emission. *EUVE* spectra of three exceptionally bright nearby stars (α Cen, Procyon, ϵ Eri) include emission lines at longer wavelengths—Ne VII (46.5 nm), Si XIII (49.9 nm), and perhaps O VI (55.4 nm) observed through the rapidly increasing interstellar absorption. Figure 6.16 shows the spectrum of the α Cen binary. However, an efficient future EUV spectrograph on a telescope of modest aperture could extend the detectable spectra of favorable stars to beyond 60 nm and perhaps to 85 nm.

There are also stellar observations of the short wavelength portion of the EUV obtained with X-ray satellites. The Wide Field Camera on the *ROSAT* satellite obtained fluxes in the 6.5–15.5 nm and 11–19.5 nm bands for a large number of nearby stars (Pounds et al. 1993; Wood et al. 1994). The Low Energy Transmission Grating Spectrometer (LETGS) on the *Chandra X-ray Observatory* is also sensitive to emission at wavelengths as long as about 17.5 nm. Wood et al. (2018) analyzed X-ray spectra of 19 main-sequence stars observed with the LETGS, obtaining emission measure distributions and coronal abundances useful for simulating EUV spectra.

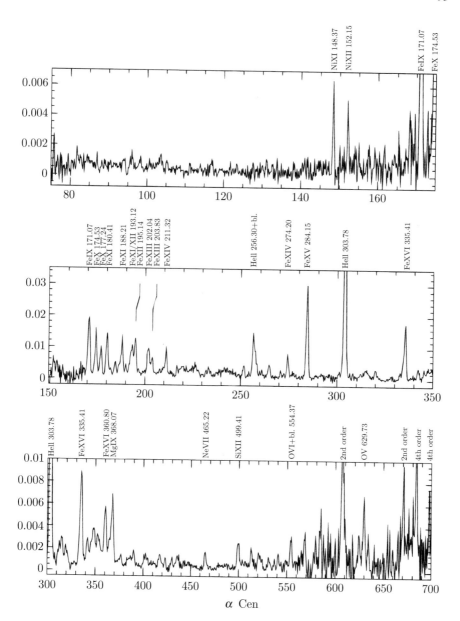

Fig. 6.16 *EUVE* spectrum of the α Cen binary with emission lines marked. Emission lines observed in second order (at half the indicated wavelength) are also marked. Figure from Craig et al. (1997). Reproduced by permission of the AAS

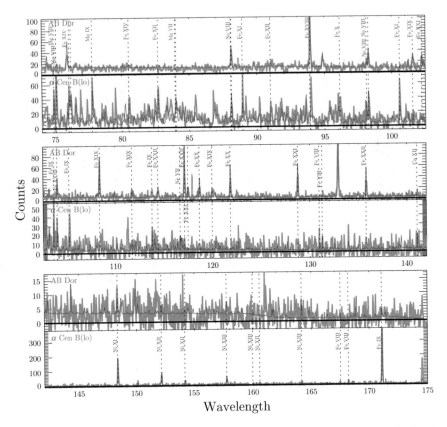

Fig. 6.17 LETGS spectra of the active K0 V star AB Dor A and the K1 V star α Cen B when it was in a low activity stage. The red lines are synthetic spectra computed from the emission measure distributions of each star. Figure from Wood et al. (2018). Reproduced by permission of the AAS

Figure 6.17 shows the long wavelength portion of the LETGS spectra of the active K0 V star AB Dor A and the K1 V star α Cen B when it was in a low activity stage. At wavelengths longer than about 14 nm the factor of 5 higher interstellar hydrogen column density N(H I) absorbs most of the AB Dor A flux compared to that of α Cen A, which has a very low column density $\log N$(H I)=17.61. Wood et al. (2018) detected 118 spectral lines from 49 ionic species with almost complete coverage of all ionization stages of Fe from Fe VIII to Fe XXIV (see Fig. 6.18). The differential emission measure distributions $Q(T)$ of the least active stars are similar to the quiet Sun ($\log L_x = 26.40$), the $Q(T)$ of moderately active stars are similar to those of solar active regions ($\log L_x = 28.66$) totally covering the stellar surfaces as previously suggested by Drake et al. (2000), and the $Q(T)$ of the most active stars are similar to coronae on the active Sun but including the high temperature emission of flares. These conclusions do not depend on stellar spectral type, only on activity levels as measured by the X-ray surface flux.

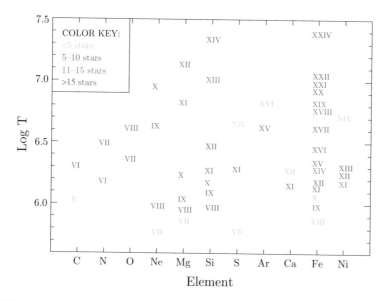

Fig. 6.18 Ionization stages detected by emission lines in the LETGS spectra of main-sequence stars. The color key indicates the number of stars detected with each ionization stage. Figure from Wood et al. (2018). Reproduced by permission of the AAS

Since the 36–91.2 nm spectral region is presently observable only for the Sun and shorter wavelengths are also unobservable for more distant stars, it is essential to develop techniques for computing, simulating or estimating the EUV fluxes and spectra of host stars. The next sections outline the various techniques that have been employed to provide the critically needed estimates of stellar EUV spectra.

6.2.2 Estimating the Solar EUV Flux over Time Using Solar Analogs

The spectra of stars with similar masses and spectral types to the Sun are useful proxies for the Sun at different ages, activity levels, and rotation rates. To estimate the solar spectrum in time, Ayres (1997) compiled X-ray to optical spectra and rotational velocities of solar-type stars ranging in age from about 60 Myr (α Persei cluster) to 4.6 Gyr (the present day Sun). He divided the observed spectral features (emission line and continuum flux) into nine temperature ranges and for each one determined power-law relations between the mean fluxes and the stellar rotation velocities. From a power-law relation of rotational velocity with age, he then determined power-law relations between flux and stellar age for the spectral features in each temperature range. He showed that the flux in high-temperature coronal emission lines decay with age much faster than lower temperature emission. To predict the EUV spectral flux with age and the resulting photoionization rates of molecules in exoplanet atmospheres, he summed the flux contributed by the spectral features in each temperature range.

Ribas et al. (2005) extended this method by compiling X-ray to optical spectra of six stars and the very quiet Sun with spectral types G0 V to G2 V and ages 0.1 Gyr (EK Dra) to 6.7 Gyr (β Hyi). Instead of grouping the observed fluxes by temperature as did Ayres et al. (1997), they chose to group the observed fluxes in wavelength bins: 1–20 Å, 20–100 Å, 100–360 Å, and 920–1200 Å. Figure 6.19 shows the resulting dependence of flux with stellar age in each wavelength bin. Since there are no data in the 360–920 Å bin, they estimated a power-law slope of -1.0, the mean of the power-law slopes of the 100–300 Å and 920–1200 Å bins. They also determined that the power-law for the reconstructed Lyman-α line flux is -0.72, similar to that for the 920–1180 Å bin. These power laws pertain only to stars with very similar masses to the Sun, but the method is being extended to M stars by Guinan and colleagues (cf. Engle et al. 2009).

Using the observed X-ray to optical spectra of the same solar analog stars, Claire et al. (2012) extended the Ribas et al. (2005) method by separating the age dependence of the strongest emission lines such as Lyman-α from the age dependence of the remaining flux in each wavelength bin. This produced multiplicative factors,

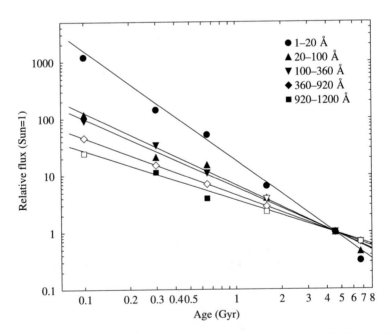

Fig. 6.19 Relative flux of solar-type stars in several bandpasses as a function of stellar age. Figure from Ribas et al. (2005). Reproduced by permission of the AAS

called flux multipliers, by which each spectral region of today's Sun should be multiplied to simulate the Sun at earlier times. This method produced complete spectra not just fluxes in broad passbands. The resulting EUV spectrum of the Sun in time shows all of the emission lines observed in today's Sun.

6.2.3 Estimating the Stellar EUV Flux from the Observed Short Wavelength Portion of the EUV Spectrum and the Solar Ratio

One approach is to multiply the observed stellar flux in a portion of the EUV band $f(\Delta\lambda)$ by the solar $f(EUV)/f(\Delta\lambda)$ ratio, where $f(EUV)$ is the flux in the complete EUV wavelength band. Lecavelier des Etangs (2007) used the S2 band (11–20 nm) stellar flux observed by the Wide Field Camera on the ROSAT satellite multiplied by the solar f(10–91.2 nm)/f(11–20 nm) flux ratio. This method assumes that the stellar flux in the total EUV band is proportional to the flux in the short wavelength portion of the EUV band with the same flux ratio as for the Sun.

One problem with this approach is that the 11–20 nm portion of the EUV spectrum is dominated by high temperature emission lines, mostly Fe XVIII to Fe XXIV formed in the corona at temperatures of 10^7 K or hotter, whereas the long wavelength portion of the EUV band is dominated by the low temperature Lyman continuum and emission lines formed in the chromosphere and transition region. Table 6.1 (adapted from Monsignori Fossi et al. 1996) shows the emission line fluxes observed by *EUVE* during a flare on the dM0 V star AU Microscopii. The fluxes are ratioed to the He II resonance line (30.39 nm), which is formed in the lower transition region. The flux ratios of all emission lines shortward of the He II line are lower limits as interstellar absorption reduces the observed flux of the He II line much more than for the emission lines at shorter wavelengths. This approach should be reliable for estimating the EUV flux of inactive early-G stars that likely have EUV spectra with the relative amounts of high- and low-temperature emission features similar to the Sun.

In the outer atmospheres of more active stars and especially M-type stars, there is relatively more high-temperature coronal plasma than low-temperature chromospheric and transition region plasma. This is shown in Fig. 6.20 where spectra of the quiet and active Sun are compared to the computed spectrum of the M1.5 V star GJ 832 (Fontenla et al. 2016). Note that the chromospheric emission that dominates the FUV spectrum is much fainter for GJ 832 than for the quiet Sun, but the X-ray emission dominated by high-temperature coronal emission lines is similar for GJ 832 and the quiet Sun.

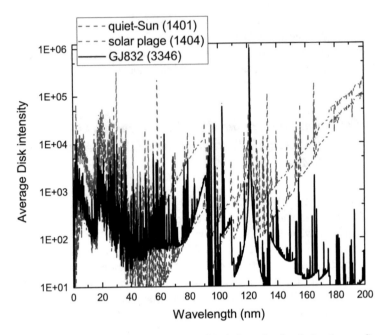

Fig. 6.20 Comparison on spectra of the quiet Sun (blue), the active Sun (solar plage, red), and the M1.5 V star GJ 832 (black). Figure from Fontenla et al. (2016). Reproduced by permission of the AAS

6.2.4 Computing the EUV Spectrum from an Emission Measure Distribution Analysis

The observed flux F in an optically thin emission line from a stellar corona can be written as

$$F = (1/4\pi d^2) \int_{T_1}^{T_2} A\phi(T)Q(T)d(lnT), \tag{6.1}$$

where d is distance to the star, A the element's abundance, ϕ the emissivity of the line per unit emission measure, and the differential emission measure

$$Q(T) = n_e n_H dV/dlnT. \tag{6.2}$$

The differential emission measure includes the collisional excitation rate for the transition, which is proportional to the electron density n_e, and the amount of plasma in the temperature range (T_1 to T_2) where the ion is abundant, which is proportional to the total amount of hydrogen in this column, $n_H dV$. $Q(T)$ is a convenient way of describing the amount of plasma responsible for the emission in a given spectral line. The analysis of many emission lines formed over a range of temperatures leads to the temperature dependence of $Q(T)$. $Q(T)$ is independent of the structure and inhomogeneity of a stellar corona, but it assumes that the emission lines are optically thin and that the plasma is in collisional ionization equilibrium. The development and application of the emission measure technique for inferring the thermal structure of stellar coronae are described by Brown and Jordan (1981) and references therein.

The measurement of $Q(T)$ from the X-ray spectrum of a star permits one to compute the fluxes of emission lines of highly ionized elements in the EUV range and thus a synthetic EUV spectrum. This technique provides reliable EUV spectra when the temperature range over which $Q(T)$ is assembled includes the important emission lines and continua in the EUV spectrum and accurate and complete atomic parameters (e.g., collisional excitation rates, spontaneous de-excitation rates, and ionization equilibria) are available

The first use of this technique was by Sanz-Forcada et al. (2011). They fitted X-ray spectra of 82 exoplanet host stars observed by the *XMM-Newton* and *Chandra* X-ray Observatories with coronal models. Figure 6.21 shows the resulting emission measure distributions for three K2 V stars. This figure shows that with increasing activity the high-temperature portion of $Q(T)$ increases more rapidly than the low-temperature portion, consistent with the relative enhancement of high-temperature emission lines compared to low-temperature emission lines in the more active stars. Since X-ray spectra provide the information needed to construct $Q(T)$ only for temperatures above 10^6 K, it is necessary to estimate the low temperature portion of $Q(T)$ from host stars with available FUV spectra. They found that for the few host stars that have high signal/noise FUV spectra, the low-temperature portion of

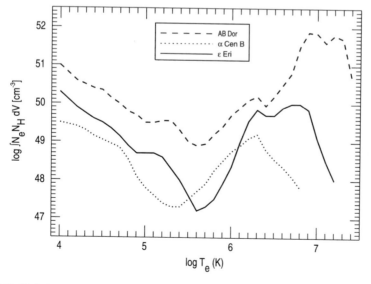

Fig. 6.21 Emission measure distributions for three K2 V stars with a range of activity levels. α Cen B is the least active star, ϵ Eri is moderately active, and AB Dor is the most active of the three stars. Figure from Sanz-Forcada et al. (2011). Reproduced with permission of ESO

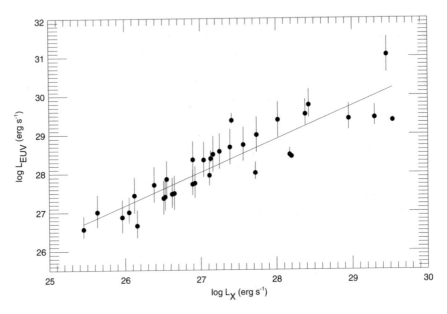

Fig. 6.22 Relation of EUV to X-ray luminosity for exoplanet host stars. Figure from Sanz-Forcada et al. (2011). Reproduced with permission of ESO

$Q(T)$ is approximately proportional to $Q(T)$ in the temperature range $\log T = 6.0$–6.3. With this scaling relation, they found a simple power law relation between the EUV (10–92 nm) and X-ray (0.5–10 nm) luminosities (see Fig. 6.22). Given the temperature sensitivity of $Q(T)$, it is not surprising that the EUV flux of the inactive star α Cen B represent 91% of the total XUV flux, whereas for the active star AB Dor the EUV flux is only 23% of the total XUV flux. These estimates of the relative importance of the EUV portion of the XUV flux may be lower limits as this analysis may not fully account for the emission by chromospheric and transition region features, in particular the Lyman continuum and perhaps the strong He II line.

Chadney et al. (2015) extended the emission measure distribution method by showing the good agreement of synthetic 8–35 nm spectra with the observed *EUVE* spectra for three stars—ϵ Eri (K2 V), AU Mic (M1 Ve) at quiescent and flaring times, and AD Leo (M4.5 Ve) at quiescent and flaring times. Figure 6.23 shows their interesting result that the EUV/X-ray flux ratio fits a power law relation with the X-ray flux. This ratio decreases with activity from 5–10 for the inactive Sun and α Cen B to about 0.3 for the stars with highest activity, in particular flaring times. They use the ROSAT X-ray flux which is for the 5–12.4 nm wavelength interval. More recently, King et al. (2018) used solar 0.5–91.2 nm fluxes from the *TIMED/SEE* solar mission to derive power-law relations between the EUV and *XMM-Newton* X-ray fluxes for K and M dwarfs. With both FUV and X-ray spectra, Louden et al. (2017) used a differential emission measure technique to compute the EUV emission spectrum from the host star of HD 209458b. They find that for this star 93% of the XUV emission is in the EUV band.

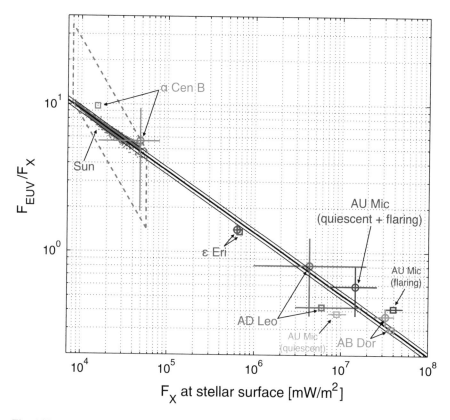

Fig. 6.23 Dependence of the EUV/X-ray flux ratio on the X-ray surface flux. The central solid line is the mean relation for the Sun and stars at quiescent and flaring times. Figure from Chadney et al. (2015)

6.2.5   Simulating the EUV Spectrum from Lyman-α and EUVE Data

Linsky et al. (2014) found that the ratio of EUV flux to the intrinsic Lyman-α flux is nearly constant over the full range of solar activity from the quietest regions to very bright plages. This is seen in both observations and semi-empirical non-LTE models (Fontenla et al. 2014) computed to fit spectra from the quietest to the most active regions on the Sun. This result led Linsky et al. (2014) to compute scaling relations for the EUV flux in 10 nm wide bands as a function of the intrinsic Lyman-α flux.

Figure 6.24 plots the f(10–20 nm)/f(Lyman-α) flux ratio vs the intrinsic Lyman-α flux multiplied by the ratio of stellar radii, $(R_{Sun}/R_{star})^2$. The seven diamond symbols connected by a solid black line are the flux ratios for the seven Fontenla et al. (2014) models computed for the faintest solar regions (on the left) to the brightest solar regions (on the right) as measured by the Ca II K line and EUV

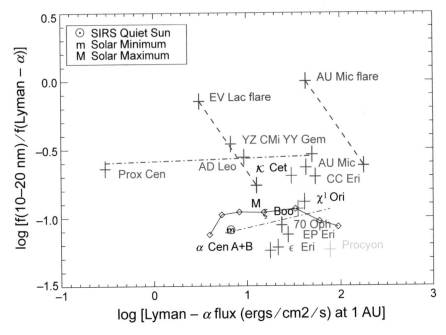

Fig. 6.24 Flux ratios for solar models (diamond symbols), the observed quiet Sun (⊙ and m symbols), active Sun (M symbol), F star (light blue), G stars (black), K stars (orange), and M stars (purple) observed by *EUVE*. Also see text. Figure from Linsky et al. (2014). Reproduced by permission of the AAS

flux. Note that the ratio is essentially constant over a factor of 30 in Lyman-α flux. The symbols m and ⊙ refer to quiet Sun ratio measurements, and the symbol M refers to an active Sun ratio. These ratios and the *EUVE* flux ratio for α Cen A+B are consistent with the ratio for the quiet Sun model. The *EUVE* ratios for the other G-type stars (black pluses), and K-type stars (orange pluses) indicate a weak trend of increasing flux ratios with increasing Lyman-α flux. The M stars (purple pluses) show higher flux ratios. The *EUVE* flux ratios of the two flaring stars (EV Lac and AU Mic), assuming likely values of the Lyman-α flux during the flares (lower right plus symbols) are consistent with the other M stars. The corresponding data for the 20–30 nm and 30–40 nm bandpasses are similar to Fig. 6.24, except that the solar flux ratios increase with increasing Lyman-α flux and are consistent with the *EUVE* data. The increase in the flux ratios with increasing Lyman-α flux for both the solar models and the *EUVE* stellar observations results from the increasing importance of high-temperature emission lines compared to lower temperature lines with increasing activity.

Extending the predictions of the EUV flux to the 40–91.2 nm wavelength region is based only on the solar models. In the 40–91.5 nm spectral region, the emission is primarily from the Lyman continuum and transition region lines that have fluxes nearly proportional the Lyman-α. The predicted EUV fluxes from the Lyman-α scaling and emission measure distribution models are consistent with each other

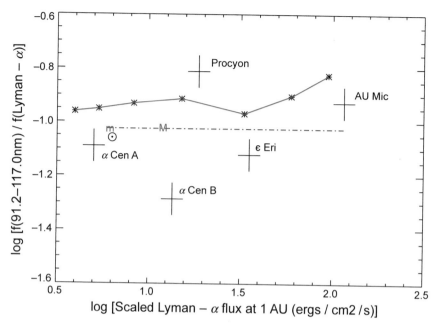

Fig. 6.25 Flux ratios for solar models (asterisk symbols), the observed quiet Sun (⊙ and m symbols), active Sun (M symbol), and stars observed by the *FUSE* satellite. The dashed line is the mean flux ratio for the *FUSE* stars. Figure from Linsky et al. (2014). Reproduced by permission of the AAS

and the *EUVE* data for the few stars in common. However, the Lyman-α scaling predictions should probably be more accurate for relatively inactive stars and the emission measure distribution method should be more accurate for the more active stars.

In the LUV (91.2–117 nm) spectral region, the model flux ratio also scales with Lyman-α and is consistent with the observations of five dwarf stars observed by the *Far Ultraviolet Spectroscopic Explorer (FUSE)* satellite (see Fig. 6.25).

The warm mini-Neptune exoplanet GJ 436b provides an important test case for the comparison of different methods for computing the EUV emission from its M3 V host star. Ehrenreich et al. (2015) computed the EUV flux incident on this exoplanet to be $1.0 \, \text{erg s}^{-1}$ from the star's X-ray emission measure distribution (Chadney et al. 2015) and nearly the same flux $1.06 \, \text{erg s}^{-1}$ based on correlations with the Lyman-α and EUVE data (Linsky et al. 2014).

6.2.6   Estimating Stellar EUV Spectra by Rescaling the Solar Irradiance Spectrum

The exoplanet host star WASP-18 (F6 V) has very faint C II, C IV, and Si IV emission lines and a very low X-ray luminosity upper limit ($\log L_x < 26.65$)

(Pillitteri et al. 2014). Compared to other stars of similar spectral type and young age (< 1 Gyr), its far-UV emission line fluxes are 10 times fainter, its X-ray emission is more than 100 times fainter, and the chromospheric emission parameter $R'_{HK} = -5.15$ is just below the basal flux limit. The likely cause of this extremely low activity level is tidal interactions by a high-mass hot Jupiter exoplanet as discussed in Chap. 14. To estimate the EUV spectrum of this very inactive host star, Fossati et al. (2018) rescaled the solar irradiance reference spectrum (Woods et al. 2009) to match the observed Si IV 139.4 nm line flux. The resulting low value of the EUV and X-ray irradiance received by the star's $10.4 M_J$ exoplanet is insufficient to drive a hydrodynamic wind, but it does stimulate a weak Jeans-escape wind (see Sect. 10.1). The extremely low value of the chromospheric activity parameter R'_{HK} for WASP-13 (G1 V) could be explained by interstellar absorption or circumstellar absorption. Fossati et al. (2015) estimated the star's EUV spectrum by rescaling the solar irradiance spectrum to fit the host star's C IV line flux.

6.2.7   Computing Stellar EUV Spectra from Model Atmospheres

The EUV spectrum of a star can be computed from a model specifying the run of temperature, density, and ionization for all important atoms as a function of height or optical depth in the atmosphere. Chapter 4 describes the assumptions underlying the computation of semi-empirical models, and Linsky (2017) has reviewed the topic in more detail with lists of many models computed for the Sun and stars. The semi-empirical models of Vernazza et al. (1973, 1976, 1981) provide synthetic EUV spectra for solar regions between least active inter-network region and the brightest plage. Fontenla et al. (2014) computed a new set of seven solar models to represent the dark quiet-Sun inter-network (model 13x0), increasingly bright network elements (models 13x1 to 13x3) and increasingly bright plage regions (models 13x4 to 13x8). Figure 4.8 shows the thermal structure of an earlier set of models for the photosphere and chromosphere regions (Fontenla et al. 2011) where model A corresponds to model 13x8 and model Q corresponds to model 13x0. Figure 6.26 compares the synthetic spectrum computed from model 13x1 with the quiet Sun spectra observed by the EVE and SOLSTICE instruments. Model 13x1 matches the observed spectra very well in the 10–40 nm and 55–91.2 nm regions but not as well in the 40–55 nm region. Although these semi-empirical models are state-of-the-art in terms of computing non-LTE ionization and excitation equilibria and radiative transfer in optically thick lines, the models are one-dimensional and, therefore, cannot account for horizontal radiative transfer in inhomogeneous media. Also, incomplete and inaccurate atomic parameters are the cause of some systematic errors in the result in the model thermal distributions. The accuracy and completeness of the EUV spectra computed with such models can be assessed from

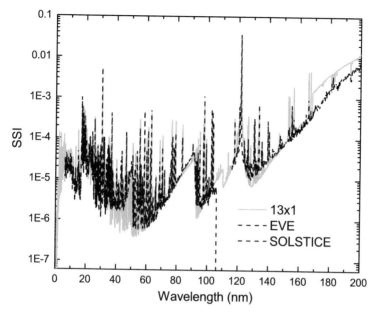

Fig. 6.26 Comparison of the solar spectral irradiance (SSI) computed from quiet Sun model 13x1 with observations obtained by the EVE and SOLSTICE instruments. Figure from Fontenla et al. (2014)

the excellent agreement of the synthetic spectrum computed from the 13x1 model with solar observations over most of the EUV region.

Figure 6.27 shows the increase in flux throughout the EUV for models constructed to match solar regions with increasing brightness. The Lyman continuum dominates the flux for all of the models below 91.2 nm to at least 60 nm. At the minimum of the sunspot cycle, the solar spectral irradiance can be fit by a mixture of models 13x0, 13x1 and 13x2 with essentially no contribution from the plage models. At sunspot maximum, the plage models can contribute about 5% of the total flux (Fontenla et al. 2011).

Extension of this method to stellar atmospheres, especially for the cooler stars, requires calculations with all of the complexity of the solar models and the inclusion of molecular opacities to match the optical and infrared portions of the spectrum. Fontenla et al. 2016) computed a model of this type for the M1.5 V star GJ 832 using the SSRPM non-LTE code. Figure 4.10 compares the thermal structure of this model to the quiet Sun, and Fig. 6.20 compares the synthetic EUV spectrum computed from this model with spectra of the quiet Sun and active Sun (plage). Models for other cool stars are being computed. For example, Peacock et al. (2019) computed a model for the M8 V star TRAPPIST-1 based on the PHOENIX non-LTE code (cf. Baron and Hauschildt 2007 and references therein) upgraded to include partial redistribution formation of strong lines. They computed EUV fluxes consistent with Lyman-α and *GALEX* FUV and NUV observations, showing that the EUV flux is

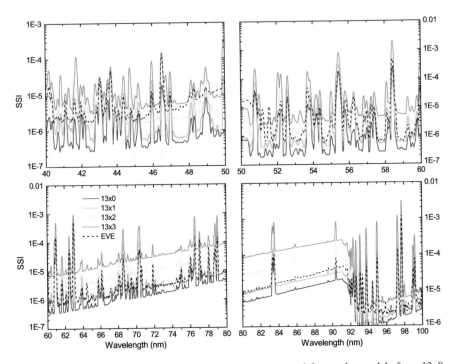

Fig. 6.27 Comparison of the solar spectral irradiance computed from solar models from 13x0 to 13x3 corresponding to a range of solar activity from very quiet levels to enhanced network emission. Figure from Fontenla et al. (2014)

highly sensitive to the assumed thermal structure of the chromosphere and transition region. The accuracy with which the EUV spectrum can be computed from such models depends upon the degree of inhomogeneity in the star's atmosphere. For relatively inactive stars, one-component models may be sufficient to explain the observed spectra, but for more active stars including stars that continuously flare at some level, multi-component models may be needed.

References

Ayres, T.R.: Evolution of the solar ionizing flux. J. Geophys. Res. **102**, 1641 (1997)

Baron, E., Hauschildt, P.H.: A 3D radiative transfer framework. II. Line transfer problems. Astron. Astrophys. **468**, 255 (2007)

Bourrier, V., Ehrenreich, D., Wheatley, P.J., Bolmont, E., Gillon, M., de Wit, J., Burgasser, A.J., Jehin, E., Queloz, D., Triaud, A.H.M.J.: Reconnaissance of the TRAPPIST-1 exoplanet system in the Lyman-α line. Astron. Astrophys. **599**, 3 (2017)

Bowyer, S., Malina, R.F.: The extreme ultraviolet explorer mission. Adv. Space Res. **11**, 205 (1991)

Brown, A., Jordan, C.: The chromosphere and corona of Procyon (α CMi, F5 IV-V). Mon. Not. R. Astron. Soc. **196**, 757 (1981)

Carpenter, K.G., Ayres, T.R., Harper, G.M., Kober, G., Nielsen, K.E., Wahlgren, G.M.: An HST COS "SNAPshot" spectrum of the K supergiant λ Vel (K4Ib-II). Astrophys. J. **794**, 41 (2014)

Chadney, J.M., Galand, M., Unruh, Y.C., Koskinen, T.T., Sanz-Forcada, J.: XUV-driven mass loss from extrasolar giant planets orbiting active stars. Icarus **250**, 357 (2015)

Claire, M.W., Sheets, J., Cohen, M., Ribas, I., Meadows, V.S., Catling, D.C.: The evolution of solar flux from 0.1 nm to 160 μm: quantitative estimates for planetary studies. Astrophys. J. **757**, 95 (2012)

Craig, N., Abbott, M., Finley, D., Jessop, H., Howell, S.B., Mathioudakis, M., Sommers, J., Vallerga, J.V., Malina, R.F.: The extreme ultraviolet explorer stellar spectral atlas. Astrophys. J. Suppl. Ser. **113**, 131 (1997)

Drake, J.J., Peres, G., Orlando, S., Laming, J.M., Maggio, A.: On stellar coronae and solar active regions. Astrophys. J. **545**, 1074 (2000)

Dring, A.R., Linsky, J., Murthy, J., Henry, R.C., Moos, W., Vidal-Madjar, A., Audouze, J., Landsman, W.: Lyman-alpha absorption and the D/H ratio in the local interstellar medium. Astrophys. J. **488**, 760 (1997)

Ehrenreich, D., Bourrier, V., Wheatley, P.J., Lecavelier des Etangs, A., Hébrard, G. Udry, S., Bonfils, X., Delfosse, X., Désert, J.-M., Sing, D.K., Vidal-Madjar, A.: A giant comet-like cloud of hydrogen escaping the warm Neptune-mass exoplanet GJ 436b. Nature **522**, 459 (2015)

Engle, S.G., Guinan, E.F., Mizusawa, T.: The living with a red dwarf program: observing the decline in dM star FUV emissions with age. AIP Conf. Proc. **1135**, 221 (2009)

Fontenla, J.L., Harder, J., Livingston, W., Snow, M., Woods, T.: High-resolution solar spectral irradiance from extreme ultraviolet to infrared. J. Geophys. Res. **116**, D20108 (2011)

Fontenla, J.M., Landi, E., Snow, M., Woods, T.: Far- and extreme-UV solar spectral irradiance and radiance from simplified atmospheric physical models. Solar. Phys. **289**, 515 (2014)

Fontenla, J., Linsky, J.L., Witbrod, J., France, K., Buccino, A., Mauas, P., Vieytes, M., Walkow-icz,L.: Semi-empirical modeling of the photosphere, chromosphere, transition region, and corona of the M-dwarfs host star GJ 832. Astrophys. J. **830**, 154 (2016)

Fossati, L., France, K., Koskinen, T., Juvan, I.G., Haswell, C.A., Lendl, M.: Far-UV spectroscopy of the planet-hosting star WASP-13: high-energy irradiance, distance, age, planetary mass-loss rate, and circumstellar environment. Astrophys. J. **815**, 118 (2015)

Fossati, L., Koskinen, T., France, K., Cubillos, P.E., Haswell, C.A., Lanza, A.F., Pillitteri, I.: Suppressed far-UV stellar activity and low planetary mass loss in the WASP-18 system. Astron. J. **155**, 113 (2018)

France, K., Schindhelm, E., Burgh, E.B., Herczeg, G.J., Harper, G.M., Brown, A., Green, J.C., Linsky, J.L., Yang, H., et al.: The far-ultraviolet "continuum" in protoplanetary disk systems. II. Carbon monoxide fourth positive emission and absorption. Astrophys. J. **734**, 31 (2011)

France, K., Linsky, J.L., Tian, F., Froning, C.S. Roberge, A.: Time-resolved ultraviolet spectroscopy of the M-dwarf GJ 876 exoplanetary system. Astrophys. J. Lett. **750**, L32 (2012)

Guinan, E.F., Engle, S.G., Durbin, A.: Living with a red dwarf: rotation and X-ray and ultraviolet properties of the halo population Kapteyn's star. Astrophys. J. **821**, 81 (2016)

Herczeg, G.J., Linsky, J.L., Valenti, J.A., Johns-Krull, C.M., Wood, B.E.: The far-ultraviolet spectrum of TW Hydrae. I. Observations of H_2 fluorescence. Astrophys. J. **572**, 310 (2002)

Herczeg, G.J., Wood, B.E., Linsky, J.L., Valenti, J.A., Johns-Krull, C.M.: The far-ultraviolet spectrum of TW Hydrae. II. Models of H_2 fluorescence in a disk. Astrophys. J. **607**, 369 (2004)

King, G.W., Wheatley, P.J., Salz, M., Bourrier, V., Czesla, S., Ehrenreich, D., Kirk, J., Lecavelier des Etangs, A., Louden, T., Schmitt, J., Schneider, P.C.: The XUV environments of exoplanets from Jupiter size to super-Earth. Mon. Not. R. Astron. Soc. **478**, 1193 (2018)

Kruczek, N., France, K., Evonosky, W., Loyd, R.O.P., Youngblood, A., Roberge, A., Wittenmyer, R.A., Stocke, J.T., Fleming, B., Hoadley, K.: H_2 fluorescence in M dwarf systems: a stellar origin. Astrophys. J. **845**, 3 (2017)

Lecavelier des Etangs, A.: A diagram to determine the evaporation status of extrasolar planets. Astron. Astrophys. **461**, 1185 (2007)

Lemaire, P., Vial, J.-C., Curdt, W., Schühle, U., Wilhelm, K.: Hydrogen Ly-α and Ly-β full Sun line profiles observed with SUMER/SOHO (1996–2009). Astron. Astrophys. **581**, A26 (2015)

Linsky, J.L.: Stellar model chromospheres and spectroscopic diagnostics. Ann. Rev. Astron. Astrophys. **55**, 159 (2017)

Linsky, J.L., Brown, A., Gayley, K., Diplas, A., Savage, B.D., Ayres, T.R., Landsman, W., Shore, S.N., Heap, S.R.: Goddard high-resolution spectrograph observations of the local interstellar medium and the deuterium/hydrogen ratio along the line of sight toward Capella. Astrophys. J. **402**, 694 (1993)

Linsky, J.L., Fontenla, J., France, K.: The intrinsic extreme ultraviolet fluxes of F5 V to M5 V stars. Astrophys. J. **780**, 61 (2014)

Louden, T., Wheatley, P.J., Briggs, K.: Reconstructing the high-energy irradiation of the evaporating hot Jupiter HD 209458b. Mon. Not. R. Astron. Soc. **464**, 2396 (2017)

Milligan, R.O., Kerr, G.S., Dennis, B.R., Hudson, H.S., Fletcher, L., Allred, J.C., Chamberlin, P.C., Ireland, J., Mathioudakis, M., Keenan, F.P.: The radiated energy budget of chromospheric plasma in a major solar flare deduced from multi-wavelength observations. Astrophys. J. **793**, 70 (2014)

Monsignori Fossi, B.C., Landini, M., Del Zanna, G., Bowyer, S.: A time-resolved extreme-ultraviolet spectroscopic study of the quiescent and flaring corona of the flare star AU Microscopii. Astrophys. J. **466**, 427 (1996)

Peacock, S., Barman, T., Shkolnik, E.L., Hauschildt, P.H., Baron, E.: Predicting the extreme ultraviolet environment of exoplanets around low-mass stars: the TRAPPIST-1 system. submitted to Astrophys. J. **871**, 235 (2019)

Pillitteri, I., Wolk, S.J., Sciortino, S., Antoci, V.: No X-rays from WASP-18. Implications for its age, activity, and the influence of its massive hot Jupiter. Astron. Astrophys. **567**, 128 (2014)

Piskunov, N., Wood, B.E., Linsky, J.L., Dempsey, R.C., Ayres, T.R.: Local interstellar medium properties and deuterium abundances for the lines of sight toward HR 1099, 31 Comae, β Ceti, and β Cassiopeiae. Astrophys. J. **474**, 315 (1997)

Pounds, K.A., Allan, D.J., Barber, C., Barstow, M.A., Bertram, D., Branduardi-Raymont, G., Brebner, G.E.C., Buckley, D., Bromage, G.E., et al.: The ROSAT Wide Field Camera all-sky survey of extreme-ultraviolet sources. I - The Bright Source Catalogue. Mon. Not. R. Astron. Soc. **260**, 77 (1993)

Redfield, S., Linsky, J.L.: The structure of the local interstellar medium. IV. Dynamics, morphology, physical properties, and implications of cloud-cloud interactions. Astrophys. J. **673**, 283 (2008)

Ribas, I., Guinan, E.F., Güdel, M., Audard, M.: Evolution of the solar activity over time and effects on planetary atmospheres. I. High-energy irradiances (1–1700 Å). Astrophys. J. **622**, 680 (2005)

Sanz-Forcada, J., Micela, G., Ribas, I., Pollock, A.M.T., Eiroa, C., Velasco, A., Solano, E., Garcia-Alvarez, D.: Estimation of the XUV radiation onto close planets and their evaporation. Astron. Astrophys. **532**, A6 (2011)

Schindhelm, E., France, K., Herczeg, G.J., Bergin, E., Yang, H., Brown, A., Brown, J.M., Linsky, J.L., Valenti, J.: Lyα dominance of the classical T Tauri far-ultraviolet radiation field. Astrophys. J. Lett. **756**, L23 (2012)

Vernazza, J.E., Avrett, E.H., Loeser, R.: Structure of the solar chromosphere. Basic computations and summary of the results. Astrophys. J. **184**, 605 (1973)

Vernazza, J.E., Avrett, E.H., Loeser, R.: Structure of the solar chromosphere. II - the underlying photosphere and temperature-minimum region. Astrophys. J. Suppl. **30**, 1 (1976)

Vernazza, J.E., Avrett, E.H., Loeser, R.: Structure of the solar chromosphere. III - models of the EUV brightness components of the quiet-sun. Astrophys. J. Suppl. **45**, 635 (1981)

Wood, B.E., Brown, A., Linsky, J.L., Kellett, B.J., Bromage, G.E., Hodgkin, S.T., Pye, J.P.: A volume-limited ROSAT survey of extreme ultraviolet emission from all nondegenerate stars within 10 parsecs. Astrophys. J. Suppl. Ser. **93**, 287 (1994)

Wood, B.E., Linsky, J.L., Hébrard, G., Williger, G.M., Moos, H.W., Blair, W.P.: Two new low galactic D/H measurements from the far ultraviolet spectroscopic explorer. Astrophys. J. **609**, 838 (2004)

Wood, B.E., Redfield, S., Linsky, J.L., Müller, H.-R., Zank, G.P.: Stellar Lyα lines in the Hubble Space Telescope archive: intrinsic line fluxes and absorption from the heliosphere and astrospheres. Astrophys. J. Suppl. Ser. **159**, 118 (2005)

Wood, B.E., Laming, J.M., Warren, H.P., Poppenhaeger, K.: A Chandra/LETGS survey of main-sequence stars. Astrophys. J. **862**, 66 (2018)

Woods, T.N., Chamberlin, P.C., Harder, J.W., Hock, R.A., Snow, M., Eparvier, F.G., Fontenla, J., McClintock, W.E., Richard, E.C.: Solar irradiance reference spectra (SIRS) for the 2008 whole heliosphere interval (WHI). J. Geophys. Res. Lett. **36**, 1101 (2009)

Woods, T.N., Eparvier, F.G., Hock, R., Jones, A.R., Woodraska, D., Judge, D., Didkovsky, L., Lean, J., Mariska, J., Warren, H., McMullin, D., Chamberlin, P., Berthiaume, G., Bailey, S., Fuller-Rowell, T., Sojka, J., Tobiska, W. K., Viereck, R.: Extreme ultraviolet variability experiment (EVE) on the solar dynamics observatory (SDO): overview of science objectives, instrument design, data products, and model developments. Solar Phys. **275**, 115 (2012)

Yang, H., Linsky, J.L., France, K.: HST/COS spectra of DF Tau and V4046 Sgr: first detection of molecular hydrogen absorption against the Lyα line. Astrophys. J. Lett. **730**, L10 (2011)

Youngblood, A., France, K., Loyd, R.O.P., Linsky, J.L., Redfield, S., Schneider, P.C., Wood, B.E., Brown, A., Froning, C., Miguel, Y., Rugheimer, S., Walkowicz, L.: The MUSCLES Treasury Survey II: intrinsic Lyman alpha and extreme ultraviolet spectra of K and M dwarfs with exoplanets. Astrophys. J. **824**, 101 (2016)

Chapter 7
Panchromatic Spectra of Exoplanet Host Stars

The spectral energy distribution (SED) emitted by a host star's photosphere, chromosphere, and corona is essential input for models of the chemistry and mass loss from exoplanet atmospheres. This chapter describes typical SEDs of different types of stars including spectral line and continuum radiation from the X-ray region to the infrared.

7.1 Spectral Energy Distribution of the Sun and Other Variable G-Type Stars

The solar irradiance spectrum describing the Sun viewed as a point source like other stars is the standard against which other G-type and cooler stars are compared. However, the solar irradiance varies over many time scales: minutes to hours as active regions evolve, the 25-day rotation period as active regions rotate in and out of sight, the 11 year sunspot/activity cycle, and infrequent but longer duration changes such as the Maunder minimum discussed below. Woods and Rottman (2002) obtained high-resolution spectral solar irradiance (SSI) measurements from X-rays to the infrared at activity maximum and minimum during sunspot cycle 22 with the *Upper Atmosphere Research Satellite (UARS)*. At the March–April 2008 solar minimum, Woods et al. (2009) obtained the Solar Irradiance Reference Spectrum between 0.1 and 2400 nm using spectrometers on several spacecraft and a rocket. The differences in the total solar irradiance (TSI) and the spectral solar irradiance (SSI) between sunspot maximum and minimum is illustrated in Fig. 7.1. The TSI, referred to as bolometric in the figure, increases by about 0.1% between minimum and maximum as the increased number of bright plages dominates over the decreased emission of dark sunspots. The large percentage change in SSI is primarily at wavelengths below 300 nm and the energy change is primarily in the 200–700 nm region.

© Springer Nature Switzerland AG 2019
J. Linsky, *Host Stars and their Effects on Exoplanet Atmospheres*,
Lecture Notes in Physics 955, https://doi.org/10.1007/978-3-030-11452-7_7

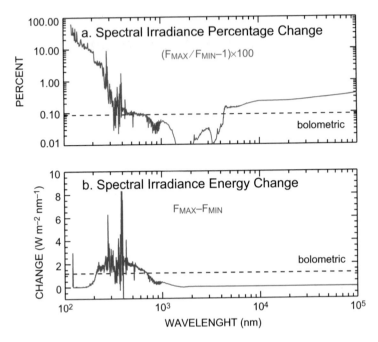

Fig. 7.1 *Top panel:* Spectral irradiance percentage change between solar minimum and maximum. The dashed line is the total solar irradiance (TSI) increase (about 0.1%) between solar minimum and maximum. The solid red line is the percent change in the spectral solar irradiance (SSI), which increases rapidly in the UV at wavelengths below 300 nm. The percent change is very small at optical wavelengths. *Bottom panel:* The dashed line marks the increase in TSI between solar minimum and maximum. The blue line marks the energy change in SSI, which is primarily in the 200–700 nm spectral region and Lyman-α. The figure does not include X-ray data. Figure from Fröhlich and Lean (2004)

After the first accurate satellite irradiance experiment launched on *NIMBUS-7* in 1978, there has a continuous record of both TSI and SSI data, which are critical input for studying climate variations on Earth. In particular, there have been searches in the solar record for periods of TSI and SSI fluxes much lower than typical solar minima, often called grand minima, which may force changes in the terrestrial climate. An example is the Maunder minimum period (1645–1715), when very few sunspots were detected and the terrestrial climate was relatively cool. Reconstructions of the TSI and SSI fluxes back in time 11,000 years to the beginning of the current warm period on Earth called the Holocene era are available based on the correlations of the TSI and SSI fluxes with activity proxies. For example, Shapiro et al. (2011) reconstructed the TSI using ice core ^{10}Be abundances that measure the shielding of cosmic rays by the heliosphere. They found that during the Maunder minimum period the TSI was 6 ± 3 W m^{-2} below the 1996 quiet Sun reference year value of 1365.5 W m^{-2}. During the last 11,000 years, they found reconstructed TSI values as low as 1356 W m^{-2} and as high as 1370 W m^{-2}, a range of -0.70%

to +0.33%. In a review of reconstructions of the solar TSI record using as proxies [10]Be and ^{14}C, Usoskin (2013) identified 27 grand minima in the Holocene era with mean durations of 70 years and the Sun relatively faint for 17% of the time. He also listed 19 grand maxima episodes (TSI values well above typical solar maxima) in the Holocene era with the Sun relatively bright 10% of the time. The Sun may presently be in a grand maxima phase. In the ultraviolet, SSI values show far larger changes than in the optical or the TSI. Figure 7.2 shows that during the Maunder minimum period, the solar 175–200 nm flux was 26.6% smaller than the 1996 quiet Sun value, whereas the 500–600 nm continuum was only 0.4% smaller.

In their search for low ultraviolet fluxes in solar-like stars, Lubin et al. (2018) found a 125–191 nm flux 23.0 ± 5.7% below the quiet Sun value for τ Cet, which is thought to be in a grand minimum state perhaps analogous to the Sun's Maunder minimum state. In their *IUE* data set of 33 dwarf stars with T_{eff} within 500 K of the solar value, no other stars showed fluxes that far below typical quiet Sun values. Searches for grand minima in distant solar-like stars with low resolution Ca II spectra is complicated by the presence of interstellar Ca II absorption that suggests low Ca II emission (Pace and Pasquini 2004). After correcting for interstellar absorption, Curtis (2017) found no Maunder minimum candidates among the solar-type stars in the 3.5–4.2 Gyr M67 cluster.

7.2 Spectral Energy Distributions of K and M Stars

Table 7.1[1] lists the spacecraft and instruments that have provided the input data for the spectral energy distributions of stars. The parameters are approximate values that may not be achieved in typical observations. The listed instruments are the *International Ultraviolet Explorer (IUE)* (Boggess et al. 1978), the *Far Ultraviolet Spectroscopic Explorer (FUSE)* (Moos et al. 2000; Sahnow et al. 2000), the *Galaxy Evolution Explorer (GALEX)* (Martin et al. 2005), the Goddard High Resolution Spectrometer (GHRS) (Brandt et al. 1994; Heap et al. 1995), the Space Telescope Imaging Spectrograph (STIS) (Kimble et al. 1998), the Advanced Camera for Surveys (ACS) (Walkowicz et al. 2008), and the Cosmic Origins Spectrograph (COS) (Green et al. 2012). The latter four instruments are on the *HST* spacecraft.

IUE obtained spectra of many bright K and M stars and the analysis of these data provided many in sights concerning stellar winds, chromospheres, and abundances. However, the quality of these data especially for faint stars was limited by *IUE's* sensitivity, instrumental scattered light, and modest dynamic range. Also, geocoronal emissions in the Lyman-α and O I lines made it difficult to extract reliable fluxes for these lines (cf. France et al. 2016). Despite these limitations, there were a few opportunities where *IUE* obtained UV spectra with coordinated observations at other wavelengths. For example, Hawley and Pettersen (1991)

---

[1]This table was originally published in Linsky (2014).

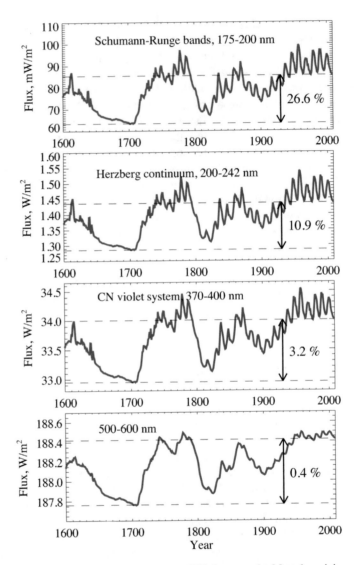

Fig. 7.2 Reconstructed spectral solar irradiance (SSI) between the Maunder minimum (1645–1715) and the 1996 solar minimum value for different wavelength bands. The minimum SSI and 1996 quiet Sun values are indicated by dashed lines. Figure from Shapiro et al. (2011). Reproduced with permission of ESO

obtained the first SED of an M star flare by observing AD Leo (dM3.5e) during an especially large flare with low-dispersion *IUE* spectra and simultaneous ground-based optical spectra. The UV emission shown in Fig. 7.3 is mostly continuum with emission lines superimposed. The results of this pioneering study will be discussed in Sect. 10.1. Other multiwavelength studies of flare stars using *IUE*

Table 7.1 Spectroscopic properties of ultraviolet instruments

Spacecraft	Timeline	Instrument mode	Spectral range (nm)	Resolving power ($\Delta\lambda/\lambda$)
IUE	1/1978–9/1996	SW-HI	115–200	∼13,000
		LW-HI	185–330	∼17,000
		SW-LO	115–200	∼280
		LW-LO	185–330	∼430
FUSE	2/1999–10/2008	SiC	91–110	17,500
		LiF	98–119	23,000
GALEX	4/2003–present	FUV	135–175	200
		NUV	175–275	90
HST/GHRS	4/1990–2/1997	Ech-A	105–173	80,000–96,000
		Ech-B	168–321	71,000–93,000
		G140M	105–173	19,000–30,000
		G160M	115–210	16,000–27,000
		G200M	160–230	20,000–32,000
		G270M	200–330	19,000–36,000
		G140L	109–190	1700–3000
HST/STIS	2/1997–8/2004	E140H	115–170	99,300–114,000
	5/2009–present	E230H	165–310	92,300–110,900
		E140M	115–170	46,000
		E230M	165–310	29,900–32,200
		G140L	115–170	950–1400
		G230L	165–310	500–960
HST/ACS	3/2002–present	PR200L	180–500	4–170
HST/COS	5/2009–present	G130M	90–125	16,000–21,000
		G160M	141–178	16,000–21,000
		G185M	170–210	22,000–28,000
		G225M	210–250	28,000–38,000
		G285M	250–320	30,000–41,000
		G140L	90–205	1500–4000
		G230L	170–320	2100–3900

include the observing campaign for Proxima Centauri (dM5e) with simultaneous X-ray observations by Haisch et al. (1983) and the YZ CMi (M4.0 Ve) observing campaign with simultaneous radio and optical data by van den Oord et al. (1996).

The STIS spectrum of AU Mic (M0 V) (see Fig. 4.4) demonstrates the ability of *HST* to obtain high-resolution spectra of bright M stars outside of flares. The inclusion of the COS instrument on *HST* with its very high throughput and low background made it feasible to observe fainter M stars, in particular interesting exoplanet host stars. With the versatile STIS and COS instruments, *HST* became the centerpiece for panchromatic observing programs involving observations with other satellites and ground-based optical, infrared, and radio observations.

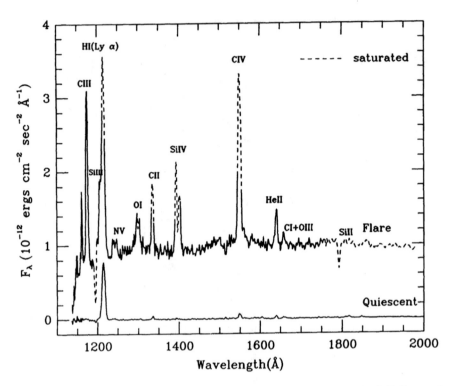

Fig. 7.3 The *IUE* spectrum of AD Leo observed during the great flare of 1985 April 12. Saturated emission lines are dashed and the quiescent spectrum is shown for comparison. Figure from Hawley and Pettersen (1991). Reproduced by permission of the AAS

The objective of the MUSCLES Treasury Survey (France et al. 2016) was to obtain complete SEDs from X-rays to the infrared of a representative sample of K and M exoplanet host stars. To the extent feasible, the data from a variety of observatories were obtained either simultaneous or within a short time period to minimize the effects of stellar variability that could frustrate the goal of obtaining intercomparable data. Figure 7.4 shows the various data sets that were compiled to form the SEDs of one of these 11 stars (Loyd et al. 2016). The X-ray data were obtained either from the *XMM-Newton* or *Chandra* Observatories. In the 7–10 nm region, where the X-ray observatories are less sensitive, the spectra were computed using the APEC plasma model code. Youngblood et al. (2016) computed the EUV spectrum using the prescriptions in Linsky et al. (2014) for the flux in 10 nm bandpasses as a function of the Lyman-α flux. *HST* observed the FUV and NUV spectra and the Lyman-α line was reconstructed from STIS spectra as described by Youngblood et al. (2016). The optical and near-IR spectra were computed using the PHOENIX atmospheric models (Husser et al. 2013) for the stellar effective temperatures. The SED for GJ 832 is shown in Fig. 7.5. The MUSCLES program data are available at https://archive.stsci.edu/prepds/muscles/. A successor program

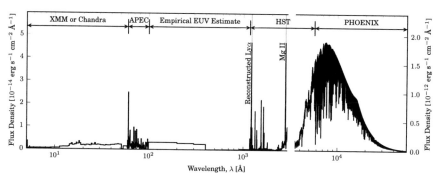

Fig. 7.4 The data sources for each spectral region used in compiling the SED of GJ 832. Spectral regions Figure from Loyd et al. (2016). Reproduced by permission of the AAS

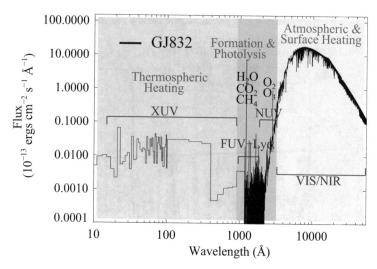

Fig. 7.5 The spectral energy distribution of the M1.5 V star GJ 832. The spectrum is a combination of near simultaneous *Chandra* X-ray spectra, *HST* UV spectra, ground-based optical and infrared spectra, and reconstructed Lyman-α and extreme ultraviolet flux. Figure from France et al. (2016). Reproduced by permission of the AAS

Mega-MUSCLES (Froning et al. 2018) will be obtaining SED's of 13 additional M dwarf host stars, and there are other programs underway to study other host stars and their time variability.

The wavelength at which chromospheric emission exceeds that from the photosphere depends on the star's effective temperature and activity. Figure 7.6 shows that for the moderately active M1.5 V star GJ 832 emission from the chromosphere becomes significant compared to the PHOENIX model photosphere near 2400 Å. At shorter wavelengths the photospheric emission decreases rapidly while emission from the chromosphere is nearly constant with wavelength. With increasing stellar

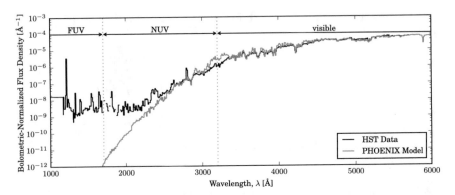

Fig. 7.6 Comparison of the observed spectrum of GJ 832 with a PHOENIX model photosphere. The increase in observed flux at wavelengths below 2400 Å is due to emission from the stellar chromosphere. Figure from Loyd et al. (2016). Reproduced by permission of the AAS

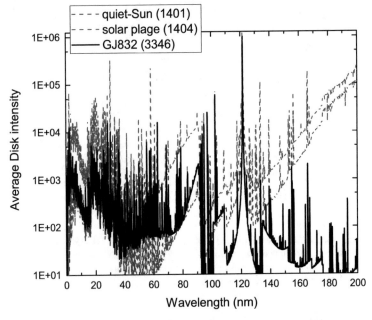

Fig. 7.7 Comparison of the synthetic spectra of the quiet Sun, a solar plage, and the M1.5 V star GJ 832. Spectra from Fontenla et al. (2016). Reproduced by permission of the AAS

effective temperature, this transition occurs at shorter wavelengths and with increasing activity it occurs at longer wavelengths.

The SEDs of M stars are qualitatively different from G-type stars and especially low activity G stars like the Sun. Figure 7.7 compares the average disk intensities, that is the flux per unit surface area, of GJ 832 with the quiet Sun and a solar plage (active Sun). The NUV intensity of the M star is 3–4 orders of magnitude

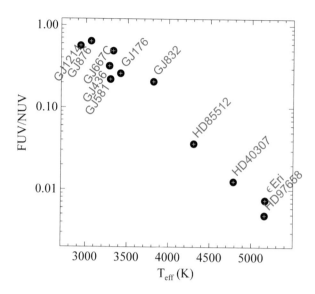

Fig. 7.8 Ratio of the 91.2–170 nm flux to the 170–320 nm flux for the stars observed in the MUSCLES Treasury Survey. Figure from France et al. (2016). Reproduced by permission of the AAS

fainter than for the Sun because the upper photosphere of an M star is cooler than for a G star and the photospheric emission is very temperature sensitive. Once the chromospheric line and continuum emission dominates the average disk intensity, the GJ 832 intensity is only one order of magnitude smaller than for the quiet Sun. In the EUV and X-ray bands the average disk intensities of GJ 832 and the quiet Sun are about the same.

The FUV/NUV flux ratio declines continuously from the late-M stars to K stars as shown in Fig. 7.8 (France et al. 2016), because the NUV emission from the photosphere increases rapidly with stellar effective temperature. The corresponding ratio for G stars is typically 10^{-3}. Plotting the same data as a function of habitable zone distance (see Fig. 7.9), one can see that the SEDs seen by exoplanets of M stars are very different from the SEDs seen from G-type stars. This change in SED shape plays an important role in the oxygen chemistry in the upper atmospheres of exoplanets as described in Chap. 12. Figure 7.9 also shows that the mean XUV flux received by exoplanets in the habitable zone is about the same for G and M stars and the mean FUV flux increases about a factor of 3 from M to G stars. Broadband photometry of a large number of stars observed by the *GALEX* satellite (e.g., Smith et al. 2014) shows that the FUV (134–177 nm) − NUV (177–283 nm) flux difference exceeds the predictions of photospheric models from G to M stars as expected because of chromospheric emission.

Fig. 7.9 *Top panel* (**a**): Stellar FUV/NUV flux ratio at the location of exoplanets in their habitable zone. M stars are on the left side of the figure. *Middle panel* (**b**): Stellar FUV flux in the habitable zones. *Bottom panel* (**c**): Stellar XUV flux in the habitable zones. Figure from France et al. (2016). Reproduced by permission of the AAS

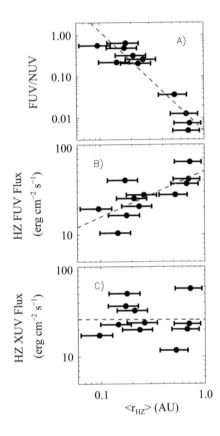

7.3   Stellar X-ray and EUV Spectra and Photometry

Table 7.2 lists the parameters of the imaging and spectroscopic instruments on X-ray and EUV satellites. The parameters are approximate values that may not be achieved in typical observations. The imaging instruments, for example the IPC and HRI instruments on *Einstein* and the PSPC and HRI instruments on *ROSAT*, have broadband sensitivity with very limited energy resolution, often only the ability to separate high energy from low energy emission. The early spectrometers, for example the FPCS and OGS on *Einstein*, had limited sensitivity. The Solid State Spectrometer (SSS) on *Einstein* provided important low-resolution spectra of many stellar sources. The Low Energy Transmission Grating (cf. Wood et al. 2018) and High Energy Transmission Grating (LETGS and HETGS) spectrometers on *Chandra* and the EPIC and RGS spectrometers on *XMM-Newton* are the prime instruments for stellar X-ray spectroscopy. Nevertheless, even the $R = 2000$ resolution at the long wavelength end of the LETGS, cannot provide accurate spectral line widths and Doppler shifts that are feasible with UV instruments. Proposed instruments may provide resolution up to 10,000 at X-ray wavelengths.

Table 7.2 Imaging and spectroscopic parameters of X-ray instruments

Spacecraft	Timeline	Instrument mode	Energy range (keV)	Resolving power $(\Delta\lambda/\lambda)$
ANS	8/1974–6/1976	SXX	1–7	
HEAO-1	8/1977–1/1979	LASS	0.25–3.0	
		CXE	0.15–60	
		MC	0.9–13.3	
Einstein	11/1978–4/1981	IPC	0.4–4	
		HRI	0.15–3.0	
		SSS	0.5–4.5	3–25
		FPCS	0.42–2.6	50–1000
		MPC	1.5–20	5
		OGS	0.15–3.0	50
EXOSAT	5/1983–5/1986	CMA	0.05–2	
		PSD		
		TGS		
		ME	1–50	
		GS	2–20	
ROSAT	6/1990–10/2011	PSPC	0.1–2.4	
		HRI		
		WFC	0.042–0.21	
EUVE	9/1992–1/2001	Imager	0.056–0.56	
		GS	0.056–0.56	≈ 200
ASCA	2/1993–3/2001	GIS	0.8–12	
		SIS	0.4–12	≤ 50
Chandra	7/1999–present	LETGS	0.071–2.5	40–2000
		HETGS	0.4–10	60–1000
		ACIS		
		HRC		
XMM-Newton	12/1999–present	EPIC	0.15–15	20–50
		RGS	0.35–2.48	200–700
		PN	0.15–15	20–50
		OM	0.019–0.069	

The spectrometer on the *Extreme Ultraviolet Explorer (EUVE)* (spectral range 7–76 nm) and the LETGS on *Chandra* (spectral range 0.5–17.5 nm) provide the only capability for spectroscopy at EUV wavelengths (see Sect. 6.2.1).

As described in Chap. 5 there are two comprehensive reviews of stellar X-ray astronomy (Güdel 2004; Güdel and Nazé 2009) that present many examples of stellar X-ray spectra. I list in Table 7.2 the X-ray instruments that have and are now providing broad band fluxes and spectra at X-ray and EUV wavelengths. These satellites have three types of instruments: imagers with very little energy resolution, spectrometers including gratings and crystals or solid state detectors with modest energy resolution, and the Optical Monitor (OM) on *XMM-Newton* to obtain UV and optical fluxes simultaneous with the X-ray and EUV observations.

Despite their low sensitivity and energy resolution, the first X-ray missions identified several of the brightest stellar X-ray sources. For example, Capella (G8 III+G1 III) was detected first by a rocket experiment (Catura et al. 1975) and then confirmed with the *ANS* satellite by Mewe et al. (1975). Subsequent detections of M dwarfs and RS CVn binary systems with *ANS* and *HEAO-1* showed that stars can have X-ray emitting coronae with X-ray luminosities orders of magnitude larger than the Sun. The increased sensitivity of the IPC and HRI imagers compared to previous missions and the modest resolution of the solid state spectrometer (SSS) on the *Einstein* satellite detected a wide variety of stars across the Hertzsprung-Russell diagram (e.g., Vaiana et al. 1981) bringing X-ray stellar observations into the mainstream of modern astrophysics. The all-sky survey by the *ROSAT* telescope provided a large number of detections that form the basis of many surveys and statistical studies of stellar X-ray properties long after the end of the mission in 2011. For example, the survey of X-ray and UV emission from M dwarfs based on *ROSAT* and more recent satellite data (Stelzer et al. 2013). Robrade and Schmitt (2005) survey the emission of active M dwarfs using the MOS, PN and RGS instruments on *XMM-Newton*.

The 0.5 arcsec imaging capability of *Chandra* allows spatial resolution of close binaries and reliable identification of sources in crowded fields, for example the identification of more than 1600 pre-main sequence stars in the Orion Molecular Cloud (Getman et al. 2005) and the identification of 1613 sources in the young cluster NGC 6231 (Damiani et al. 2016). In their survey of the Taurus Molecular Cloud, Güdel et al. (2007) identified 136 pre-main sequence stars using the EPIC, RGS, and OM instruments on *XMM-Newton*.

The spectral resolution of the LETGS and HETGS instruments on *Chandra* and the RGS on *XMM-Newton* support the construction of differential emission measure distributions $Q(T)$ described Sect. 5.3 and studies of coronal abundances based on emission lines from ions formed over a wide range of temperatures. Examples of such studies are the survey of solar analog stars with the RGS (Telleschi et al. 2005) and the $Q(T)$ analysis of 19 stars with spectral types F1 V to M4.5 V by Wood et al. (2018) using the LETGS.

7.4 Stellar UV, Optical, and Infrared Spectra

There are a number of libraries of flux-calibrated stellar spectra covering ultraviolet and longer wavelengths for a wide range of stars. For example, the Stellar Spectral Flux Library compiled by Pickles (1998) includes flux-calibrated spectra of 131 stars of all spectral types and luminosity classes with wavelengths between 115 nm and 2.5 μm. The UV fluxes are from the *IUE* satellite. The complete archive of stellar observations with the *IUE, EUVE, FUSE, GALEX*, and *HST* is available at the Mikulski Archive for Space Telescopes (MAST) at http://archive.stsci.edu.

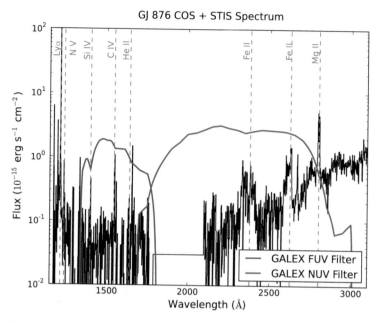

Fig. 7.10 The red and blue solid lines are the relative throughputs of the two *GALEX* passbands. The black line shows the spectrum of GJ 876 (M4 V) observed with COS by France et al. (2013). Prominent emission lines are marked. Figure from Jones and West (2016). Reproduced by permission of the AAS

To achieve the highest quality spectra, Ayres (2010) created the StarCAT[2] catalog of postprocessed STIS echelle spectra to achieve photometric accuracy of about 4% and absolute velocity accuracy of $1 \mathrm{km} \ \mathrm{s}^{-1}$. StarCAT contains FUV and NUV echelle spectra of 545 stars observed by STIS between 1997 and 2004. The Advanced Spectral Library (ASTRAL) contains accurately processed high-resolution spectra of eight stars in the spectral range F5 IV-V (Procyon) to M3.5 III (γ Cru) observed by STIS (see Ayres 2013). These data are available at http://casa.colorado.edu/~ayres/ASTRAL/.

The *Galaxy Evolution Explorer(GALEX)* surveyed two-thirds of the sky in two broadbands: FUV (135–175 nm) and NUV (175–250 nm). The overlap of these filters in Fig. 7.10 shows that the FUV filter measures primarily emission lines formed in the chromosphere and transition region, whereas the NUV band samples primarily the photospheric continuum and chromospheric emission lines of Mg II and Fe II. The FUV/NUV flux ratio thereby samples emission in the upper atmosphere but not the corona. Shkolnik and Barman (2014) extracted from the *GALEX* archive M1–M4 dwarf stars within about 50 pc of the Sun with a range of ages to study the decline in FUV and NUV flux with stellar age (cf. Jones and West 2016).

---

[2]http://casa.colorado.edu/~ayres/StarCAT/.

References

Ayres, T.R.: StarCAT: a catalog of space telescope imaging spectrograph ultraviolet echelle spectra of stars. Astrophys. J. Suppl. Ser. **187**, 149 (2010)

Ayres, T.R.: Advanced spectral library (ASTRAL): cool stars edition. Astron. Nachr. **334**, 105 (2013)

Boggess, A., Bohlin, R.C., Evans, D.C., Freeman, H.R., Gull, T.R., Heap, S.R., Klinglesmith, D.A., Longanecker, G.R., Sparks, W., West, D.K.: In-flight performance of the IUE. Nature **275**, 377 (1978)

Brandt, J.C., Heap, S.R., Beaver, E.A., Boggess, A., Carpenter, K.G., Ebbets, D.C., Hutchings, J.B., Jura, M., Leckrone, D.S., Linsky, J.L., et al.: The Goddard high resolution spectrograph: instrument, goals, and science results. Publ. Astron. Soc. Pac. **106**, 890 (1994)

Catura, R.C., Acton, L.W., Johnson, H.M.: Evidence for X-ray emission from Capella. Astrophys. J. Lett. **196**, L47 (1975)

Curtis, J.L.: No Maunder minimum candidates in M67: mitigating interstellar contamination of chromospheric emission lines. Astron. J. **153**, 275 (2017)

Damiani, F., Micela, G., Sciortino, S.: A Chandra X-ray study of the young star cluster NGC 6231: low-mass population and initial mass function. Astron. Astrophys. **596**, A82 (2016)

Fontenla, J., Linsky, J.L., Witbrod, J., France, K., Buccino, A., Mauas, P., Vieytes, M., Walkowicz, L.: Semi-empirical modeling of the photosphere, chromosphere, transition region, and corona of the M-dwarfs host star GJ 832. Astrophys. J. **830**, 154 (2016)

France, K., Froning, C.S., Linsky, J.L., Roberge, A., Stocke, J.T., Tian, F., Bushinsky, R., Désert, J.-M., Mauas, P., Vietes, M., Walkowicz, L.: The ultraviolet radiation environment around M dwarf exoplanet host stars. Astrophys. J. **763**, 149 (2013)

France, K., Loyd, R.O.P., Youngblood, A., Brown, A., Schneider, P.C., Hawley, S.L., Froning, C.S., Linsky, J.L., Roberge, A., et al.: The MUSCLES Treasury Survey I: motivation and overview. Astrophys. J. **820**, 89 (2016)

Fröhlich, C., Lean, J.: Solar radiative output and its variability: evidence and mechanisms. Astron. Astrophys. Rev. **12**, 273 (2004)

Froning, C.S., France, K., Loyd, R.O.P., Youngblood, A., Brown, A., Schneider, C., Berta-Thompson, Z., Kowalski, A.: The mega-MUSCLES HST Treasury Survey. AAS Meeting #231, id. 111.05 (2018)

Getman, K.V., Flaccomio, E., Broos, P.S., Grosso, N., Tsujimoto, M., Townsley, L., Garmire, G.P., Kastner, J., Li, J., et al.: Chandra Orion ultradeep project: observations and source lists. Astrophys. J. Suppl. Ser. **160**, 319 (2005)

Green, J.C., Froning, C.S., Osterman, S., Ebbets, D., Heap, S.H., Leitherer, C., Linsky, J.L., Savage, B.D., Sembach, K., Shull, J.M., et al.: The cosmic origins spectrograph. Astrophys. J. **744**, 60 (2012)

Güdel, M.: X-ray astronomy of stellar coronae. Astron. Astrophys. Rev. **12**, 71 (2004)

Güdel, M., Nazé, Y.: X-ray spectroscopy of stars. Astron. Astrophys. Rev. **17**, 309 (2009)

Güdel, M., Briggs, K.R., Arzner, K., Audard, M., Bouvier, J., Feigelson, E.D., Franciosini, E., Glauser, A., Grosso, N., Micela, G., et al.: The XMM-Newton extended survey of the Taurus molecular cloud (XEST). Astron. Astrophys. **468**, 353 (2007)

Haisch, B.M., Linsky, J.L., Bornmann, P.L., Stencel, R.E., Antiochos, S.K., Golub, L., Vaiana, G.S.: Coordinated Einstein and IUE observations of a disparitions brusques type flare event and quiescent emission from Proxima Centauri. Astrophys. J. **267**, 280 (1983)

Hawley, S.L., Pettersen, B.R.: The great flare of 1985 April 12 on AD Leonis. Astrophys. J. **378**, 725 (1991)

Heap, S.R., Brandt, J.C., Randall, C.E., Carpenter, K.G., Leckrone, D.S., Maran, S.P., Smith, A.M., Beaver, E.A., Boggess, A., Ebbets, D.C., et al.: The Goddard high resolution spectrograph: in-orbit performance. Publ. Astron. Soc. Pac. **107**, 871 (1995)

Husser, T.-O., Wende-von Berg, S., Dreizler, S., Homeier, D., Reiners, A., Barman, T., Hauschildt, P.H.: A new extensive library of PHOENIX stellar atmospheres and synthetic spectra. Astron. Astrophys. **553**, A6 (2013)

Jones, D.O., West, A.A.: A catalog of GALEX ultraviolet emission from spectroscopically confirmed M dwarfs. Astrophys. J. **817**, 1 (2016)

Kimble, R.A., Woodgate, B.E., Bowers, C.W., Kraemer, S.B., Kaiser, M.E., Gull, T.R., Heap, S.R., Danks, A.C., Boggess, A., Green, R.F., et al.: The on-orbit performance of the space telescope imaging spectrograph. Astrophys. J. **492**, 83 (1998)

Linsky, J.L.: The radiation environment of exoplanet atmospheres. Challenges **5**, 351 (2014)

Linsky, J.L., Fontenla, J., France, K.: The intrinsic extreme ultraviolet fluxes of F5 V to M5 V stars. Astrophys. J. **780**, 61 (2014)

Loyd, R.O.P., France, K., Youngblood, A., Schneider, C., Brown, A., Hu, R., Linsky, J., Froning, C.S., Redfield, S., Rugheimer, S., Tian, F.: The MUSCLES Treasury Survey III: X-ray to infrared spectra of 11 M and K stars hosting planets. Astrophys. J. **824**, 102 (2016)

Lubin, D., Melis, C., Tytler, D.: Ultraviolet flux decrease under a grand minimum from IUE short-wavelength observation of solar analogs. Astrophys. J. Lett. **852**, L4 (2018)

Martin, D.C., Fanson, J., Schiminovich, D., Morrissey, P., Friedman, P.G., Barlow, T.A., Conrow, T., Grange, R., Jelinsky, P.N., Milliard, B., et al.: The Galaxy Evolution Explorer: a space ultraviolet survey mission. Astrophys. J. Lett. **619**, L1 (2005)

Mewe, R., Heise, J., Gronenschild, E.H.B.M., Brinkman, A.C., Schrijver, J., den Boggende, A.J.F.: Detection of X-ray emission from stellar coronae with ANS. Astrophys. J. Lett. **202**, L67 (1975)

Moos, H.W., Cash, W.C., Cowie, L.L., Davidsen, A.F., Dupree, A.K., Feldman, P.D., Friedman, S.D., Green, J.C., Green, R.F., Gry, C., et al.: Overview of the far ultraviolet spectroscopic explorer mission. Astrophys. J. Lett. **538**, L1 (2000)

Pace, G., Pasquini, L.: The age-activity-rotation relationship in solar-type stars. Astron. Astrophys. **426**, 1021 (2004)

Pickles, A.J.: A stellar spectral flux library: 1150–25000 Å. Publ. Astron. Soc. Pac. **110**, 863 (1998)

Robrade, J., Schmitt, J.H.M.M.: X-ray properties of active M dwarfs as observed by XMM-Newton. Astron. Astrophys. **435**, 1073 (2005)

Sahnow, D.J., Moos, H.W., Ake, T.B., Andersen, J., Andersson, B.-G., Andre, M., Artis, D., Berman, A.F., Blair, W.P., Brownsberger, K.R., et al.: On-orbit performance of the Far Ultraviolet Spectroscopic Explorer satellite. Astrophys. J. Lett. **538**, L7 (2000)

Shapiro, A.I., Schmutz, W., Rozanov, E., Schoell, M., Haberreiter, M., Shapiro, A.V., Nyeki, S.: A new approach to the long-term reconstruction of the solar irradiance leads to large historical solar forcing. Astron. Astrophys. **529**, 67 (2011)

Shkolnik, E.L., Barman, T.S.: HAZMAT. I. The evolution of far-UV and near-UV emission from early M stars. Astron. J. **148**, 64 (2014)

Smith, M.A., Bianchi, L., Shiao, B.: Interesting features in the combined GALEX and Sloan color diagrams of solar-like Galactic populations. Astron. J. **147**, 159 (2014)

Stelzer, B., Marino, A., Micela, G., López-Santiago, J., Liefke, C.: The UV and X-ray activity of the M dwarfs within 10 pc of the Sun. Mon. Not. R. Astron. Soc. **431**, 2063 (2013)

Telleschi, A., Güdel, M., Briggs, K., Audard, M., Ness, J.-U., Skinner, S.L.: Coronal evolution of the Sun in time: high-resolution X-ray spectroscopy of solar analogs with different ages. Astrophys. J. **622**, 653 (2005)

Usoskin, I.G.: A history of solar activity over millennia. Living Rev. Sol. Phys. **10**, 1 (2013)

Vaiana, G.S., Cassinelli, J.P., Fabbiano, G., Giacconi, R., Golub, L., Gorenstein, P., Haisch, B.M., Harnden Jr., F.R., Johnson, H.M., Linsky, J.L., et al.: Results from an extensive Einstein stellar survey. Astrophys. J. **245**, 163 (1981)

van den Oord, G.H.J., Doyle, J.G., Rodonò, M., Gary, D.E., Henry, G.W., Byrne, P.B., Linsky, J.L., Haisch, B.M., Pagano, I., Leto, G.: Flare energetics: analysis of a large flare on YZ Canis Minoris observed simultaneously in the ultraviolet, optical and radio. Astron. Astrophys. **310**, 908 (1996)

Walkowicz, L.M., Johns-Krull, C.M., Hawley, S.L.: Characterizing the near-UV environment of M dwarfs. Astrophys. J. **677**, 593 (2008)

Wood, B.E., Laming, J., Warren, H.P., Poppenhaeger, K.: A Chandra/LETGS survey of main-sequence stars. Astrophys. J. **862**, 66 (2018)

Woods, T., Rottman, G.: Solar ultraviolet variability over time periods of aeronomic interest. In: Mendilloi, M., Nagy, A., White, H. (eds.) Comparative Aeronomy in the Solar System. Geophysical Monograph Series, vol. 221. American Geophysical Union, Washington, D.C. (2002)

Woods, T.N., Chamberlin, P.C., Harder, J.W., Hock, R.A., Snow, M., Eparvier, F.G., Fontenla, J., McClintock, W.E., Richard, E.C.: Solar irradiance reference spectra (SIRS) for the 2008 whole heliosphere interval (WHI). Geophys. Res. Lett. **36**, L01101 (2009)

Youngblood, A., France, K., Loyd, R.O.P., Brown, A., Mason, J.P., Schneider, P.C., Tilley, M.A., Berta-Thompson, Z.K., Buccino, A., Froning, C.S., et al.: The MUSCLES Treasury Survey. IV. Scaling relations for ultraviolet, Ca II K, and energetic particle fluxes from M dwarfs. Astrophys. J. **843**, 31 (2016)

Chapter 8
Stellar Winds

Mass-loss rates of cool stars are notoriously difficult to observe.—Vidotto et al. (2014)

In addition to photons at all wavelengths, stars eject ionized and neutral particles to space. These typically magnetized winds interact with the upper layers of exoplanet atmospheres producing charge-exchange and other reactions that facilitate mass loss. In this chapter, I describe the different types of winds observed in hot and cool stars. For some star types, it is relatively easy to measure diagnostics of mass loss but in solar type stars it is more difficult. I describe the different techniques for measuring or estimating stellar mass-loss rates and the diverse models developed to explain stellar mass loss.

8.1 Stellar Winds Across the H-R Diagram

In a seminal paper, Deutsch (1956) observed 10 km s^{-1} blue-shifted absorption lines in the spectrum of the α Her triple system, and concluded that the M5 Ib supergiant is losing mass at the prodigious rate of $\dot{M} > 3 \times 10^{-8} M_\odot \text{ yr}^{-1}$. He concluded that the Doppler-shifted resonance lines of all neutral and singly-ionized atoms metal lines are formed in a highly inhomogeneous circumstellar envelope with a radius of about 1000 au and that other late-type supergiants also show high mass-loss rates. Although the physical cause for this mass loss was unknown at that time, he argued that radiation pressure was far too weak to be the cause and that magnetic fields could play a role. In a prescient statement, he called attention to the possibility of "non-adiabatic compressible hydrodynamic outflows having the character of a slowly accelerated expansion starting from the reversing layer with a very small velocity."

© Springer Nature Switzerland AG 2019
J. Linsky, *Host Stars and their Effects on Exoplanet Atmospheres*,
Lecture Notes in Physics 955, https://doi.org/10.1007/978-3-030-11452-7_8

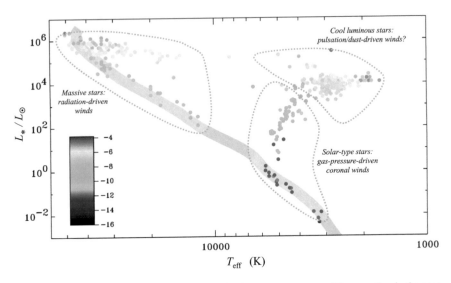

Fig. 8.1 Stellar mass loss rates color in units of solar masses per year. The grey line is the zero age main sequence. Figure courtesy of Steven Cranmer

Deutsch's discovery of high mass-loss rates in M supergiants and the likely effects of the mass loss on stellar evolution stimulated many spectroscopic searches for winds from a variety of stars leading to the discovery of winds from three types of stars (see Fig. 8.1):

Hot star winds: Stars with spectral types O and early-B, corresponding to effective temperatures in excess of 20,000 K, typically have mass-loss rates of about 10^{-7} $M_\odot$ yr^{-1}. These large mass-loss rates are measured from P Cygni-like profiles (see Fig. 8.2) and free-free continuum radio emission. Since hot stars have lifetimes of order 1 or several million years, these large mass-loss rates can effect their evolution. There is now agreement that these winds are driven by radiation pressure on multiply ionized metals (cf., Castor et al. 1975; Lamers and Cassinelli 1999).

Cool Giants and supergiants: Powerful winds of cool luminous stars and especially supergiants like Betelgeuse (α Ori) can have mass-loss rates as large as $10^{-6} M_\odot$ yr^{-1}. Relatively low velocity line shifts, P Cygni line profiles, and free-free radio emission are useful observing techniques. Luminosity class giants cooler than about spectral type K2 III have winds observable by blue-shifted absorption features in the Mg II and other chromospheric lines (Linsky and Haisch 1979; Harper et al. 1995). The acceleration mechanisms for these cool winds can be radiation pressure on dust, pulsations, and Alfvén waves, or a combination of these processes. Warm winds are also observed in the transition regions lines of warm giants and supergiants such as β Dra (G2 Ib-IIa) and α Aqr (G2 Ib) (Dupree et al. 2005). Lammers and Cassinelli (1999) provide a theoretical background for these types of winds, and Willson (2000) reviews both the theory and observing techniques.

Cool dwarf stars: Winds of main sequence stars similar to and cooler than the Sun
had not been observed prior to 1960 and were thought to be too weak to be
detected spectroscopically. In one of the few examples of theoretical prediction
preceeding observations, Parker (1958, 1960) predicted that the Sun must have a
wind on the basis that the only sensible solution to the hydrodynamic equations
describing an atmosphere with a hot lower boundary condition (the corona) and
a zero pressure upper boundary condition (vacuum in space) was a flow pattern
starting from a very low speed and passing though a critical point where the
flow speed equals the escape speed from the Sun (Fig. 8.2). Above this critical
point, the flow is supersonic increasing, asymptotically, to its terminal speed. The
figure shows the properties of the solar wind for an assumed coronal temperature
of 2×10^6 K. The location of the critical point and the mass-loss rate depend
critically on this temperature.

Solar wind properties were first systematically studied by the *Mariner 2* spacecraft on its travels to and past Venus in 1962 (Neugebauer and Snyder 1966),
although the existence of the solar wind was suspected previously from the
antisolar direction of comet tails (e.g., Biermann 1957) and the auroral emission
seen near the Earth's magnetic poles. Since the Sun has a wind based on sound
physical principles, it is safe to say that all main-sequence stars with hot coronae
like the Sun should also have winds. However, the tiny solar mass-loss rate
of 2×10^{-14} $M_\odot$ yr^{-1} (Feldman et al. 1977; Cohen 2011) provides a severe
challenge for actually detecting winds of cool dwarf stars. Another challenge is
that the solar mass-loss rate varies only weakly with magnetic activity with a
value of $2 \times 10^{-14} M_\odot$ yr^{-1} at solar minimum increasing to only 3×10^{-14} at
solar maximum (Wang 1998).

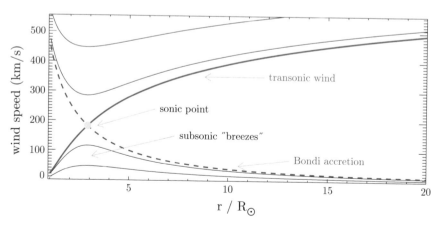

Fig. 8.2 Thermal pressure (Parker-type) winds. The transonic wind model is for an isothermal
$T = 2 \times 10^6$ K solar corona. The sonic point moves inward (outward) with higher (lower) coronal
temperatures. Figure courtesy of Steven Cranmer

Most exoplanets detected so far have host stars with effective temperatures between F stars, which are somewhat hotter than the Sun, and the late-M dwarfs, which are much cooler. The rest of this chapter concerns observations and predictions of winds in these F through M stars collectively called cool dwarf winds for which the solar wind may be a useful prototype.

I concentrate on stellar mass loss, although simulations and theoretical models of mass loss predict both mass and angular momentum loss rates at the same time. Stellar rotation evolution is a major topic in astrophysics as it also plays an important role in the dynamo generation of magnetic fields, age determination through gyrochronology (e.g., Barnes 2003), and stellar internal structure. However, this large topic is beyond the scope of this book. The study of M dwarf rotation rates by Newton et al. (2016) contains references to papers on rotation measurements and angular momentum evolution of cool stars.

Even though mass-loss rates for cool dwarfs except the Sun were unknown until recently, estimates of mass-loss rates through observations and simulations have made great progress in the last 20 years. However, our understanding of mass-loss rates remains inadequate because systematic errors can be very important both for the simulations and and the analysis of the few observations presently available.

8.2   Observations of Mass Loss in Solar-Type and Cooler Stars

Many decades of solar research with numerous satellites and theoretical studies now provide a detailed picture of the solar wind that is the likely prototype for other cool main-sequence stars. The solar wind generally has two flow regimes (e.g., McComas et al. 2007). The high-speed wind with a typical speed of 800 km s^{-1} originates in open magnetic field structures located in coronal holes, which are regions of lower coronal temperature and X-ray flux generally located near the solar poles. The low-speed solar wind with typical speed of 400 km s^{-1} originates in magnetically complex regions located near the solar equator that are bright sources of X-ray emission. The mass-loss rate of these two components is about the same because the factor of two higher speed of the wind from open field regions is compensated by about a factor of two lower density. Thus the solar mass-loss rate $\dot{M}$, which is proportional to the product of wind speed and density, does not change by a large amount with the phase of the solar magnetic cycle or the X-ray flux. The physical processes responsible for the solar wind are more complicated than for the Parker-type hydrodynamic wind because magnetic processes (Alfvèn wave pressure, MHD shocks, and magnetic field divergence with height) can play important roles in controlling $\dot{M}$ and the wind speed. The addition of magnetic energy dissipation in the subsonic regime interior to the critical point increases $\dot{M}$, while the addition of magnetic energy in the supersonic flow exterior to the critical point increases the wind speed but not $\dot{M}$. One would expect, therefore, that the critical parameters for winds of main-sequence stars would be the star's gravity and coronal temperature, which are the important parameters for hydrodynamic winds, and the coronal magnetic field strength and structure that can modify the coronal wind speed and $\dot{M}$ and the dependence of these parameters on location across the stellar surface.

How might winds of main-sequence stars be detected despite their likely very small mass-loss rates? Since it is not yet feasible to fly a robotic instrument inside of the astrosphere of a nearby star, one must use some very sensitive spectroscopic or imaging technique. There have been searches for spatially extended emission using the Lyman-α line, which have not been successful in part because of the high geocoronal background emission in the Lyman-α line (Wood et al. 2003). Other emission lines are much fainter than Lyman-α and would be less likely probes of extended emission indicative of mass outflow. An alternative technique would be to search for blue-shifted absorption or P Cygni line profiles that identify mass loss in hot stars. However, with mass-loss rates anticipated to be about six orders of magnitude smaller than for typical hot stars, these spectroscopic techniques are unlikely to be successful and such line profile shapes have not yet been detected in cool main-sequence stars. Another more sensitive spectroscopic observational technique is needed to detect mass loss from cool main sequence stars.

8.2.1 The Astrosphere Method for Estimating Stellar Mass Loss

In a seminal theoretical paper, Baranov (1990) solved the gas dynamical equations describing the interaction of outflowing protons in the solar wind with neutral hydrogen atoms in the interstellar medium that the Sun is moving through with a relative speed of 23.84 ± 0.90 km s^{-1} (see Redfield and Linsky 2008). An essential component of their interaction is charge exchange between the solar wind protons and interstellar neutral hydrogen atoms. In heliospheric models, the solar wind expands supersonically until becoming subsonic at a termination shock (TS). *Voyager 1* passed through the TS located at 94.01 au on 16 December 2004 (Stone et al. 2005), and *Voyager 2* passed through the TS at 83.7 au on 30 August 2007 (Stone et al. 2008). The different distances to the TS discovered by the two *Voyager* spacecraft traveling in different directions from the Sun highlight the asymmetric shape of the TS produced by the Sun's trajectory through the LISM. Beyond the TS near 140 au is located the heliopause, the contact surface separating the solar wind plasma from the interstellar plasma. In the heliopause, the subsonic outflowing solar wind protons charge exchange with the inflowing interstellar hydrogen atoms producing an increase the the neutral hydrogen density that peaks near 300 au according to models computed by Zank et al. (2013). Charge exchange reactions decelerate, heat, and compress hydrogen atoms forming a "hydrogen wall" around the Sun. In Zank et al. (2013) plasma models, there is either a bow shock or a bow wave depending on whether the interstellar magnetic field is weaker or stronger than about 3 μG. Near 600 au the neutral hydrogen density is predicted to reach values near 0.2 cm^{-3} typical of the LIC, which provides a sensible demarcation for the beginning of the pristine interstellar medium largely unchanged by the heliosphere.

Realistic plasma models of the interaction between the ionized solar wind and interstellar hydrogen now include charge exchange, the orientation of the Sun's motion relative to the LISM flow vector, the inflow of neutral hydrogen atoms that have not charge exchanged through kinetic calculations, and the physical processes within the multi-component plasma. The models treat the ions as a multi-fluid plasma but the neutral hydrogen atoms with their long path lengths require special kinetic modeling often by a Monte Carlo code. An important aspect of all of these calculations (e.g., Baranov and Malama 1993, Zank et al. 1996, 2013, Izmodenov et al. 2002) is the presence of the hydrogen wall surrounding the Sun. The heliospheric absorption on the red side of the Lyman-α interstellar absorption feature was first observed in HST/GHRS observations of α Cen by Linsky and Wood (1996). They determined that the properties of the absorbing hydrogen atoms in the solar hydrogen wall are, $T = 29,000 \pm 5,000$ K, $\log N(HI) = 14.74 \pm 0.24$ and a red-shifted velocity of 2–4 km s^{-1} relative to the LISM flow for this line of sight. These parameters are very different from the temperature (7500 ± 1300 K), neutral hydrogen column density in the LISM to the nearest stars ($\log N$(H I)=17.2–18.0), and upwind speed of the LISM gas (23.84 km s^{-1}) (Redfield and Linsky 2008). If the hydrogen wall gas around stars has properties similar to the solar hydrogen wall, then astrospheres (stellar analogs to the heliosphere) should be spectrally observable. The best and perhaps only way of detecting stellar hydrogen walls is by observing high resolution spectra of the strongest spectral line of the most abundant element, the Lyman-α line of atomic hydrogen.

Figure 8.3 portrays the journey of Lyman-α photons emitted in the chomosphere of a star that first pass through a star's hydrogen wall, then the interstellar medium, and finally the Sun's hydrogen wall before being observed by HST. Using the Sun as a guide, the Lyman-α photons emitted in the star's chromosphere are assumed to be distributed in wavelength as a broad emission line centered at the star's radial velocity with a small self-reversal near line center. Hydrogen atoms in the star's hydrogen wall absorb the core of the original emission line centered at the velocity of the hydrogen wall gas. As seen from the star, this velocity is slower and thus red-shifted compared to the interstellar flow velocity because of the charge-exchange reactions. As seen from an external perspective by *HST*, the absorption in the star's hydrogen wall is at a faster speed and thus blue-shifted relative to the interstellar flow velocity. Absortion by interstellar hydrogen (and deuterium) then produces a wider absorption feature centered on the interstellar flow velocity. Finally, there is absorption by gas in the solar hydrogen wall which is slowed down and therefore redshifted relative to the interstellar flow velocity. The effect of these three absorptions is an observed Lyman-α profile in which most of the absorption is centered at the interstellar velocity but there is additional absorption on the short wavelength side of the interstellar absorption produced by hydrogen atoms in the stellar hydrogen wall, and additional absorption on the long wavelength side of the interstellar absorption produced by hydrogen atoms in the solar hydrogen wall. A very useful result of the Gayley et al. (1997) calculations of the interaction between

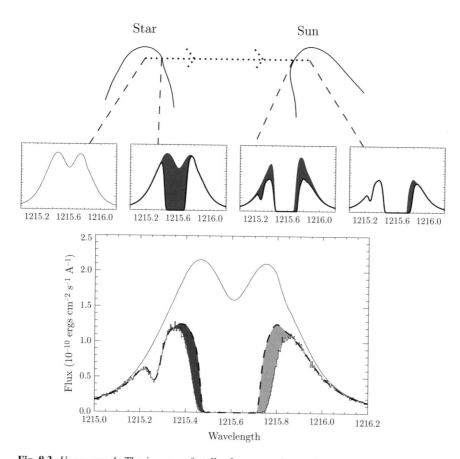

Fig. 8.3 *Upper panel:* The journey of stellar Lyman-α photons from a stellar chromosphere through the stellar hydrogen wall (the curved line around the star), the interstellar medium, and the hydrogen wall around the Sun. The shapes of the hydrogen walls are produced by the flow of interstellar gas from the upper part of the diagram. *Middle panel:* The shape of the Lyman-α emission line emitted by the star (left profile), absorption (purple) by the stellar hydrogen wall (second profile), additional absorption (purple) by interstellar hydrogen and deuterium (third profile), and additional absorption (purple) by the solar hydrogen wall (right profile). *Bottom panel:* The stellar Lyman-α profile before (upper solid line) and after the three absorptions. The absorption produced only by interstellar hydrogen is outlined by the dashed line, the extra absorption produced by the stellar hydrogen wall is red, and the extra absorpion produced by the solar hydrogen wall is green. Figure from Wood (2004)

the solar wind and interstellar hydrogen is that the extra absorption on the red side in the interstellar absorption is almost entirely produced in the solar hydrogen wall and that the extra absorption on the blue side is almost entirely produced in the stellar hydrogen wall. Both effects are true for almost all orientations of the line of sight to a star relative to the interstellar flow direction.

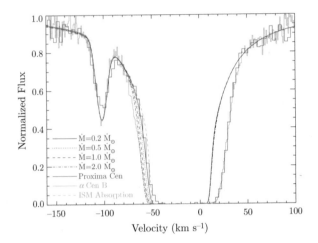

Fig. 8.4 Normalized Lyman-α profiles of Proxima (red) and α Cen B (green). The dashed green line shows absorption by interstellar hydrogen. The extra absorption at positive velocities for both stars is produced by hydrogen in the Sun's hydrogen wall. The extra absorption seen on the blue side of the interstellar absorption is produced by hydrogen in the astrosphere of the α Cen AB binary. The various dashed lines on the blue side of the interstellar absorption are the additional absorption predicted by models of the α Cen AB astrosphere with different assumed mass-loss rates. Figure from Wood et al. (2001). Reproduced by permission of the AAS

The clean separation of absorption by the interstellar medium, the stellar hydrogen wall (astrospheric absorption), and the solar hydrogen wall (heliospheric absorption) is shown by high-resolution Lyman-α spectra of α Cen B (K2 V) and α Cen C (M6 V) often called Proxima Centauri. Since the two stars are separated only 2.2 degrees on the sky, the heliospheric absorption should be the same for the two stars while the astrospheric absorption could very different. Figure 8.4 shows the observed Lyman-α profiles of Proxima Cen (red histogram) and α Cen B (green histogram). The dashed line is the interstellar hydrogen absorption inferred from the shapes of the interstellar deuterium line (at -100 km s^{-1}) and the Mg II and Fe II lines (not shown). Note that the extra absorption on the red side of the interstellar absorption is the same for both stars as predicted, while the extra absorption on the blue side is different for the two stars. To proceed from a measurement of the astrospheric absorption measured in a Lyman-α line profile to a plausible stellar mass-loss rate requires a plasma physics model that incorporates the physical processes just described. Wood (2004) has reviewed these models up to that time, and the heliospheric absorption feature observed in spectra along the line of sight to at least 13 stars is generally well matched by the models.

Beginning with the first detection of astrospheric absorption in the α Cen system by Linsky and Wood (1996), there are now detections of astrospheric absorption in the Lyman-α spectra of 15 single and binary stars (Wood et al. 2002, 2005, 2014) including 12 dwarf stars (spectral types G2–M5.5 V) and 4 subgiant or giant stars (G4 III-IV to K0 IV). Figure 8.4 shows the mass-loss rates for α Cen B obtained by comparing the observed astrospheric absorption with hydrodynamic models scaled from a solar model in which different mass-loss rates are produced by changing the proton density in the wind (Wood et al. 2001). The best match to the observed astrospheric absorption is for $\dot{M} = 2.0\dot{M}_\odot$. This likely represents the mass-loss rate of the α Cen AB binary as both stars are located in a common astrosphere. The Lyman-α spectrum of Proxima Cen shows no obvious astrospheric absorption consistent with an upper limit to the mass-loss rate of $\dot{M} \leq 0.2\dot{M}_\odot$. The very different mass-loss rates for these two stars located close together in the same interstellar cloud provides strong empirical evidence that the excess absorption on the short wavelength side of the interstellar absorption is produced by neutral hydrogen in the astrospheres of these stars. Nevertheless, the estimates of $\dot{M}$ are subject to systematic errors in the models and the assumption that the wind speed is the same as for the Sun (400 km s^{-1}) (Wood et al. 2001). $\dot{M}$ should vary inversely with the assumed wind speed, which could be larger cooler dwarf stars that have larger escape speeds.

Additional astrosphere absorption measurements and mass-loss estimates have been obtained by Wood et al. (2002, 2005) using the STIS instrument on *HST*. Figure 8.5 shows examples of the short wavelength portions of the Lyman-α lines

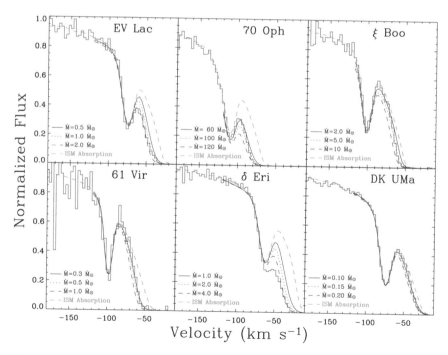

Fig. 8.5 Lyman-α line profiles of six stars observed by the HRS and STIS instruments. Shown are the short wavelength portions of the line including the blue side of the interstellar hydrogen absorption and the interstellar deuterium absorption line. Dashed lines indicate additional absorption due to astrospheric hydrogen for different assumed mass loss rates (Wood et al. 2005). Reproduced by permission of the AAS

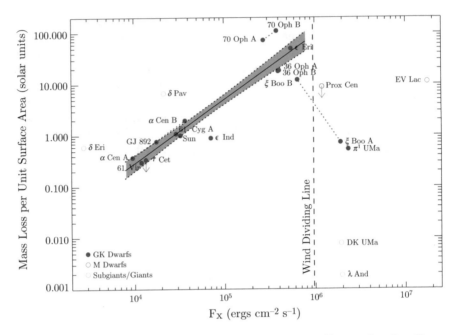

Fig. 8.6 Measured mass loss rates per unit stellar surface area vs X-ray surface flux. Figure courtesy of Brian Wood

in the spectra of six nearby stars and model fits for different values of $\dot{M}$ (Wood et al. 2005). The inferred mass-loss rates for the stars studied to date are in the range 0.15–100 $\dot{M}_\odot$. Figure 8.6 shows that the correlation of mass-loss rate with X-ray flux fails for stars with X-ray surface fluxes greater than 10^6 ergs cm^{-2} s^{-1} roughly corresponding to a stellar age of about 600 Myr. The mass-loss rates of the active stars π^1 UMa (G1.5 V) (Wood et al. 2014) and ξ Boo A (K4 V) (Wood and Linsky 2010) both lie far below the extraolation of the correlation and the M dwarfs Prox Cen (M5.5 V) and EV Lac (M4 V) also lie below the correlation but less so. Wood et al. (2014) suggest that a change in magnetic topology occurs near the "wind dividing line" at $F_x = 10^6$ ergs cm^{-2} s^{-1} to dramatically decrease the mass-loss rate, but what might this be and does the Sun provide any clues?

The question of the relation of wind mass flux with X-ray flux is more complex as indicated by solar observations. On the basis of monitoring the wind mass flux by many spacecraft and the X-ray flux measured by the *GOES* spacecraft over a period of more than 30 years, three cycles of solar magnetic activity, Cohen (2011) found no significant change in the ambient (i.e., nonflare) mass-loss rate despite a more than order of magnitude change in the solar X-ray flux between maximum and minimum. How could this occur? Since mass loss occurs in regions of open magnetic field lines and the X-ray emission is mainly from closed magnetic loops, Cohen (2011) argued that the fraction of the low corona with open magnetic fields must be roughly the same independent of the activity level as measured by the X-ray flux. This explanation is consistent with the theoretical models of Garraffo et al. (2015a,b) in which the mass-loss rate depends on the fraction of a stellar surface with open field lines. The correlation of mass-loss rate with X-ray flux for stars more active than the Sun and the absence of this correlation in the very active stars shown in Fig. 8.6 could be explained by the increase and then decrease in fraction of a stellar surface with open field lines. This tentative conclusion requires further study.

8.2.2   Upper Limits from Radio Observations

There have been several attempts to infer mass-loss rates for cool dwarf stars from the free-free continuum emission at radio wavelengths (e.g., Gaidos et al. 2000). Free-free emission from a fully ionized hot wind is predicted to vary with frequency as $\nu^{-0.6}$, but there are other sources of radio emission that can complicate the analysis—gyro-synchrotron emission at cm wavelengths (Güdel 2002) and thermal emission from stellar chromospheres at centimeter and millimeter wavelengths (Villadsen et al. 2014; Liseau et al. 2016). Since free-free emission from winds is predicted to be weak, it is essential to determine whether a detection is from the stellar wind or from another emission mechanism.

The most recent study by Fichtinger et al. (2017) used the Karl G. Jansky Very Large Array (VLA) with its upgraded sensitivity and angular resolution to observe four young G1 V to G5 V stars at 4–8 GHz (6 cm) and 12–18 GHz (2 cm). The stars (EK Dra, π^1 UMa, χ^1 Ori, and κ^1 Cet) are proxies for the young Sun at ages of 100–650 million years. Figure 8.7 compares the upper limits for the derived mass-loss rates with several theoretical models (see Sect. 8.2) and the measurements using the astrosphere technique (see Sect. 8.4). In particular, the upper limit for the radio emission from π^1 UMa implies a mass-loss rate 250 times larger than $\dot{M} = 1.0 \times 10^{-14} M_\odot$ yr^{-1} measured by the astrospheres technique (Wood et al. 2014). Clearly, more sensitive centimeter-wave measurements are needed to test the models and the astrospheric measurements.

According to Fichtinger et al. (2017), the radio upper limits are already sufficient low to constrain the mass of the young Sun to be $<1.004 M_\odot$, which is much less than the 1.03–1.07 $M_\odot$ that is needed to keep the young Earth warm enough to avoid its surface from being fully glaciated, which is commonly called the "faint young Sun paradox" (Sagan and Mullen 1972; Kasting and Catling 2003).

Radio emission can also occur when the strong magnetic winds of a host star interact with a planet's magnetic field leading to reconnection and energetic electrons that propagate along the planet's magnetic field lines and emit cyclotron radiation. Vidotto et al. (2010) investigated this scenario and found that a close-in hot Jupiter at 0.05 AU from a young active host star can emit low frequency radio power 10^5 times larger than now emitted by Jupiter. Vidotto et al. (2012) also computed the radio emission from τ Boo b, a hot Jupiter located at 0.0462 AU from an F7 V host star.

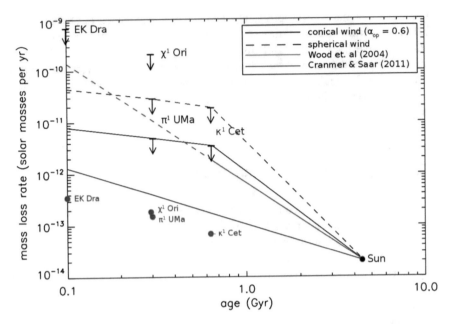

Fig. 8.7 Mass-loss rate upper limits obtained from VLA radio observations by Fichtinger et al. (2017). For π^1 UMa and κ^1 Cet the upper limits are for 6 cm (upper symbol) and for 2 cm (lower symbol). The other two stars had detected emission but the source was not deemed to be from the stellar winds. The upper limits are for wind emission alone. The red dots are the predictions of MHD models for these stars (Johnstone et al. 2015a,b) and the blue line is the time evolution of mass-loss rates predicted by the MHD models of Cranmer and Saar (2011). The red line (and its dotted extension) are the predicted evolution that Wood (2004) proposed based on the analysis of Ly-α astrospheric absorption. The black dashed line is the time evolution predicted for a spherical wind and the black solid line is for a conical wind. Reproduced with permission of ESO

8.2.3   Upper Limits from X-ray Observations

The detection of X-ray emission from comets that is explained by charge-exchange reactions between solar wind ions and neutral atoms in the comet's atmosphere (Lisse et al. 1996) stimulated Wargelin and Drake (2001) to propose that charge-exchange X-rays could measure the mass-loss rates of nearby dwarf stars. They argued that stellar wind ions, mainly protons, would interact with neutral hydrogen in the local interstellar medium to produce charge-exchange X-rays surrounding the star that would measure the mass-loss rate and other properties of the stellar wind. They calculated that emission in the O VII ion at 570 eV from a stellar wind 100 times stronger than solar should be detectable at a distance of 3 pc and perhaps 10 pc by an X-ray observatory like *Chandra*. However, sensitive searches have not detected charge-exchange X-ray emission near a star. For example, the 3σ upper limit of the charge-exchange X-ray emission from the nearest star, Proxima Cen (1.3 pc), is $<14\dot{M}_\odot$ (Wargelin and Drake 2001, 2002), which is far above the $\leq 0.2\dot{M}_\odot$ upper limit obtained by Wood et al. (2001) using the astrosphere technique.

8.2.4 Mass-Loss Rates from Transit Observations

Another approach is to infer stellar mass-loss rates from absorption features in the stellar Lyman-α emission line as exoplanets transit their host stars. Bourrier et al. (2016) presented simulations of the interactions between the ionized wind of GJ 436 (M2.5 V) and neutral hydrogen atoms in the hydrodynamical blow-off from its close-in (0.0287 au semi-major axis) warm Neptune-like exoplanet. These simulations include the neutral hydrogen atoms abraded from the exoplanet's large exosphere by the stellar wind and the stellar wind protons that are neutralized by charge exchange with hydrogen atoms in the exoplanet's exosphere. These two populations of neutral hydrogen atoms can be distinguished observationally because the planetary neutrals escape with relatively slow speeds, whereas the neutralized stellar wind protons retain the high speed of the stellar wind and are, therefore, observed in the blue wing (–120 to –40 km s^{-1}) of the Lyman-α line.

Lavie et al. (2017) summarized the extensive set of observations of absorption in the Lyman-α line during transits of hot Jupiters, beginning with observations of HD 209458 by Vidal-Madjar et al. (2003), and of mini-Neptunes like GJ 436b. For this star, transits begin about 3 h prior to first contact and extend to at least 10 h after egress (Lavie et al. 2017). They also detected absorption in the red wing of the Lyman-α line and in the Si III with uncertain causes. Bourrier et al. (2016) explained the pre-transit absorption seen in the Lyman-α line as produced by planetary neutrals abrased from the exoplanet's coma by the stellar wind. The post-transit absorption is produced by an extended comet-like tail consisting mostly of neutralized stellar wind protons (see Fig. 8.8).

The identification of the neutralized stellar wind protons in the exoplanet's cometary tail from blue-shifted absorption in the Lyman-α line after transit allowed Bourrier et al. (2016) to estimate the density and outflow speed of the stellar wind. With these parameters, Vidotto and Bourrier (2017) computed a spherically symmetric steady-state stellar wind model with a mass-loss rate $\dot{M} = 1.2^{+1.3}_{-0.75} \times 10^{-15} M_\odot$ yr^{-1}, corresponding to 0.02–0.13 times the solar mass-loss rate. In their model the astrosphere size is 25 au, about four times smaller than for the Sun. The Wood et al. (2005) scaling relation between mass-loss rate and X-ray surface flux predicts for GJ 436 $\dot{M} \approx 4.8 \times 10^{-15} M_\odot$ yr^{-1}, only a factor of four above the value estimated by Vidotto and Bourrier (2017). The transit method for estimating stellar mass-loss rates should be applicable to other stars like HD 189733 that have transiting exoplanets with comet-like tails.

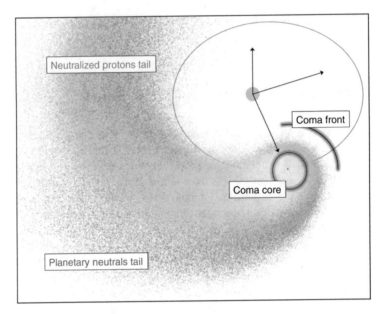

Fig. 8.8 A schematic representation of the neutral hydrogen cloud surrounding GJ 436b. The blue region represents neutral hydrogen atoms escaping from the planet and the yellow region represents the neutralized stellar wind protons. The three arrows from the star to the planet's orbit refer to the three epochs with *HST* observations. Figure courtesy of V. Bourrier. Reproduced with permission of ESO

8.2.5 Mass-Loss Rates from M Dwarf Companions of White Dwarfs

Accretion of stellar wind gas from a close-in M dwarf star onto the surface of a DAZ white dwarf companion provides another method for infering mass-loss rates. Debes (2006) observed the presence of Ca II ions in the atmospheres of three DAZ white dwarfs indicating that accretion from the companion M dwarf must be balanced against the settling rate of metal ions in white dwarf atmospheres, which should otherwise be pure hydrogen. There are a number of uncertainties in the method including the assumption of Bondi-Hoyle accretion rates, the mass-fraction in the white dwarf's convective envelope, and diffusion rates. Table 8.1 lists the inferred mass-loss rates for three M dwarfs using the WD binary technique and compares the mass-loss rates per unit surface area (last column) with those obtained by the transit and astrosphere techniques. The wide range in these results indicates no clear trend with M dwarf mass. Further work is needed to determine realistic errors in the application of these techniques. Also, more observations and analysis are needed to search for the dependence of mass-loss rates on stellar mass, rotation rate, age, activity, and magnetic field structure. Until we have better understanding of these issues, the best approach is to use the astrosphere measurements for EV Lac or theoretical models (see below) as a guidance for other M dwarfs.

Table 8.1 Comparison of empirical mass-loss rates for M dwarfs

Star	Technique	$M_{MD}/M_\odot$	$R_{MD}/M_\odot$	$\dot{M}_{MD}$	$\dfrac{(\dot{M}_{MD}/\dot{M}_\odot)}{(R_{MD}/R_\odot)^2}$
GJ 436	Transit	0.45	0.44	1.2×10^{-15}	0.31
1213+528	WD binary	0.36	0.32	1×10^{-16}	0.049
EV Lac	Astrosphere	0.35	0.36	2×10^{-14}	7.5
1026+002	WD binary	0.23	0.25	1×10^{-16}	0.080
Prox Cen	Astrosphere	0.12	0.14	$<4 \times 10^{-15}$	<10
Prox Cen	X-ray	0.12	0.14	$<3 \times 10^{-13}$	<750
0419-487	WD binary	0.095	0.189	6×10^{-15}	8.5

8.2.6 Could Coronal Mass Ejections be Important for Mass Loss?

Stellar wind models typically treat the outflowing gas as spatially and temporally uniform. Observations by the LASCO (Large Angle and Spectrometric Corona-graph) instrument on the *SOHO* spacecraft and earlier spacecraft, however, show that the solar wind is highly structured with a wide range of local densities and speeds (e.g., Crooker et al. 1996; DeForest et al. 2018). The largest events in the solar wind are coronal mass ejections (CMEs), which are eruptions of large magnetic loops following destabilization of their magnetic footprints usually associated with energetic flares.

Solar CMEs have been observed with masses of 10^{12} to above 10^{16} g and 10^{26} to more than 10^{28} ergs of X-ray emission in the *GOES* bandpass (0.1–0.8 nm) (Aarnio et al. 2011, 2012). The largest mass recorded in a solar CME is about 2×10^{17} g and the largest kinetic energy is 1.2×10^{33} erg (Gopalswamy et al. 2009). Drake et al. (2013) found that the kinetic energy in CMEs is typically 200 times the flare X-ray emission in the *GOES* bandpass, but the X-ray emission in this bandpass is typically only 1% of the flare's total radiation including UV and optical emission. As a result, the kinetic enegy of CMEs is at least twice the total radiative emission. Typical CME outflow speeds are 200–1000 km s^{-1}, and the estimated mass-flux rate for the Sun is small, in the range $(6.6 - 22) \times 10^{-16} M_\odot$ yr^{-1}, depending on a number of assumptions. Thus, CMEs account for only 3–10% of the measured solar wind mass-loss rate of $2 \times 10^{-14} M_\odot$ yr^{-1}. Could CMEs be far more important and perhaps dominate the mass-loss rate of more active stars which have several orders of magnitude higher flare rates and energies than the Sun?

Aarnio et al. (2012) found an empirical relation between solar CME mass (M) and total X-ray energy (E) of the associated flare in the form $M = KE^\beta$, where K is a constant and $\beta = 0.63 \pm 0.04$. To estimate the mass-loss rate of pre-main sequence (PMS) stars in the Orion star forming complex, they extrapolated this relation by six orders of magnitude from 10^{30} ergs observed in very large solar flares to 10^{36} ergs seen in Orion PMS star large flares. Extrapolations over six orders of magnitude may be far off, but let's proceed anyway. The extrapolation of the mass-energy relation to very active PMS stars yielded CME mass-loss rates at least 30 times solar to an upper limit of 10^5 times solar.

In a similar analysis, Drake et al. (2013) found that an extrapolation of the solar CME masses vs flare X-ray energy correlation to active stars with $L_x = 10^{30}$ erg s^{-1} (only 10^3 times that of the quiet Sun) results in CME mass-loss rates of $10^{-10} \dot{M}_\odot$, about 5000 times that of the Sun. Also the kinetic energy in these powerful CMEs would be 1% of their star's total luminosity. Drake et al. (2013) argued that for such mass-loss rates the X-ray and kinetic energy fluxes would be highly unrealistic. The highest mass-loss rate found by the astrosphere technique is only 100 times solar and the active M dwarf flare star EV Lac has a solar-like mass-loss rate even though its X-ray luminosity is $\log L_x = 28.99$ (Wood et al. 2002).

A further test of whether the correlation of CME mass with flare energy might be valid at high flare energies is provided by the events that occurred on the Sun in October 2014. At that time, active region AR12192 was the most energetic active region recorded in the last 24 years. There were 6 X-class flares, the most energetic of X-ray flares seen on the Sun, but only one CME and that was associated not with an X-class flare but with a much weaker M-class flare at the edge of the active region. The events on this very well studied active region severely question the putative correlation of flare X-ray flux with CME mass loss. Sun et al. (2015) proposed that the unexpected behavior seen in AR12192 could be explained by relatively weak non-potential magnetic fields, strong overlying fields, and small changes in the magnetic field occuring during the flares. Could this scenario also occur on very active stars with very energetic flares?

CMEs have not yet been definitively detected from stars, but there are two possible detections. During a 4 min period in the decay phase of the 1980 August 20 flare on Proxima Cen, the X-ray flux decreased and neutral hydrogen column density increased. Haisch et al. (1983) suggested that the passage of a CME (or possibly an ejected cool prominence) in front of the X-ray emiting plasma could explain this observation. The 1977 August 30 superflare on the so-called "Demon Star" Algol (B8 V + K2 IV) showed a large increase in the hydrogen column density at the onset of the X-ray flare consistent with the evolution and ejection of a CME (Moschou et al. 2017).

There are stellar searches underway to detect Type II radio emission that is usually associated with solar CME events (e.g., Crosley and Osten 2018). Since flares on active stars can be 10,000 times as X-ray luminous as major solar flares, extrapolations of solar CME mass-loss rates to active stars predict that CME mass-loss could be more important than steady-state mass loss. This prediction needs to be tested. Using 3D MHD simulations, Alvarado-Gómez et al. 2018) found that large-scale dipolar magnetic fields of 75 G aligned with the rotation axis of the star can confine CMEs for flare energies less than about 3×10^{32} ergs, about X20 on the *GOES* scale (see Chap. 12), and reduce the outflow velocities of more energetic CMEs that can escape. This numerical simulation suggests that active stars with strong magnetic fields may produce far fewer CMEs that predicted on the basis of the correlation of solar CME properties with flare X-ray emission. The range of stellar CME properties as a function of energy in the X-ray band and the strength and orientation of the coronal magnetic field should be explored further, because Segura et al. (2010) have shown that the impact of a large CME on a planetary atmosphere can lead to chemical reactions that destroy ozone for a long period of time (months to years), allowing stellar UV photons to penetrate to the surface of a planet and thereby inhibiting habitability on the planet's surface.

8.2.7 Mass-Loss Rates from Slingshot Prominences

Rapidly rotating young main-sequence stars often show transient absorptions in Hα and Ca II K lines, which have been interpreted as low-temperature gas temporarily trapped in magnetic loops near the Keplerian co-rotation radius where the effective gravity is zero. The prototype example of this slingshot phenomenon is AB Dor (K0 V) first studied by Robinson and Collier Cameron (1986) and other examples have been studied by Jardine and Collier Cameron (2019). They noted that when the co-rotation radius lies outside of the sonic point radius (where the outflow speed equals the sound speed) but inside of the Alfvénic point radius, the gas outflow along the magnetic loops can become supersonic before reaching the co-rotation radius. In this "limit cycle regime", the gas at the loop base cannot know what is happening near the co-rotation radius and there is no way of readjusting the outflow. Eventually the mass constrained by the loop near the co-rotation radius becomes too large to be supported by the magnetic field against centrifugal forces and must escape as a transient slingshot wind.

Jardine and Collier Cameron (2019) quantified this scenario to estimate mass-loss rates from the typical recurrence times for prominence disruption, typical gas masses in the co-rotating loops measured from absorption lines, and the likely footpoint areas of these loops on the stellar surface. The mass-loss rates per unit stellar surface area are 10^2 to 10^4 times larger than the corresponding solar mass-loss rate for stars with X-ray surface fluxes 10^6 to more than 10^8 erg cm^{-2} s^{-1}. These results extend the trend of increasing mass-loss rates seen in Fig. 8.6 to far higher values for stars with X-ray surface fluxes well beyond the wind dividing line proposed by Wood et al. (2014). It is important to test this new wind model and its assumptions by future observations and simulations.

8.3 Simulations and Models

Modelling the winds of low-mass stars is a tricky business. Johnstone et al. (2015a)

Parker's hydrodynamic outflow model in which thermal pressure of the hot coronal plasma is the driving force provides the basis for understanding cool star winds. However, this model assumes that the magnetic field only heats the corona and constrains the outflow direction to be radial but does not participate in accelerating the wind. More complete models of the solar and stellar winds should include magnetic fields, especially for active stars with magnetic field strengths and fluxes far larger than for the Sun.

One approach is to extend the Parker-type thermally driven wind model to stars by including the magnetic field geometry. With this approach, See et al. (2015, 2017) developed wind models for 66 stars with wind temperatures, speeds, and magnetic field divergence scaled from solar values and assuming that the source surface where the wind becomes radial has a potential magnetic field obtained from observed Zeeman-Doppler images (ZDIs). The mass-loss rates computed with this approach are proportional to the open magnetic flux.

The initial rationale for including magnetic fields in stellar wind models (cf., Weber and Davis 1967) was to understand the effect of torques exerted by magnetic fields on the rotational evolution of cool stars. Subsequent models (e.g., Matt et al. 2012a) considered how the mass-loss rate could depend on stellar parameters and rotation rate. For example, Holzwarth and Jardine (2007) developed a polytropic model for thermally driven winds in which the coronal temperature and magnetic flux depend on the stellar rotation rate with power law indices set to fit empirical constraints for the Sun and cool stars. For slowly rotating stars, their model predicts that the mass-loss rate is proportional to the X-ray flux with a small power-law index, $\dot{M} \propto F_x^{0.5}$, such that $\dot{M}$ could be as large as ten times solar for very active stars with large X-ray surface fluxes. These predicted mass-loss rates for such stars as ϵ Eri, 70 Oph, and 36 Oph are much lower than determined by the astrosphere method.

A different polytropic wind model developed by Johnstone et al. (2015a,b) uses observed properties of the solar wind and stellar rotation evolution to predict $\dot{M}$ for stars as a function of mass (0.4–$1.1 M_\odot$) and age (0.1–5 Gyr). The heating of the wind plasma with distance from the star is not explicitly computed but is implicit in the value of the polytropic index α for the equation of state $P = K\rho^\alpha$, where P is the gas pressure, K is a constant, and ρ is the gas density. In their solar model, α and other parameters vary with radial position to fit spacecraft measurements. The temperature and density at the base of the wind driven by thermal pressure gradients control the $\dot{M}$ and are different for the solar high and low speed winds. If Alfvén wave pressure is significant, then the required base temperature need not be as high to produce the same mass-loss rate.

Extension of their solar model to stars follows from measured stellar rotation rates as a function of age and mass and the decrease in rotation rate with time due to the torque exerted by the wind flux on the magnetic field inside of Alfvén radius and thus on the star's rotation rate. With assumptions concerning the temperature or sound speed at the base of the wind, they predict $\dot{M}$ and asymptotic wind speeds. Figure 8.9 shows their predicted mass-loss rate for the Sun in time and comparison with observations and other theoretical models. The age dependences, $\dot{M} \propto t^\beta$, of the Cranmer and Saar (2011) ($\beta = -1.1$) and Suzuki et al. (2013) ($\beta = -1.23$) models are consistent with the Johnstone et al. model ($\beta = -0.75$ for $t > 0.7$ Gyr), but much less steep than the $\beta = -2.33 \pm 0.55$ dependence predicted by Wood et al. (2005) (see Sect. 8.2). As a result, Johnstone et al. predict that the maximum mass-loss rate that the young Sun had if initially a fast rotator should be about 37 times

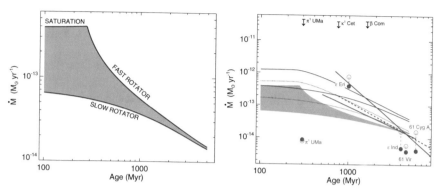

Fig. 8.9 *Left:* Mass-loss rates computed for the Sun in time between the limits of initially slow or fast rotation. Saturation of the mass-loss rate at high rotation rates is described in the text. *Right:* Comparison of the range of solar mass-loss rates (green) with radio observation upper limits (Gaidos et al. 2000) and astrosphere technique measurements (red dots) by Wood et al. (2005). The blue line is the prediction of Suzuki et al. (2013), the dashed red line is the scaling law of Wood et al. (2005), and the thin solid and dashed lines are the models of Cranmer and Saar (2011). Figure from Johnstone et al. (2015b). Reproduced with permission of ESO

its present value but only about 7 times larger if it were initially a slow rotator. They pointed out that the maximum value of $\dot{M}_{sat}$ (the saturation mass-loss rate) occurs because with increasing driving force (i.e., higher temperatures at the base of the wind), the critical point moves inward where higher densities radiate more energy leaving less energy available to drive the wind. Therefore, slowly rotating low-mass stars spin down faster than high-mass stars rotating at the same rate; a $0.5M_\odot$ star has a mass-loss rate per unit area ten times larger than for a $1.0M_\odot$ star. Johnstone et al. (2015b) pointed out that the Wood et al. scaling relation includes stars with a range of masses and that for the Sun and four stars with masses similar to the Sun, $\beta = -1.33$, which is consistent with the other values of β. O'Fionnagáin and Vidotto (2018) called attention to the decrease in coronal temperatures in stars older than about 2 Gyr that leads to a sharp decline in wind temperatures, spin down rate and mass-loss rate.

Whether young solar-type stars have modest mass-loss rates or the much larger rates predicted by Wood et al. (2005) is critically important for exoplanet habitability (see Chap. 11) and must be tested by further observations and models based on realistic MHD calculations.

The physical cause of saturated mass-loss rates and the dependence of $\dot{M}_{sat}$ on the photospheric input energy, magnetic flux and morphology, and stellar mass were the subjects of MHD numerical experiments by Suzuki et al. (2013). Two important physical processes in and above the chromosphere are the reflection of Alfvén waves down to lower layers and radiation losses; both processes absorb energy that would otherwise drive the mass loss. With increasing input energy, magnetic waves levitate and increase chromospheric and coronal densities. Since radiation losses in these optically thin layers are proportional to ρ^2, where ρ is the density, radiation losses increase faster than the input wave energy, leading to a maximum in the mass-loss rate with $\dot{M}_{sat} \propto (B_o f_o)^{1.82}$. These numerical experiments predict that in the unsaturated regime $\dot{M} \propto t^{-1.1}$, but in the saturated regime the dependence on $(B_o f_o)^{1.82}$ predicts mass-loss rates that can be 1000 times larger than today's Sun for young active stars.

Cranmer and Saar (2011) developed a model for stellar winds in which MHD turbulence generated in the subphotospheric convection zone propagates upward with increasing amplitude until shocks heat the outer atmosphere and drive mass loss where the magnetic field is open. The driving force for the winds of cool dwarf stars is primarily thermal from the hot coronal plasma (as in Parker's model), because densities in the 10^4 K chromosphere are too low to radiate most of the input energy. For evolved giant stars, however, densities in the chromosphere are sufficiently large to radiate much of the input wave energy. For these stars, there is no hot corona and wave pressure gradients are an important driving source for the wind. The mass loss rate for both cool and hot outer atmospheres is primarily determined by the fraction f of the stellar surface covered by open magnetic regions (small for inactive stars like the Sun and large for active stars), but also by the field strength in the photosphere and the scale length for the divergence of the field with height. Turbulent energy propagates upward in the diverging flux tubes as transverse kink-mode Alfvén waves and dissipates energy as the wave speed approaches the Alfvén speed ($v_A = B(4\pi\rho)^{-1/2}$). Cranmer and Saar (2011) predicted mass-loss rates using observations that in the photosphere the magnetic and gas pressures are about equal, $B_{\text{photo}}^2/8\pi = 1.13 P_{\text{photo}}$, and that the product of the photospheric field strength and filling factor, $B_{\text{photo}} f$, scales with the stellar rotation period. In the absence of magnetic fields, the basal mass-loss rates shown in Fig. 8.10 depend only on the stellar effective temperature that controls the energy in the subphotospheric turbulent waves (see Sect. 2.2). The rapid decrease in the basal mass-loss rate at $T_{\text{eff}} > 9000$ K occurs because convection becomes an unimportant process for

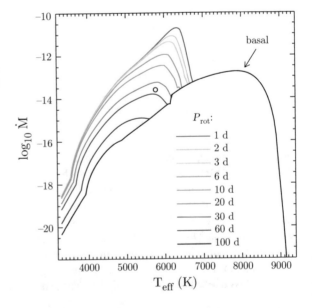

Fig. 8.10 Predicted mass loss rates as a function of effective temperature and rotation period. The term "basal" refers to nonmagnetic atmospheres. The small circle indicates the location of the Sun. Figure from Cranmer and Saar (2011). Reproduced by permission of the AAS

energy transport. The inclusion of magnetic fields greatly enhances $\dot{M}$ for cool dwarf stars as the upwardly propagating kink-mode waves are far more energetic than nonmagnetic acoustic waves leading to hotter coronae and larger mass-loss rates. Their model predicts that $\dot{M} \propto t^{-1.1}$ for cool stars on the main sequence with ages between 0.2 and 7 Gyr and is consistent with the solar mass-loss rate. An important point is that wind acceleration and coronal plasma heating occur for magnetic waves only when they have a compressional component. Pure Alfvén cannot transfer energy to the plasma without conversion to compressional waves.

For very rapidly-rotating stars, magneto-centrifugally driven winds can be the major source of mass-loss. An example of these models is the study of the very rapidly rotating M4 V star V374 Peg with a rotation period of 0.44 days. Vidotto et al. (2011) used the 3D MHD numerical code BATS-R-US to compute the mass-loss rate in the magnetic geometry obtained from a Zeeman Doppler image (ZDI, see Sect. 3.3). They found that at the base of the stellar corona the ratio of gas to magnetic pressure, $\beta = P_{gas}/P_{mag} \approx 10^{-5}$, or about five orders of magnitude smaller than for the corresponding layer in the Sun. For V374 Peg, magnetic pressure and centrifugal forces rather than thermal pressure drive the much denser stellar wind with a terminal speed much higher than in the solar wind. The resulting mass-loss rate $\dot{M} = (4\text{–}40) \times 10^{-11} M_\odot \, \mathrm{yr}^{-1}$ is 2000–20,000 times solar. This very different wind regime from the Sun has not yet been confirmed empirically.

Recent stellar wind models computed with the BATS-R-US code, which solves the MHD equations in a three dimensional geometry for a specified magnetic field morphology, provide insights into the importance of the magnetic field morphology. For example, Garraffo et al. (2015a) computed a set of models for solar-type stars in which the morphology is assumed to be a spherical harmonic with multipole order n between 1 and 10. They found that $\dot{M}$ decreases by a factor of 50 when n increases from 1 to 10. The reason for this dramatic decrease in $\dot{M}$ and also in angular momentum loss is that the fraction of the stellar surface area with open field lines permitting mass outflow decreases rapidly with increasing magnetic complexity (cf., Réville et al. 2015). Counteracting this effect is an increase in mass-loss rate with increasing magnetic field strength (see Fig. 8.11). Garraffo et al. (2016a) next computed mass-loss rates for magnetic geometries with different values of the spherical harmonic distribution parameter m but they found that m has no significant effect on $\dot{M}$. They proposed analytic scaling factors for estimating mass-loss and angular momentum loss rates based on a complexity parameter computed from the stellar ZDI. Garraffo et al. (2015b) and previously Holzwarth and Jardine 2005) found that the latitude of starspots and their associated strong magnetic fields can modify the mass and angular momentum loss rates by a factor of two. Both rates decrease when spots are located at higher latitudes as is typically seen in active stars. Since magnetic field strength and morphology can be very different for stars with similar mass, one should expect a wide range of mass-loss rates for stars with similar masses but different ages, rotation rates, and activity levels. It is important to note that these calculations assume that a star with a complex field geometry does not also have a dipolar field component. The presence of a dipole, for which the magnetic field strength decreases with height far more slowly than for complex fields, will dominate the rates of mass loss and angular momentum loss.

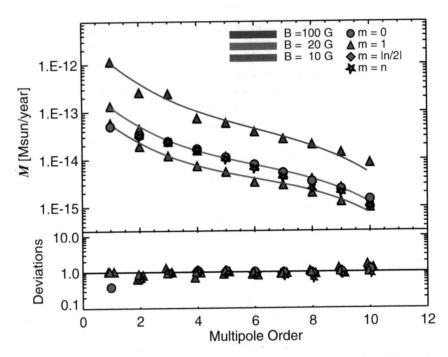

Fig. 8.11 Mass-loss rates computed for solar-type stars with different magnetic field strengths and morphologies as a function of multipole order. The cyan points are for models with a constant magnetic flux normalized to a 10 G dipole. Figure from Garraffo et al. (2015a). Reproduced by permission of the AAS

Simulations of the mass-loss and angular momentum loss rates for specific stars are now been computed using 3D MHD codes like BATS-R-US to compute heating and wind acceleration by Alfvén waves in the observed magnetic structure obtained with ZDIs. For example, Alvarado-Gómez et al. (2016) simulated the winds of four planet-hosting stars: HD 22049 (ϵ Eri K2 V), HD 1237 (G8 V), and HD 147513 (G5 V), and the Sun at both activity maximum and minimum. Each of these simulations shows regions of fast, low density winds and slow, higher density winds as well as current sheets. The computed mass-loss rates and wind properties depend on whether the magnetic fields were obtained directly from the ZDIs or from spherical harmonic representations of the ZDIs. They obtained a relation between mass-loss rate and X-ray flux, $\dot{M} \propto F_x^{\gamma}$, where $\gamma = 0.79^{+0.19}_{-0.15}$. The value of γ is smaller than the empirical value $\gamma = 1.34 \pm 0.18$ found by Wood et al. (2005). This relation is based on simulations of $\dot{M}$ and F_x using the MHD code rather than on scaling relations.

Using the same MHD code and ZDIs, do Nascimento et al. (2016) obtained $\dot{M} = 9.7 \times 10^{-13} M_{\odot}$ yr^{-1} for the young (0.4–0.6 Gyr) Sun proxy star κ^1 Cet (HD 20630, G5 V). This mass-loss rate for a solar-type star with $F_x \approx 10^6$ erg cm^{-2} s^{-1} is about 50 times the present day solar rate and similar to the 60–130 times rate predicted by Wood et al. (2005). A similar simulation by Garraffo et al. (2016b) for the M5.5 Ve star Prox Cen, the Sun's nearest neighbor, found a mass-loss rate $\dot{M} \sim 1.5 \times 10^{-14} M_\odot \, yr^{-1}$, nearly a factor of 10 higher than the upper limit that Wood et al. (2001) obtained from their astrosphere analysis. Garraffo et al. (2016b) mentioned that the difference may result from the astrosphere result being model dependent. A similar simulation for the host star Trappist-1, which has at least 7 exoplanets, resulted in $\dot{M} \sim 3 \times 10^{-14} M_\odot \, yr^{-1}$ (Garraffo et al. 2017).

Using a different 3D MHD code (ALF3D) that includes three fluids (electrons, protons, and interstellar pickup protons), Airapetian and Usmanov (2016) computed $\dot{M}$ for the Sun at age 700 Myr, which is consistent with the Wood et al. (2005) scaling law and similar to $\dot{M}$ for κ^1 Cet (do Nascimento et al. (2016), which has a similar age. More recently, Alvarado-Gómez et al. (2018) used two different 3D MHD codes to compute $\dot{M} \approx 2.5 \times 10^{-13} M_\odot \, yr^{-1}$ for an active solar-type star. This mass loss rate about 12.5 times the solar value is similar to that of 36 Oph (K1 V+K1 V) (Wood et al. 2005).

Vidotto et al. (2014) simulated the winds of three rapidly rotating active M dwarfs and three slowly rotating less active M dwarfs (P_{rot}=14.0–18.6 days) with very different mass-loss rates. They called attention to the asymmetric properties of the wind that occur when the magnetic field distribution is far from axisymmetric and that the mass-loss rate is proportional to unsigned open magnetic surface flux, $\dot{M} \propto \Phi^{0.89\pm0.19}$.

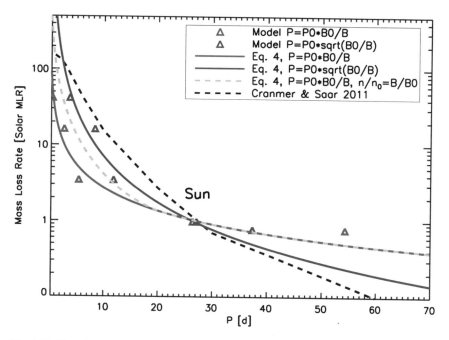

Fig. 8.12 Mass-loss rates as a function of stellar rotation period. The red triangles and dashed line assume that the dipole magnetic field strength B_{polar} is proportional to the stellar angular velocity $B_{polar} \propto \Omega$, and the blue triangles and solid line assume that $B_{polar} \propto \Omega^2$. The black dashed line is for the simulations of Cranmer and Saar (2011). Figure from Cohen and Drake (2014). Reproduced by permission of the AAS

From a large grid of simulations computed with the BATS-R-US code and an assumed magnetic dipolar geometry, Cohen and Drake (2014) obtained a set of scaling laws relating $\dot{M}$ to the dipole magnetic field strength, stellar rotation rate, and coronal base density (Fig. 8.12). These models are bi-modal with faster low-density winds emerging from coronal holes and slower high-density winds emerging from helmut streamers like in the solar case. With the assumption of coronal base densities similar to the Sun, they inferred mass-loss rates much lower than the values assumed by Matt et al. (2012b) and Vidotto et al. (2011).

Included in Table 8.2 are simulations of stellar winds and the solar wind at different ages (100 Myr and 700 My) and various conditions (for minimum and

Table 8.2 MHD simulations of winds based on magnetic images

Star	Spectral type	$\dot{M}$ $(10^{-14} M_\odot \text{ yr}^{-1})$	$\dot{M}/\dot{M}_\odot$	Thermal distribution	Reference
τ Boo	F7 V	270	135	Polytrope	1
HD 179949	F8 V	80	40	Polytrope	14
Solar min	G2 V	2.76	1.38	Alfvén waves	2
Solar max	G2 V	4.80	2.40	Alfvén waves	2
100 Myr	G2 V	7–40	3.5–20	Polytrope	3
700 Myr	G2 V		26–50	Alfvén waves	4
$F_x = 10^6$	G2 V	15–50	30–100	Alfvén waves	5
Saturation	G2 V	64	37	Polytrope	3
κ^1 Cet	G5 V	97	48.5	Polytrope	6
HD 147513	G5 V	11.4	5.7	Alfvén waves	2
HD 1237	G8 V	4.7–13.9	2.35–6.95	Alfvén waves	2
HD 73256	G8	21	10.5	Polytrope	14
Active star		16	12.5	Alfvén waves	15
HD 189733	K0 V	8.3	4.15	Alfvén waves	7
AB Dor	K0 V	10–500	5–250	Alfvén waves	8
HD 102195	K0 V	32	16	Polytrope	14
HD 130322	K0 V	58	29	Polytrope	14
HD 46375	K1 IV	19	9.5	Polytrope	14
ϵ Eri	K2 V	2.77–10.2	1.38–5.1	Alfvén waves	2
Less active	M1 V	700–1000	350–500	Polytrope	9
More active	M1 V	1200–7000	600–3500	Polytrope	9
V374 Peg	M4 V	4000–40,000	2000–20,000	Polytrope	10
EV Lac	M4 V	3	1.5	Alfvén waves	11
Prox Cen	M5.5 Ve	~ 1.5	~ 0.75	Alfvén waves	12
Trappist-1	M8 V	~ 3	~ 1.5	Alfvén waves	13

References: (1) Vidotto et al. (2012), (2) Alvarado-Gómez et al. (2016), (3) Johnstone et al. (2015b), (4) Airapetian and Usmanov (2016), (5) Suzuki et al. (2013), (6) do Nascimento et al. (2016), (7) Cohen et al. (2011), (8) Cohen et al. (2010), (9) Vidotto et al. (2014), (10) Vidotto et al. (2011), (11) Cohen et al. (2014), (12) Garraffo et al. (2016a), (13) Garraffo et al. (2017), (14) Vidotto et al. (2015), (15) Alvarado-Gómez et al. (2018)

maximum activity today, X-ray surface flux $F_x = 10^6$ erg cm^{-2} s^{-1}, and for saturated mass-loss rate). These simulations assume that the thermal distribution where the winds are driven is either computed from the dissipation of Alfvén waves or approximated through the value of the assumed polytrope index α. There are systematic errors associated with both types of models. Both types must assume a value for the density at the base of the wind and are based upon observed magnetic field images with low spatial resolution. The Alfvén wave driven simulations assume a specific type of magnetic wave, whereas there are a variety of such waves and energy conversion among these waves. The polytropic simulations assume a coronal temperature based on solar analogy, but $\dot{M}$ depends sensitively on the coronal temperature (cf., Airapetian and Usmanov 2016). The orbital speeds of close-in Jupiters are supermagnetosonic with bow shocks surrounding the exoplanets as they plow through the stellar wind (Vidotto et al. 2015).

The very high mass-loss rate of V374 Peg can be understood in terms of magneto-centrifugal acceleration for this very rapidly rotating star. However, the mass-loss rates for the three more slowly rotating early-M stars simulated with a polytropic thermal structure are 400 times larger than for the later M dwarfs (Proxima Cen and Trappist-1) simulated with Alfvén wave thermal structures. This large difference and the systematic errors associated with both types of simulations suggests that further work is needed to determine the best way to determine the thermal distributions that drive stellar winds of cool dwarfs. In addition, the strong dependence of mass-loss rates on the uncertain density at the base of the corona is demonstrated by the large range in $\dot{M}$ in the models for AB Dor for which the only variable is the coronal base density (Cohen et al. 2010). At this point in time, observational and theoretical estimates of mass-loss rates should be viewed with caution.

In Sect. 10.2, I will discuss the effects of stellar winds on the magnetospheres and atmospheres of exoplanets.

8.4 For the Future

While there are many promising simulations of stellar mass-loss rates as a function of stellar properties (e.g., stellar rotation rate, age, mass, and magnetic-field properties), it is essential to test and verify the predictions of these simulations with observations. Except for inferences of mass-loss rates of three M dwarfs from accretion onto their white dwarf companions, the astrosphere technique is the only available tool that is now providing mass-loss rates rather than upper limits. However, the astrosphere technique has an important limitations. The Lyman-α absorption in an astrosphere occurs only when the star is embedded in interstellar matter that includes substantial neutral hydrogen. Otherwise there is no charge exchange between the outflowing stellar wind protons and interstellar neutral hydrogen to produce a decelerated hydrogen wall and the extra absorption on the short-wavelength side of the interstellar absorption feature in the Lyman-α

line profile. Thus the non-detection of the astrospheric mass-loss signal could either mean that the stellar mass-loss rate is too small to detect or that the star is embedded in a completely ionized interstellar gas. However, the detection of stellar hydrogen wall absorption clearly indicates that a star has measurable mass loss. New observations of nearby stars are needed to determine the dependence of mass-loss rates on stellar parameters, and STIS is the only available instrument that can obtain the required high spectral resolution through a narrow slit to minimize geocoronal emission. Of particular interest is the mass-loss rates of M dwarfs for which we have very few data. Are the properties of the winds of M dwarfs very different from F-K stars? How do mass-loss rates depend on the star's magnetic field properties? ZDI's now provide information on the complexity of the large-scale magnetic field structure, which may (Garraffo et al. 2015a,b) or may not (Jardine et al. 2017) control the mass-loss rates. This topic should be tested observationally by various techniques including more sensitive observations of radio emission, X-ray flux, and stellar CMEs to provide realistic stellar mass-loss rates.

References

Aarnio, A.N., Stassun, K.G., Hughes, W.J., McGregor, S.L.: Solar flares and coronal mass ejections: a statistically determined flare flux - CME mass correlation. Solar Phys. **268**, 195 (2011)

Aarnio, A.N., Matt, S.P., Stassun, K.G.: Mass loss in pre-main-sequence stars via coronal mass ejections and implications for angular momentum loss. Astrophys. J. **760**, 9 (2012)

Airapetian, V.S., Usmanov, A.V.: Reconstructing the solar wind from its early history to current epoch. Astrophys. J. Lett. **817**, L24 (2016)

Alvarado-Gómez, J.D., Hussain, G.A.J., Cohen, O., Drake, J.J., Garraffo, C., Grunhut, J., Gombosi, T.I.: Simulating the environment around planet-hosting stars. II. Stellar winds and inner astrospheres. Astron. Astrophys. **594**, A95 (2016)

Alvarado-Gómez, J.D., Drake, J.J., Cohen, O., Moschou, S.P., Garraffo, C.: Suppression of coronal mass ejections in active stars by an overlying large-scale magnetic field: a numerical study. Astrophys. J. **862**, 93 (2018)

Baranov, V.B.: Gasdynamics of the solar wind interaction with the interstellar medium. Space Sci. Rev. **52**, 89 (1990)

Baranov, V.B., Malama, Y.G.: Model of the solar wind interaction with the local interstellar medium - numerical solution of self-consistent problem. J. Geophys. Res. **98**, A9, 15157 (1993)

Barnes, S.A.: On the rotational evolution of solar- and late-type stars, its magnetic origins, and the possibility of stellar gyrochronology. Astrophys. J. **586**, 464 (2003)

Biermann, L.: Solar corpuscular radiation and the interplanetary gas. Observatory **77**, 109 (1957)

Bourrier, V., Lecavelier des Etangs, A., Ehrenreich, D., Tanaka, Y.A., Vidotto, A.A.: An evaporating planet in the wind: stellar wind interactions with the radiatively braked exosphere of GJ 436 b. Astron. Astrophys. **591**, A121 (2016)

Castor, J.I., Abbott, D.C., Klein, R.I.: Radiation-driven winds in of stars. Astrophys. J. **195**, 157 (1975)

Cohen, O.: The independency of stellar mass-loss rates on stellar X-ray luminosity and activity level based on solar X-ray flux and solar wind observations. Mon. Not. R. Astron. Soc. **417**, 2592 (2011)

Cohen, O., Drake, J.J.: A grid of MHD models for stellar mass loss and spin-down rates of solar analogs. Astrophys. J. **783**, 55 (2014)

Cohen, O., Drake, J.J., Kashyap, V.L., Hussain, G.A.J., Gombosi, T.I.: The coronal structure of AB Doradus. Astrophys. J. **721**, 80 (2010)

Cohen, O., Kashyap, V.L., Drake, J.J., Sokolov, I.V., Garraffo, C., Gombosi, T.I.: The dynamics of stellar coronae harboring hot Jupiters. I. A time-dependent magnetohydrodynamic simulation of the interplanetary environment in the HD 189733 planetary system. Astrophys. J. **733**, 67 (2011)

Cohen, O., Drake, J.J., Glocer, A., Garraffo, C., Poppenhaeger, K., Bell, J.M., Ridley, A.J., Gombosi, T.I.: Magnetospheric structure and atmospheric Joule heating of habitable planets orbiting M-dwarf stars. Astrophys. J. **790**, 57 (2014)

Cranmer, S.R., Saar, S.H.: Testing a predictive theoretical model for the mass loss rates of cool stars. Astrophys. J. **741**, 54 (2011)

Crooker, N.U., Burton, M.E., Siscoe, G.L., Kahler, S.W., Gosling, J.T., Smith, E.J.: Solar wind streamer belt structure. J. Geophys. Res. **101**, A11, 24331 (1996)

Crosley, M.K., Osten, R.A.: Constraining stellar coronal mass ejections through multi-wavelength analysis of the active M dwarf EQ Peg. Astrophys. J. **856**, 39 (2018)

Debes, J.: Measuring M dwarf winds with DAZ white dwarfs. Astrophys. J. **652**, 652 (2006)

DeForest, C.E., Howard, R.A., Velli, M., Viall, N., Vourlidas, A.: The highly structured outer solar corona. Astrophys. J. **862**, 18 (2018)

Deutsch, A.J.: The circumstellar envelope of Alpha Herculis. Astrophys. J. **123**, 210 (1956)

do Nascimento Jr., J.-D., Vidotto, A.A., Petit, P., Folsom, C., Castro, M., Marsden, S.C., Morin, J., Porto de Mello, G.F., Meibom, S., Jeffers, S.V., Guinan, E., Ribas, I.: Magnetic field and wind of Kappa Ceti: toward the planetary habitability of the young Sun when life arose on Earth. Astrophys. J. Lett. **820**, L15 (2016)

Drake, J.J., Cohen, O., Yashiro, S., Gopalswamy, N.: Implications of mass and energy loss due to coronal mass ejections on magnetically active stars. Astrophys. J. **764**, 170 (2013)

Dupree, A.K., Lobel, A., Young, P.R., Ake, T.B., Linsky, J.L., Redfield, S.: The far-ultraviolet sopectroscopic survey of luminous cool stars. Astrophys. J. **622**, 629 (2005)

Feldman, W.C., Asbridge, J.R., Bame, S.J., Gosling, J.T.: The solar output and its variation. In: White, O.R. (ed.) Proceedings of a Workshop, held in Boulder, Colorado, April 26–28, 1976, p. 351. Colorado Associated University Press, Boulder (1977)

Fichtinger, B., Güdel, M., Mutel, R.L., Hallinan, G., Gaidos, E., Skinner, S.L., Lynch, C., Gayley, K.G.: Radio emission and mass loss rate limits of four young solar-type stars. Astron. Astrophys. **599**, 127 (2017)

Gaidos, E.J., Güdel, M., Blake, G.A.: The faint young Sun paradox: an observational test of an alternative solar model. J. Geophys. Res. Lett. **27**, 501 (2000)

Garraffo, C., Drake, J.J., Cohen, O.: Magnetic complexity as an explanation for bimodal rotation populations among young stars. Astrophys. J. Lett. **807**, L6 (2015a)

Garraffo, C., Drake, J.J., Cohen, O.: The dependence of stellar mass and angular momentum losses on latitude and the interaction of active region and dipolar magnetic fields. Astrophys. J. **813**, 40 (2015b)

Garraffo, C., Drake, J.J., Cohen, O.: The missing magnetic morphology term in stellar rotation evolution (Research Note). Astron. Astrophys. **595**, A110 (2016a)

Garraffo, C., Drake, J.J., Cohen, O.: The space weather of Proxima Centauri b. Astrophys. J. Lett. **833**, L4 (2016b)

Garraffo, C., Drake, J.J., Cohen, O., Alvarado-Gómez, J.D., Moschou, S.P.: The threatening magnetic and plasma environment of Trappist-1 planets. Astrophys. Let. **843**, L33 (2017)

Gayley, K.G., Zank, G.P., Pauls, H.L., Frisch, P.C., Welty, D.E.: One- versus two-shock heliosphere: constraining models with Goddard high resolution spectrograph Ly-α spectra toward α centauri. Astrophys. J. **487**, 259 (1997)

Gopalswamy, N., Yashiro, S., Michalek, G., Stenborg, G., Vourlidas, A., Freeland, S., Howard, R.: The SOHO/LASCO CME Catalog. EM&P **104**, 295 (2009)

Güdel, M.: Stellar radio astronomy: probing stellar atmospheres from protostars to giants. Ann. Rev. Astron. Astrophys. **40**, 217 (2002)

Haisch, B.M., Linsky, J.L., Bornmann, P.L., Stencel, R.E., Antiochos, S.K., Golub, L., Vaiana, G.S.: Coordinated Einstein and IUE observations of a disparitions brusques type flare event and quiescent emission from Proxima Centauri. Astrophys. J. **267**, 280 (1983)

Harper, G.M., Wood, B.E., Linsky, J.L., Bennett, P.D., Ayres, T.R., Brown, A.: A semiempirical determination of the wind velocity structure for the hybrid-chromosphere star α Trianguli Australis. Astrophys. J. **452**, 407 (1995)

Holzwarth, V., Jardine, M.: A further "degree of freedom" in the rotational evolution of star. Astron. Astrophys, **444**, 661 (2005)

Holzwarth, V., Jardine, M.: Theoretical mass loss rates of cool main-sequence stars. Astron. Astrophys. **463**, 11 (2007)

Izmodenov, V., Wood, B.E., Lallement, R.: Hydrogen wall and heliosheath Lyα absorption toward nearby stars: possible constraints on the heliospheric interface plasma flow. J. Geophys. Res. **107**, 1308 (2002)

Jardine, M., Collier Cameron, A.: Slingshot prominences: nature's wind gauges. Mon. Not. R. Astron. Soc. **482**, 2853 (2019)

Jardine, M., Vidotto, A.A., See, V.: Estimating stellar wind parameters from low-resolution magnetograms. Mon. Not. R. Astron. Soc. **465**, L25 (2017)

Johnstone, C.P., Güdel, M., Lüftinger, T., Toth, G., Brott, I.: Stellar winds on the main-sequence. I. Wind model. Astron. Astrophys. **577**, A27 (2015a)

Johnstone, C.P., Güdel, M., Brott, I., Lüftinger, T.: Stellar winds on the main-sequence. II. The evolution of rotation and winds. Astron. Astrophys. **577**, A28 (2015b)

Kasting, J.F., Catling, D.: Evolution of a habitable planet. Ann. Rev. Astron. Astrophys. **41**, 429 (2003)

Lamers, J.G.L.M., Cassinelli, J.P.: Introduction to Stellar Winds. Cambridge University Press, Cambridge (1999)

Lavie, B., Ehrenreich, D., Bourrier, V., Lecavelier des Etangs, A., Vidal-Madjar, A., Delfosse, X., Gracia Berna, A., Heng, K., Thomas, N., Udry, S., Wheatley, P.J.: The long egress of GJ 436b's giant exosphere. Astron. Astrophys. **605**, L7 (2017)

Linsky, J.L., Haisch, B.M.: Outer atmospheres of cool stars. I - The sharp division into solar-type and non-solar-type stars. Astrophys. J. Lett. **229**, 27 (1979)

Linsky, J.L., Wood, B.E.: The α Centauri line of sight: D/H ratio, physical properties of local interstellar gas, and measurement of heated hydrogen (the 'hydrogen wall') near the heliopause. Astrophys. J. **463**, 254 (1996)

Liseau, R., De la Luz, V., O'Gorman, E., Bertone, E., Chavez, M., Tapia, F.: ALMA's view of the nearest neighbors to the Sun. The submm/mm SEDs of the α Centauri binary and a new source. Astron. Astrophys. **594**, A109 (2016)

Lisse, C.M., Dennerl, K., Englhauser, J., Harden, M., Marshall, F.E., Mumma, M.J., Petre, R., Pye, J.P., Ricketts, M.J., Schmitt, J., Trumper, J., West, R.G.: Discovery of X-ray and extreme ultraviolet emission from comet C/Hyakutake 1996 B2. Science **274**, 205 (1996)

Matt, S.P., MacGregor, K.B., Pinsonneault, M.H., Greene, T.P.: Magnetic braking formulation for Sun-like stars: dependence on dipole field strength and rotation rate. Astrophys. J. Let. **754**, L26 (2012a)

Matt, S.P. Pinzón, G., Greene, T.P., Pudritz, R.E.: Spin evolution of accreting young stars. II. Effect of accretion-powered stellar winds. Astrophys. J. **745**, 101 (2012b)

McComas, D.J., Velli, M., Lewis, W.S., Acton, L.W., Balat-Pichelin, M., Bothmer, V., Dirling Jr., R.B., Feldman, W.C., Gloeckler, G., Habbal, S.R., et al.: Understanding coronal heating and solar wind acceleration: case for in situ near-Sun measurements. Rev. Geophys. **45**, 1004 (2007)

Moschou, S.-P., Drake, J.J., Cohen, O., Alvarado-Gomez, J.D., Garraffo, C.: A monster CME obscuring a demon star flare. Astrophys. J. **850**, 191 (2017)

Neugebauer, M., Snyder, C.W.: Mariner 2 observations of the solar wind, 1, average properties. J. Geophys. Res. **71**, 4469 (1966)

Newton, E.R., Irwin, J., Charbonneau, D., Berta-Thompson, Z.K., Dittmann, J.A., West, A.A.: The rotation and Galactic kinematics of mid M dwarfs in the solar neighborhood. Astrophys. J. **821**, 93 (2016)

O'Fionnagáin, D., Vidotto, A.A.: The solar wind in time: a change in the behaviour of older winds? Mon. Not. R. Astron. Soc. **476**, 2465 (2018)

Parker, E.N.: Dynamics of the interplanetary gas and magnetic fields. Astrophys. J.**128**, 664 (1958)

Parker, E.N.: The hydrodynamic theory of solar corpuscular radiation and stellar winds. Astrophys. J. **132**, 821 (1960)

Redfield, S., Linsky, J.L.: The structure of the local interstellar medium. IV. Dynamics, morphology, physical properties, and implications of cloud-cloud interactions. Astrophys. J. **673**, 283 (2008)

Réville, V., Brun, A.S., Matt, S.P., Strugarek, A., Pinto, R.F.: The effect of magnetic topology on thermally driven wind: toward a general formulation of the braking law. Astrophys. J. **798**, 116 (2015)

Robinson, R.D., Collier Cameron, A.: Fast H-α variations on a rapidly rotating spotted star. Proc. Astron. Soc. Aust. **6**, 308 (1986)

Sagan, C., Mullen, G.: Earth and Mars: evolution of atmospheres and surface temperatures. Science **177**, 52 (1972)

See, V., Jardine, M., Fares, R., Donati, J.-F., Moutou, C.: Time-scales of close-in exoplanet radio emission variability. Mon. Not. R. Astron. Soc. **450**, 4323 (2015)

See, V., Jardine, M., Vidotto, A.A., Donati, J.-F., Boro Saikia, S., Fares, R., Folsom, C.P., Hébrard, E.M., Jeffers, S.V., Marsden, S.C., Morin, J., Petit, P., Waite, I.A., BCool Collaboration: studying stellar spin-down with Zeeman-Doppler magnetograms. Mon. Not. R. Astron. Soc. **466**, 1542 (2017)

Segura, A., Walkowicz, L.M., Meadows, V., Kasting, J., Hawley, S.: The effect of a strong stellar flare on the atmospheric chemistry of an Earth-like planet orbiting an M dwarf. Astrobiology **10**, 751 (2010)

Stone, E.C., Cummings, A.C., McDonald, F.B., Heikkila, B.C., Lal, N., Webber, W.R.: Voyager 1 explores the termination shock region and the heliosheath beyond. Science **309**, 5743 (2005)

Stone, E.C., Cummings, A.C., McDonald, F.B., Heikkila, B.C., Lal, N., Webber, W.R.: An asymmetric solar wind termination shock. Nature **454**, 71 (2008)

Sun, X., Bobra, M.G., Hoeksema, J.T., Liu, Y., Li, Y., Shen, C., Couvidat, S., Norton, A.A., Fisher, G.H.: Why Is the great solar active region 12192 flare-rich but CME-poor? Astrophys. J. **804**, 28 (2015)

Suzuki, T.K., Imada, S., Kataoka, R., Kato, Y., Matsumoto, T., Miyahara, H., Tsuneta, S.: Saturation of Stellar Winds from Young Suns. Publ. Astron. Soc. Jpn. **65**, 98 (2013)

Vidal-Madjar, A., Lecavelier des Etangs, A., Désert, J.-M., Ballester, G.E., Ferlet, R., Hébrard, G., Mayor, M.: An extended upper atmosphere around the extrasolar planet HD209458b. Nature **422**, 143 (2003)

Vidotto, A.A., Bourrier, V.: Exoplanets as probes of the winds of host stars: the case of the M dwarf GJ 436. Mon. Not. R. Astron. Soc. **470**, 4026 (2017)

Vidotto, A.A., Opher, M., Jatenco-Pereira, V., Gombosi, T.I.: Simulations of winds of weak-lined T Tauri stars. II. The effects of a tilted magnetosphere and planetary interactions. Astrophys. J. **720**, 1262 (2010)

Vidotto, A.A., Jardine, M., Opher, M., Donati, J.F., Gombosi, T.I.: Powerful winds from low-mass stars: V374 Peg. Mon. Not. R. Astron. Soc. **412**, 351 (2011)

Vidotto, A.A., Fares, R., Jardine, M., Donati, J.-F., Opher, M., Moutou, C., Catala, C., Gombosi, T.I.: The stellar wind cycles and planetary radio emission of the τ Boo system. Mon. Not. R. Astron. Soc. **423**, 3285 (2012)

Vidotto, A.A., Jardine, M., Morin, J., Donati, J.F., Opher, M., Gombosi, T.I.: M-dwarf stellar winds: the effects of realistic magnetic geometry on rotational evolution and planets. Mon. Not. R. Astron. Soc. **438**, 1162 (2014)

Vidotto, A.A., Fares, R., Jardine, M., Moutou, C., Donati, J.-F.: On the environment surrounding close-in exoplanets. Mon. Not. R. Astron. Soc. **449**, 4117 (2015)

Villadsen, J., Hallinan, G., Bourke, S., Güdel, M., Rupen, M.: First detection of thermal radio emission from solar-type stars with the Karl G. Jansky Very Large Array. Astrophys. J. **788**, 112 (2014)

Wang, Y.-M.: Cyclic magnetic variations of the Sun. In: Donahue, R.A., Bookbinder, J.A. (eds.) ASP Conf. Ser. 154, The Tenth Cambridge Workshop on Cool Stars, Stellar Systems and the Sun, p.131 (1998)

Wargelin, B.J., Drake, J.J.: Observability of stellar winds from late-type dwarfs via charge exchange X-Ray emission. Astrophys. J. Lett. **546**, L57 (2001)

Wargelin, B.J., Drake, J.J.: Stringent X-ray constraints on mass loss from Proxima Centauri. Astrophys. J. **578**, 503 (2002)

Weber, E.J., Davis Jr., L.: The angular momentum of the solar wind. Astrophys. J. **148**, 217 (1967)

Willson, L.A.: Mass loss from cool stars: impact on the evolution of stars and stellar populations. Ann. Rev. Astron. Astrophys. **38**, 573 (2000)

Wood, B.E.: Astrospheres and solar-like stellar winds. Living Rev. Solar Phys. **1**, 2 (2004)

Wood, B.E., Linsky, J.L.: Resolving the ξ Boo binary with Chandra, and revealing the spectral type dependence of the coronal "FIP Effect". Astrophys. J. **717**, 1279 (2010)

Wood, B.E., Linsky, J.L., Müller, H.-R., Zank, G.P.: Observational estimates for the mass-loss rates of α Centauri and Proxima Centauri using Hubble Space Telescope Lyα spectra. Astrophys. J. Lett. **547**, L49 (2001)

Wood, B.E., Müller, H.-R., Zank, G.P., Linsky, J.L.: Measured mass-loss rates of solar-like stars as a function of age and activity. Astrophys. J. **574**, 412 (2002)

Wood, B.E., Linsky, J.L., Müller, H.-R., Zank, G.P.: A search for Lyα emission from the astrosphere of 40 Eridani A. Astrophys. J. **591**, 1210 (2003)

Wood, B.E., Müller, H.-R., Zank, G.P., Linsky, J.L., Redfield, S.: New mass-loss measurements from astrospheric Lyα absorption. Astrophys. J. Lett. **628**, L143 (2005)

Wood, B.E., Müller, H.-R., Redfield, S. Edelman, E.: Evidence for a weak wind from the young Sun. Astrophys. J. Lett. **781**, L33 (2014)

Zank, G.P., Pauls, H.L., Williams, L.L., Hall, D.T.: Interaction of the solar wind with the local interstellar medium: a multifluid approach. J. Geophys. Res. **101**, 21639 (1996)

Zank, G.P., Heerikhuisen, J., Wood, B.E., Pogorelov, N.V., Zirnstein, E., McComas, D.J.: Heliospheric structure: the bow wave and the hydrogen wall. Astrophys. J. **763**, 20 (2013)

Chapter 9
Activity Indicator Correlations

Measurements of activity indicators including stellar X-ray and UV emission and wind flux are essential for calculating the photochemistry and hydrodynamical outflow rate in exoplanet atmospheres at the present time, and it is also important to estimate what the stellar fluxes were at earlier times and will likely become in the future. Assessing the habitability of an exoplanet requires knowledge of the host star's emission back to the protoplanetary stage. We therefore need to know how emission depends on stellar properties including mass, age, and rotation rate, which depends on a star's age and mass. This chapter reviews the systematic studies of stars in different mass ranges that have led to correlations and scaling laws for the dependence of stellar emission on age and rotation rate. Christensen-Dalsgaard and Aguirre (2018) compare the uncertainties in the commonly used dating techniques for host stars (e.g., membership in clusters, gyrochronology, evolution) with the most precise asteroseismology method presently used only for sufficiently bright stars. The studies cited in this chapter, generally based on ages obtained before the analysis of Kepler photometry using asteroseismology, are a representative sample of the many studies directed to understanding how activity parameters depend on stellar parameters. These studies have also identified correlations among different activity indicators that facilitate predictions of the strength of an activity indicator from observations of other activity indicators.

9.1 Correlations of Activity Indicators with Stellar Parameters: Mass, Age, and Rotation Period

Astronomers have known for a long time that emission in the cores of the Ca II H and K lines occurs only in stars of late spectral type (F, G, K, and M), that stars in young clusters show far brighter emission than older stars, and that the Ca II emission is stronger in rapidly rotating stars (e.g., Kraft 1967). The first study to quantify

© Springer Nature Switzerland AG 2019
J. Linsky, *Host Stars and their Effects on Exoplanet Atmospheres*,
Lecture Notes in Physics 955, https://doi.org/10.1007/978-3-030-11452-7_9

the age and rotation dependence of Ca II emission was that of Skumanich (1972), who showed that this emission feature and the stellar rotation period decay with age as $t^{-1/2}$. Subsequent studies have characterized the age and rotational period dependence of emission in the Ca II H and K lines, the corresponding Mg II h and k lines, and the Ca II infrared triplet lines (e.g., Hartmann et al. 1984; Pace and Pasquini 2004), Busà et al. 2007). The recent study of Ca II H and K emission for solar twins ($T_{eff} = 5777 \pm 100$ K and solar like gravity and abundances) between ages 0.6–9 Gyr shows a power law slope $t^{-0.52}$ (Lorenzo-Oliveira et al. 2018).

Another indicator of activity in stellar chromospheres is the Hα line, which becomes a deeper absorption line in active G and K stars and less active M stars but becomes an emission line as seen against the faint continuum of the more active M and cooler stars (e.g., Walkowicz and Hawley 2009). In their survey of ultracool dwarf stars, Schmidt et al. (2015) detected Hα emission in M7 to L8 stars with the fraction of active stars, identified by Hα emission, peaking at 90% for L0 dwarfs.

Simon et al. (1985) extended the study of activity indicators to those formed above the chromosphere by analyzing ultraviolet spectra obtained with the *IUE* spacecraft. They fitted the fluxes of low chromosphere lines (e.g., Ca II, Mg II, O I) and transition region lines (e.g. C II, Si IV, C IV, N V) with power laws (see Table 9.1) for F7–G2 V stars in clusters as young as 300 Myr. They also showed that the data can be fit with exponential distributions in which significant decay is not apparent until near 1 Gyr. A major question is whether the decrease in flux of an activity indicator begins at an early phase in a star's evolution, or whether the activity indicator has a saturated value until a time t_0 and decays thereafter.

There have been a number of studies of stellar X-ray emission as a function of stellar age and rotational period. Güdel et al. (1997) analyzed *ASCA* and *ROSAT* spectra of eleven stars with spectral types similar to the Sun (G0 V to G5 V), ages 0.07 to 9 Gyr, and rotational periods $P_{rot} = 2.75$ days to about 30 days. The X-ray spectra of the younger stars consist of a hot (5–30 MK) component and a cooler component similar to the nonflaring Sun. The temperature of the hot component decreases rapidly with decreasing rotational period (proportional to $P_{rot}^{-0.55}$) and age ($t^{-0.3}$). The hot component, which is produced by continuous flaring and accelerates high-energy electrons responsible for gyrosynchrotron radio emission, is essentially gone by about age 500 Myr except for occasional flares. Older solar-like stars, therefore, have X-ray spectral energy distributions shifted to longer wavelengths than young stars. The total X-ray luminosity from both components is saturated at a value near $L_x/L_{bol} \approx 10^{-3}$ until an age of about 70 Myr ($P_{rot} = 2.75$ days), the parameters for the star EK Dra (G0 V). As solar-type stars age and their rotation decelerates, their X-ray luminosity decays as $P_{rot}^{-2.64\pm0.12}$ and $t^{-1.5}$ (see Fig. 9.1). The X-ray luminosity of the nonflaring Sun is presently at the level $L_X/L_{bol} \approx 10^{-6.3}$, a factor of 2000 below the saturation level.

Ribas et al. (2005) have assembled fluxes from a variety of X-ray and UV instruments for stars similar in spectral type to the Sun (G0–G2 V) and ages $t = 0.1 - 7$ Gyr. As shown in Fig. 9.2 and listed in Table 9.1, there is a trend of more rapid decrease with time at shorter wavelengths that correspond to higher temperatures from the transition region to the corona. This age dependence is

Table 9.1 Age dependence of activity parameters

Spectral feature	Mass range	Number of stars	t_0 (Myr)	α	Source	Ref.
Ca II H+K	F4–K5			$-1/2$	Optical	1
Ca II H+K	Solar twins	60	600	-0.52	Optical	10
Ca II, Mg II	F7–G2 V	29		-0.50 ± 0.07	*IUE*	3
NUV	0.3–0.65	215	≈ 300	-0.84 ± 0.009	*GALEX*	2
NUV	M0–M5 V	72		-0.90 ± 0.01	*GALEX*	6
FUV	0.3–0.65	215	≈ 300	-0.99 ± 0.19	*GALEX*	2
FUV	M0–M5 V	72		-0.72 ± 0.05	*GALEX*	6
Lyman-α	M1–M5 V	9		-0.647	*HST*	8
C II–N V	F7–G2 V	31		-1.01 ± 0.12	*IUE*	3
92–118 nm	G0–G2 V	7		-0.85	Many	7
10–36 nm	G0–G2 V	7		-1.20	Many	7
2–10 nm	G0–G2 V	7		-1.27	Many	7
0.1–10 nm	G0–G2 V	7		-1.92	Many	7
X-ray	G0–G5 V	12	≈ 70	-1.5	*ROSAT, ASCA*	4
X-ray	0.3–0.65	215	≈ 300	-1.36 ± 0.32	*ROSAT*	2
X-ray	F7–G9 V	61		$-1.5^{0.3}_{-0.2}$	*Einstein*	5
X-ray	M0–M5 V	93		-1.10 ± 0.02	*ROSAT, XMM*	6
X-ray	M0–M5 V	12		-1.424	*Chandra*	8
X-ray SF	F4–M2 V	14	1000	-2.8 ± 0.72	*Chandra, XMM*	9

References: (1) Skumanich (1972); (2) Shkolnik and Barman (2014); (3) Simon et al. (1985); (4) Güdel et al. (1997); (5) Maggio et al. 1987; (6) Stelzer et al. (2014); (7) Ribas et al. (2005); (8) Guinan et al. (2016); (9) Booth et al. (2017); Lorenzo-Oliveira et al. (2018)

usually fitted by a power-law t^{α}. For these stars, which are likely good prototypes for the Sun in time, the X-ray emission decreases a factor of 1000 from $t = 0.1$ Gyr to the present age of the Sun, while the transition-region emission in the 92–120 nm bandpass decreases by only a factor of 30.

In their near volume-limited sample of M0–M5 V dwarf stars within 10 pc, Stelzer et al. (2014) obtained power-law relations for the decay with age of the activity ratios L_{band}/L_{bol} for the nonphotospheric fluxes in the *GALEX* NUV and FUV bands and for the X-ray flux observed by the *ROSAT* and *ASCA* satellites.

Guinan et al. (2016) assembled *Chandra* and *HST/COS* spectra of nine M1.5–M5.5 V stars with ages from 0.2 Gyr to 11.5 Gyr to study the evolution of X-ray and UV emission from M stars. The oldest star in the group is Kapteyn's star (M1.5 V) which is metal poor high radial velocity star. Guinan et al. (2016) found that the X-ray emission of these M stars decays with a steeper slope of $\alpha = -1.424$ (see Fig. 9.3) than the reconstructed Lyman-α flux with its slope $\alpha = -0.647$ (see Fig. 9.4). Over the course of 11.5 Gyr, the star's X-ray emission decreased a factor of 71, whereas its Lyman-α emission, indicative of chromospheric emission, decreased by only a factor of 7 from the very strong fluxes it had at age 0.2 Gyr. Despite this decay in emission and the star's very old age, the X-ray emission received by a

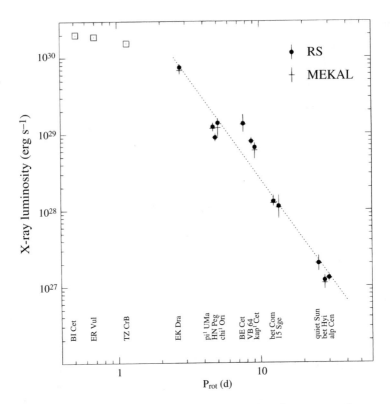

Fig. 9.1 Dependence of X-ray luminosity on rotational period for solar-mass stars. Star names are at the bottom of the figure. RS and MEKAL refer to different plasma spectroscopy codes used to infer the X-ray luminosity. Figure from Güdel et al. (1997). Reproduced by permission of the AAS

putative exoplanet in the habitable zone (at 0.17 AU) is 20 times larger than what the Earth receives from the Sun, and the Lyman-α flux is 1.7 times larger than that received by the Earth. These fluxes from this very old star and the far larger fluxes when the star was young are essential input for evaluating whether the exoplanet could be habitable. For the ultracool stars, there is evidence for a steep decline in L_x/L_{bol} from the mid-M stars to the mid-L dwarfs (Berger et al. 2010).

Shkolnik and Barman (2014) studied 215 stars observed by the *GALEX* spacecraft in the FUV and NUV spectral bands. These stars are members of a variety of moving groups with ages from 10 Myr (TW Hydra Association) to 650 My (Hyades Cluster) and older field stars. After subtraction of photospheric emission, the fluxes in the FUV and NUV passbands are nearly constant until 300 My and then decay as $t^{-\alpha}$ (see Fig. 9.5).

More recently, Booth et al. (2017) investigated the previously uncertain X-ray-age relation for a sample of dwarf stars older than 1 Gyr with precise ages obtained primarily from asteroseismology. Their plot of the X-ray surface flux, $L_x/(4\pi R^2)$,

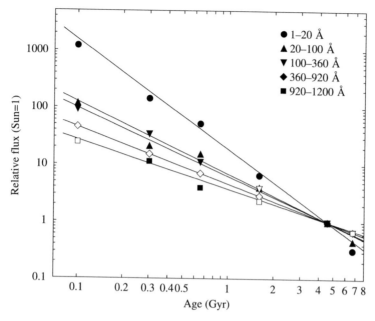

Fig. 9.2 Fluxes in different passbands relative to the Sun as a function of age. The stars are similar to the Sun with spectral types in the range G0–G2 V. Figure from Ribas et al. (2005). Reproduced by permission of the AAS

as a function of age for stars without hot Jupiter companions shows a power-law relation with a slope -2.80 ± 0.72, much steeper than previous determinations (e.g., Jackson et al. 2012; Guinan et al. 2016). Further study is needed to confirm this relation, which is based on only 14 stars, to search for any dependence on T_{eff} and thus on the depth of the convective zone and to identify the cause of this apparent very rapid decrease in X-ray surface flux with age.

9.1.1 Rotation Evolution and Activity Indicators

Reiners and Mohanty (2012) developed a model for the rotation evolution of stars including spin-up as a star contracts during the premain sequence phase and spin-down due to angular momentum loss by magnetic winds. The age t_{crit} at which a star spins down from the saturated to the unsaturated magnetic field regime is very long for low mass stars because their small radii result in small wind torques and contraction of fully convective stars produces spin-up. The increase in t_{crit} with decreasing mass shown in Fig. 9.6 is similar to the empirical activity lifetimes measured by the stellar age at which Hα line emission is no longer observed (West et al. 2008). The close agreement of these two parameters shown in the figure likely results from the rapid spin-down time scale and the resulting decrease in the mean

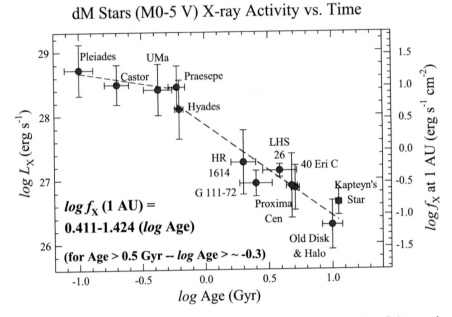

Fig. 9.3 X-ray luminosities and fluxes at 1 AU for M0–M5 V stars. Figure from Guinan et al. (2016). Reproduced by permission of the AAS

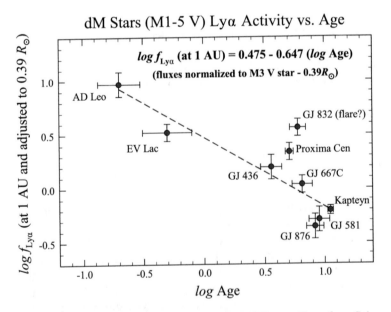

Fig. 9.4 Lyman-α fluxes at a distance of 1 AU from M1–M5 V stars. Figure from Guinan et al. (2016). Reproduced by permission of the AAS

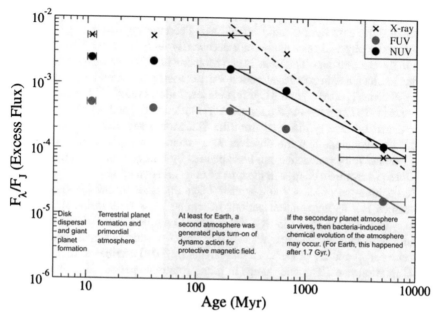

Fig. 9.5 Median X-ray and *GALEX* FUV and NUV excess fluxes as a function of stellar age. Excess fluxes are observed fluxes minus the photospheric component. Figure from Shkolnik and Barman (2014). Reproduced by permission of the AAS

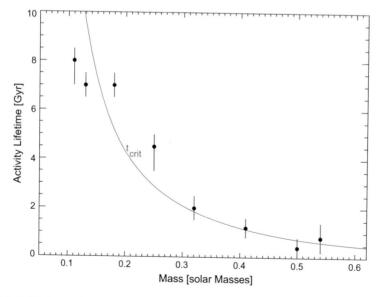

Fig. 9.6 Solid dots and error bars are activity lifetimes estimated by West et al. (2008) on the basis of the detection limit of Hα emission and ages from Galactic dynamics. The red line is the time scale for rapid rotational breaking. Figure from Reiners and Mohanty (2012). Reproduced by permission of the AAS

magnetic fields in low-mass stars that occurs for stars older than t_{crit}. The observed fluxes of other activity indicators should also decrease with stellar age and mass for the same physical reasons, but not necessarily with the same time scale as is seen in the Hα line data. The Reiners and Mohanty (2012) rotation evolution model predicts a decrease in rotational period for 1.0 solar mass stars between the ages of the Pleiades (100 Myr) and 8 Gyr that closely follows the $t^{-1/2}$ law proposed by Skumanich (1972), however, the rotational evolution of 0.5 and 0.1 solar mass stars is predicted to be very different from the Skumanich law. Reiners and Mohanty (2012) also conclude that the observed steep change in t_{crit} near $0.35 M_\odot$ where stars become fully convective can be explained by a large change in stellar radius and does not require a change in the type of magnetic dynamo.

The spin-down of stars younger than 1 Gyr does not follow the simple $t^{-0.5}$ Skumanich law as the rotation periods of similar mass stars in the same young cluster show a wide range in P_{rot} despite their common age. Solar-like stars in clusters with known ages less than 1 Gyr exhibit a wide range in their activity indicators R'_{HK} and L_x at each age that Gondoin (2018) ascribes to their different initial rotation rates. The wide range in initial rotational periods as stars reach the main sequence should damp out to a uniform rotation rates within 100 Myr and follow the Skumanich law, but such damping is not observed to occur until 1 Gyr for solar mass stars. Garraffo et al. (2018) reviewed the various attempts to explain this unexpected behavior and proposed that magnetic field complexity may be the cause. Zeeman Doppler images of young stars show complex magnetic morphologies that simplify into more nearly dipolar configurations as stars age, P_{rot} increases, and R_0 increases from < 0.1 in the saturated regime to larger values in the unsaturated regime. With MHD simulations, Garraffo et al. (2015) showed that the angular momentum loss of a star is much smaller for complex magnetic geometries than for simple dipole geometries. Thus stars with initially high rotation rates will stay in the saturated regime with much less angular momentum loss for a longer time than stars that were initially slow rotators. Garraffo et al. (2018) argued that this diversity in angular momentum loss rates produces the wide range in rotational periods for stars of similar mass and age until about 1 Gyr for solar mass stars.

The MEarth project is providing a wealth of rotational periods for M dwarf stars from observations of the periodic variations in optical brightness as starspots rotate onto and off of the star's observed hemisphere. With measured rotational periods and Hα spectra of a sample of 164 stars, West et al. (2015) investigated the fraction of these stars that are active using for an activity criterion the presence of emission in the Hα line. By this criterion, all of the M1–M4 V stars in their sample with rotational periods $P_{rot} < 26$ days are active and all of the M5–M8 V stars with $P_{rot} < 86$ days are active. The M1–M4 V stars show a clear decrease in activity with $P_{rot} > 26$ days, but the M5–M8 V stars maintain the same level of activity until $P_{rot} > 86$ days. They argued that rotation indicates age because the faster rotating M dwarfs are kinematically young and the slow rotators are kinematically old. Newton et al. (2016, 2018) extended this work with a larger sample of stars, concluding that M dwarfs maintain their rapid rotation for several Gyr and then rapidly spin down to rotation periods longer than 100 days with the lowest mass stars taking the longest time to spin down. Thus the combination of rotation period, mass and kinematics leads to rough age estimates for M dwarfs in the field.

Analysis of Kepler's systematic optical monitoring of a large number of stars allowed McQuillan et al. (2013, 2014) to create a data base of 34,030 stars with $P_{rot} = 0.2 - 70$ days. Although these stars do not have ages from $H\alpha$ or other spectroscopic activity indicators, the rotation periods of the slowest rotators are consistent with an age of about 4.5 Gyr. An interesting result is the detection of two peaks at 19 and 33 days in the rotation period distribution with a definite gap in between that suggests that star formation in the solar neighborhood occurred at two different times producing two stellar populations with different mean ages. A similar result was suggested by Vaughan and Preston (1980) on the basis of a deficiency in intermediate Ca II H and K line strength for F and G stars within 25 pc of the Sun.

9.1.2 Which Is the Better Parameter: Rossby Number or Rotation Period?

Using a group of 41 F2–K5 V stars with well-determined rotation periods, Noyes et al. (1984) computed the chromospheric emission in the Ca II H and K lines and compared these data to different rotation parameters. For their chromospheric Ca II emission parameter, they used R'_{HK}, the observed fluxes in 0.1 nm passbands centered on the H and K lines corrected for photospheric emission and divided by the total bolometric luminosity of the star. They found that their data fit a smooth curve with minimal scatter when R'_{HK} was plotted against the Rossby number $R_0 = P_{rot}/\tau_{conv}$. Here τ_{conv} is the convective turnover time in the convective zone, which depends on the depth of the convective zone and thus on stellar mass. While τ_{conv} plays an important role in hydrodynamical dynamo theories for creating magnetic fields, this parameter is not accurately measured as it likely varies with depth in a star's convective zone. In their analysis, Noyes et al. (1984) empirically modified theoretical estimates of τ_{conv} in order to obtain an optimal fit to the R'_{HK} data. Like other activity parameters, R'_{HK} measures the heating rate in a region of the stellar atmosphere, in this case the chromosphere, and therefore should depend on the magnetic fields responsible for the heating even though the physical mechanism for the heating remains uncertain.

Figure 9.7 shows a plot the $H\alpha$ emission normalized by the stellar bolometric luminosity vs. R_0. This figure is based on 466 M stars with measured $H\alpha$ emission and photometric rotational periods mostly measured by the MEarth program (Newton et al. 2017). For these stars, $L_{H\alpha}/L_{bol}$ is saturated at the value $(1.49 \pm 0.08) \times 10^{-4}$ for $R_0 < 0.21 \pm 0.02$. At larger values of R_0, $L_{H\alpha}/L_{bol}$ is proportional to t^β with a power-law slope of $\beta = -1.7 \pm 0.1$. Douglas et al. (2014) found similar parameters for the M and K dwarfs in the 600 Myr old Hyades and Praesepe clusters. An important and unexpected result is that the power-law slope does not vary appreciably with stellar mass and, therefore, whether the stars

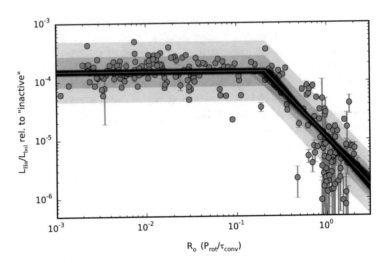

Fig. 9.7 Ratio of the relative Hα emission, $L_{H\alpha}/L_{bol}$, vs Rossby number for M stars. Figure from Newton et al. (2017). Reproduced by permission of the AAS

are fully convective when $M < 0.35M_\odot$ (Chabrier and Baraffe 1997) or have a radiative cores with convective envelopes when $M > 0.35M_\odot$. The extensive Kepler data (McQuillan et al. 2013, 2014) also show no evidence for a change in the rotation-mass distribution near the $M = 0.35M_\odot$ threshold for the onset of complete convection. Why stellar dynamos or at least their effect on activity indicators do not change appreciably across this important threshold in stellar interior structure is not understood.

The success in fitting the Ca II activity indicator with a simple parameter stimulated many studies of other activity indicators to determine whether they are well fit by plots against the Rossby number or other rotation parameters. For example, Pallavicini et al. (1981) found that the X-ray luminosities of a variety of G–M stars observed by the Einstein spacecraft are consistent with L_x proportional to $(\upsilon \sin \iota)^{1.9\pm0.5}$. Vilhu and Walter (1987) called attention to the rapidly rotating stars for which $L_x/L_{bol} \approx 10^{-3}$ and other activity indicators follow similar relations with no dependence on rotation period or stellar mass. With the much larger data set of *ROSAT* X-ray luminosities including stars in young clusters, Pizzolato et al. (2003) found that the $L_x/L_{bol} \approx 10^{-3}$ saturation regime increases from about 1.5 days for solar mass stars to about 4 days for $0.5M_\odot$ stars. For more slowly rotating stars in the unsaturated regime, the data can be fit either by $L_x/L_{bol} \alpha R_0$ or by $L_x \alpha P_{rot}^{-2}$, but which formula is a more physical way of understanding the data?

Employing a much larger data base of rotation periods and X-ray luminosities, Wright et al. (2011) quantified the saturated and unsaturated regimes in terms of plots of $\log(L_x/L_{bol})$ vs Rossby number. With an unbiased subsample of stars, they found that in the saturated regime ($R_0 < 0.13$) the best fit is $(L_x/L_{bol}) = -3.13 \pm 0.22$. For very fast rotating F, G, and K stars, there is some evidence

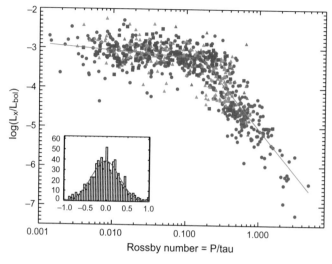

Fig. 9.8 Normalized X-ray luminosity vs Rossby number. Colors indicate stellar ages from < 50 Myr (blue), 85–150 Myr (green), 600–700 Myr (magenta), and older field stars (red). The near horizontal line is the fit to the saturated regime and the sloping line is the fit to the unsaturated regime. The insert plot indicates the number of stars as a function of Rossby number. Figure from Reiners et al. (2014). Reproduced by permission of the AAS

for a decrease in L_x/L_{bol} but its cause is uncertain. In the unsaturated regime ($R_0 > 0.13$), L_x/L_{bol} in proportional to R_0^{β}, where $\beta = 2.70 \pm 0.13$, a much larger number than some previous estimates. To extend the $\log(L_x/L_{bol})$ vs R_0 relation to fully convective stars, Wright et al. (2018) obtained *Chandra* X-ray fluxes for 19 slowly rotating M3 V to M6 V stars in the unsaturated regime. For these stars, $\beta = -2.3^{+0.4}_{-0.6}$, which is consistent with the earlier result for warmer stars that have radiative cores and convective envelopes. This remarkable agreement in slope requires that fully convective stars either have solar-like $\alpha\Omega$ magnetic dynamos without a tachocline shear layer between a radiative core and a convective envelope, or that fully convective stars have a magnetic dynamo very different from the Sun that somehow produces the same rotation activity relations as seen in solar-like stars. Wright et al. (2018) discuss these options.

Reiners et al. (2014) also analyzed a large data set of X-ray observations including stars in young clusters. Figure 9.8 shows their L_x/L_{bol} data plotted relative to the Rossby number. This figure also shows the two regimes: a saturated regime ($R_0 < 0.2$) in which L_x/L_{bol} is nearly constant when a star rotates rapidly, and an unsaturated regime ($R_0 > 0.2$) in which L_x/L_{bol} decreases with increasing R_0 and therefore slower rotation. They found, however, that there is less scatter when L_x in the unsaturated regime is plotted relative to P_{rot}^2 with no dependence on stellar mass, convective zone depth or any other stellar parameter as Pallavicini et al. (1981) originally proposed. Reiners et al. conclude that the star's rotation alone determines L_x and the coronal heating rate.

The Reiners et al. (2014) conclusion fits in with another result. By combining solar and stellar X-ray measurements with unsigned magnetic flux measurements Φ, Pevtsov et al. (2003) obtained the remarkable result that the two quantities are linearly proportional for solar regions, the solar disk average, and stars over 12 orders of magnitude as shown in Fig. 2.1. Although the data show scatter about the power-law relation $L_x \alpha \Phi^p$ where $p = 1.13 \pm 0.05$, the near-linear relation indicates that closed magnetic field structures in the corona are heated by a mechanism proportional to the mean magnetic flux in the underlying solar/stellar surface. The magnetic heating mechanism may well be the same in the Sun and stars and appears to be independent of stellar parameters such as radius and effective temperature. The combination of the two results concerning L_x implies that the magnetic flux in a stellar atmosphere depends only on the stellar rotation rate.

9.2   Correlations Among Activity Indicators

It is reasonable to expect that the emission in two activity indicators (usually emission lines) in a number of stars will correlate, that is $\log F_1 = m \log F_2 + c$ with a power-law index m of about unity, when both emission lines are formed in the same region of the stellar atmospheres with similar thermal-height distributions. Figure 4.8 shows models of different regions of the Sun with different heating rates and activity levels. An interesting aspect of this figure is that the thermal structures of these models have very similar shapes in their chromospheres. Other grids of stellar model atmospheres also show similar thermal structures. This suggests that for a group of stars with a moderate range in mass and gravity but with very different activity levels, that different emission features formed in their chromospheres should correlate with $m \approx 1.0$. The transition regions in Fig. 4.8 have thermal structures that are similar to each other but are very different from the underlying chromosphere regions. This suggests that transition region lines should correlate with slopes $m \approx 1.0$, but that the correlation of transition region lines with chromosphere lines may well have slopes significantly different from unity. The correlation of coronal X-ray emission with transition region and chromosphere emission may also have slopes different from unity. Empirical studies of power-law indices among different activity indicators are needed both to better understand the thermal structures of stellar atmospheres and to predict the fluxes of activity indicators that have not or can not be observed.

9.2.1   Correlations Among Chromosphere and Transition Region Activity Indicators

There have been a number of studies of the correlations of transition region emission lines (e.g., He II, C II, C III, Si IV, C IV, and N V) with emission lines formed

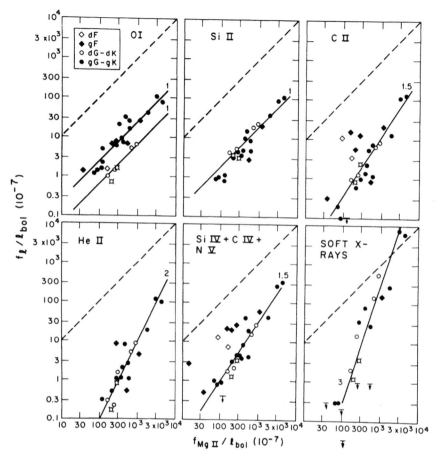

Fig. 9.9 Correlations of emission line surface fluxes divided by the bolometric flux of transition region lines with respect to the chromospheric Mg II lines. Open symbols are for dwarf stars and filled symbols are for giants. Solid lines are power-law fits with slopes 1 (chromospheric lines), 1.5 (transition region lines), and 3 (coronal X-ray emission) as noted. The dashed line is for the correlation of Mg II with itself. Values for the Sun at minimum and maximum are spiked circles. Figure from Ayres et al. (1981). Reproduced by permission of the AAS

in the chromosphere (e.g., Ca II H and K, Mg II h and k, O I, and Hα). Ayres et al. (1981) presented the first set of correlations based on early *IUE* spectra of a group of 4 dwarf and subgiant stars with spectral types F2 IV to K2 V and 9 giant and supergiant stars with spectral types A7 III to K2 III. Despite the diversity of star types and evolution, Fig. 9.9 shows that the correlations of chromospheric lines (O I and Si II) with Mg II have power-law indices of 1, whereas the correlation of transition region lines with Mg II have power-law slopes of 1.5. The power-law slope for X-rays relative the Mg II is 3, and for the He II 164 nm line the slope is 2. The intermediate slope for the He II line likely results from this line being formed

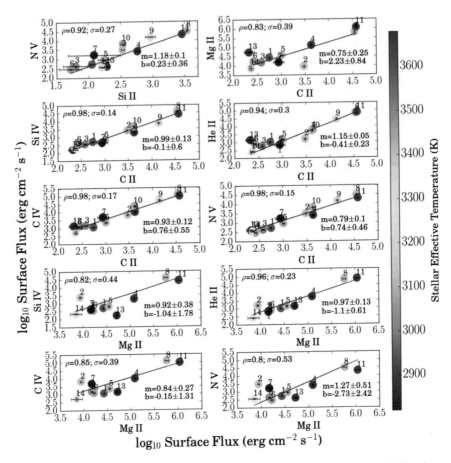

Fig. 9.10 Correlations of emission line surface fluxes of transition region lines with the chromospheric Mg II lines for K and M dwarfs observed in the MUSCLES Treasury Survey. The red lines are power law fits to F_{line1} proportional to F_{line2}^{m}. The power-law fits lie in the range $m = 0.75 - 1.27$. Figure from Youngblood et al. (2017). Reproduced by permission of the AAS

in part by recombination following X-ray photoionization and in part by collisional excitation in the transition region. Ayres et al. (1981) discussed possible causes for the large scatter among the data for giant stars and suggested that the different slopes may indicate different heating mechanisms.

The spectra of the 4 K stars and 7 M stars observed in the MUSCLES Treasury Survey allowed Youngblood et al. (2017) to correlate the surface fluxes of chromospheric and transition region lines. Figure 9.10 shows plots of these surface fluxes with power-law fits to the data in the form F_{line1} proportional to F_{line2}^{m}. The fits of the He II, C II, C IV, and N V lines with respect to the chromospheric Mg II lines have power-law indices in the range $m = 0.84$–1.27. Fits of transition region lines relative to the Ca II H and K lines have power-laws in a similar range, $m = 0.88$–1.35.

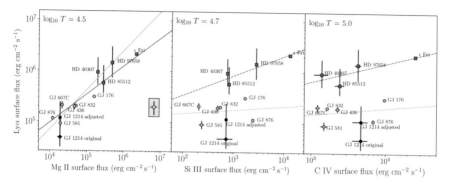

Fig. 9.11 Correlations of the Lyman-α surface flux compared with the Mg II surface flux for M dwarfs (orange circles) and K dwarfs (red squares). The orange and red solid lines are power-law fits to the M dwarfs and K dwarfs, respectively. Figure from Youngblood et al. (2016). Reproduced by permission of the AAS

Power-law fits with indices close to unity are also found when plotting the *GALEX* FUV and NUV fluxes vs. Mg II flux (Shkolnik et al. 2014) and vs. Hα flux (Jones and West 2016). The broad band FUV/NUV flux ratios observed by France et al. (2016) increase rapidly with decreasing stellar effective temperature and thus with decreasing distance of the habitable zone from the star. The FUV/NUV flux ratio plays an important role in the oxygen photochemistry of an exoplanet's atmosphere (see Sect. 11.3).

The increasing power-law slope obtained by comparing emission from hotter regions with respect to chromospheres is consistent with the increasing power-law slopes of coronal X-ray emission as a function of stellar age relative to chromospheric emission (see Fig. 9.2). However, the Lyman-α line is an exception. After reconstructing the intrinsic Lyman-α emission of 40 dwarf and giant stars, Wood et al. (2005) found that the power-law index for Lyman-α emission relative to Mg II is 0.82 ± 0.04 for F and G dwarfs and 0.89 ± 0.06 for K dwarfs not unity as suggested by the Ayres et al. (1981) correlation of chromospheric lines relative to Mg II emission. Youngblood et al. (2016) also found power law indices less than unity: 0.55 ± 0.11 for the MUSCLES K dwarfs, and 0.77 ± 0.10 for the M dwarfs (Fig. 9.11). The very large optical depth of the Lyman-α line compared to other chromospheric lines may be a cause of the different slopes. Typical ratios of the intrinsic unreddened Lyman-α to Mg II fluxes are 10 ± 3 (France et al. 2013). Linsky et al. (2013) discuss correlations of the Lyman-α flux with Ca II and Mg II fluxes for different spectral type stars.

The Hα and Ca II lines are both formed in chromospheres, but their correlation is more complex than is seen for other chromosphere lines. The reason is that for stars with very weak activity the Hα line is in absorption compared with the brighter near-IR continuum. With increasing activity as indicated by stronger Ca II K line emission, the increase in Hα opacity forces the line to first become deeper in absorption (Hα equivalent width becomes more negative) and then the line fills in

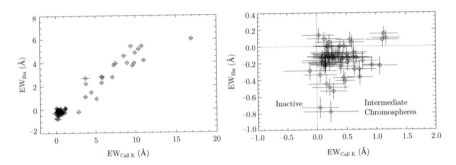

Fig. 9.12 *Left:* Correlation of equivalent widths of the Hα the Ca II K lines. Equivalent widths of the Hα line are negative when the line is in absorption and positive when the line is in emission. *Right:* An expanded version of the plot on the left for low activity stars showing that with increasing activity (measured by the Ca II K line emission) the Hα line first shows stronger absorption and then goes into emission. Figure from Walkowicz and Hawley (2009). Reproduced by permission of the AAS

to become an emission line (positive equivalent width). This is shown in radiative transfer calculations (e.g., Cram and Mullan 1985) and in the observations of many M3 dwarf stars (Walkowicz and Hawley 2009). The left hand panel of Fig. 9.12 shows that for active stars the two lines are well correlated. The right hand panel, which is an expanded view of the low activity stars, shows that the Hα equivalent width first becomes more negative and then becomes positive as the Ca II K line flux increases. The transition from Hα absorption to emission depends on stellar type because for hotter stars the near-IR continuum becomes brighter making the Hα line appear fainter. For cooler stars the near-IR continuum is weaker and the Hα line goes into emission at lower activity levels. This complex behavior of the Hα line makes it an ambiguous indicator of stellar activity for low activity stars.

9.2.2 Correlations Among Chromosphere and Coronal Activity Indicators

Figure 9.13 compares measurements of the Lyman-α and X-ray fluxes of stars from spectral type F0 V to M8 V. The data are mostly from Linsky et al. (2013) and the data for Trappist-1 is from Guinan et al. (2016). The data for the F, G, and K stars are well correlated with slopes increasing slowly for the cooler stars. The slope for the M1-M2 stars is steeper and the cooler M stars lie below the warmer M stars. At an X-ray flux level of about 0.1, the Lyman-α flux of Trappist-1 is more than 100 times fainter than for the F–K stars. This indicates the rate of heating of stellar coronae compared to chromospheres declines rapidly with decreasing effective temperature in the late M stars. During solar flares, several continua and emission lines formed in the chromosphere and transition region closely follow the evolution of the X-ray

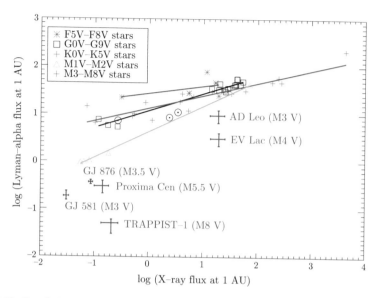

Fig. 9.13 Correlations of the Lyman-α flux with X-ray flux, both measured at 1 AU distance from the star. The Lyman-α flux becomes relatively weaker compared to the X-ray flux with later spectral type especially for the late M dwarfs. Figure from Linsky (2019)

flux (Milligan et al. 2014), demonstrating the tight correlation of chromospheric and coronal activity indicators for an individual star.

9.2.3 Correlations Among Coronal Activity Indicators

Figure 6.22 shows that EUV emission is correlated with X-ray emission for F6 V to M4 V stars observed with the *XMM-Newton, Chandra*, and *ROSAT* satellites (Sanz-Forcada et al. 2011). The power-law slope is close to unity, $m = 0.860 \pm 0.073$, as expected because much of the EUV flux is in coronal ions. In their sample of late-type dwarfs and giants, Ayres et al. (2003) found that the Fe XXI 135.4 nm line is correlated with X-ray emission with a slope $m \approx 1.0$, but the correlation of the Fe XII lines at 124.2 nm and 134.9 nm with X-rays has a slope of only 0.5. This difference may result from the Fe XXI line being formed at 10^7 K where the X-ray emission from active stars is formed, but the Fe XII line is formed near 10^6 K at the top of the transition region or in cool magnetic loops where the thermal structure is different from that of a hot corona. The Fe XVIII 97.4 nm coronal emission line formed near 6×10^6 K is also correlated with the X-ray flux with slope of about unity (Redfield et al. 2003).

Güdel and Benz (1993) and Benz and Guedel (1994) described their surprising result that the soft X-ray and radio emissions (typically at 6 cm wavelength) are

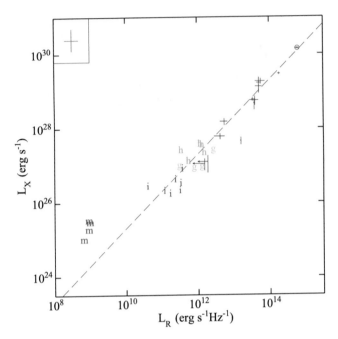

Fig. 9.14 Correlation of X-ray and radio (4.9 and 8.5 GHz) luminosities. Symbols are for solar microflares (m), impulsive solar flares (i), gradual solar flares (h), active M dwarfs (error bars), and a long duration flare on the M dwarf EQ Peg B (circle with error bar). The dashed line is has a power-law slope of 1. Figure from Benz and Guedel (1994). Reproduced by permission of the AAS

correlated with $\log L_x = \log L_R^m + 15.5 \pm 0.5$. This result is surprising for two reasons: (1) soft X-ray emission is produced by thermal electrons, whereas the cm-wavelength radio emission is gyrosynchrotron emission by nonthermal electrons, and (2) the correlation over many orders of magnitude fits most solar flares and a wide variety of active stars and binary systems (see Fig. 9.14). The authors explained the correlation of X-ray and radio emission as likely resulting from the heating of thermal electrons and acceleration of nonthermal electrons being parts of the same flaring process. Excluding solar microflares, the power-law index is $m = 1.02 \pm 0.17$. The quiet Sun, which has no measurable gyrosynchrotron radio emission, and solar microwave flares are considered to be radio-poor as they lack or have minimal nonthermal electrons perhaps because the plasma is too cool to accelerate many nonthermal electrons. Benz and Guedel (1994) point out that the correlation includes stars with stronger magnetic fields covering a much larger fraction of their stellar surfaces than the Sun and also includes lower gravity active stars with much larger magnetic loop structures confining the radiating nonthermal electrons. Ultracool stars with spectral types M7 to L3.5 violate the Güdel-Benz relation in the opposite sense. They have observed radio emission, roughly the same as warmer M stars, but very weak undetected X-ray emission (e.g., Berger et al. 2010). These X-ray poor stars have heated chromospheres based on $H\alpha$ emission but weak or infrequent heating to coronal temperatures. This was previous shown by the weak X-ray emission relative to Lyman-α emission of the coolest M stars in Fig. 9.13.

References

Ayres, T.R., Marstad, N.C., Linsky, J.L.: Outer atmospheres of cool stars. IX - A survey of ultraviolet emission from F-K dwarfs and giants with IUE. Astrophys. J. **247**, 545 (1981)

Ayres, T.R., Brown, A., Harper, G.M., Osten, R.A., Linsky, J.L., Wood, B.E., Redfield, S.: Space Telescope Imaging Spectrograph survey of far-ultraviolet coronal forbidden lines in late-type stars. Astrophys. J. **583**, 963 (2003)

Benz, A.O., Guedel, M.: X-ray/microwave ratio of flares and coronae. Astron. Astrophys. **285**, 621 (1994)

Berger, E., Basri, G., Fleming, T.A., Giampapa, M.S., Gizis, J.E., Liebert, J., Martn, E., Phan-Bao, N., Rutledge, R.E.: Simultaneous multi-wavelength observations of magnetic activity in ultracool dwarfs. III. X-ray, radio, and H activity trends in M and L dwarfs. Astrophys. J. **709**, 332 (2010)

Booth, R.S., Poppenhaeger, K., Watson, C.A., Silva Aguirre, V., Wolk, S.J.: An improved age-activity relationship for cool stars older than a gigayear. Mon. Not. R. Astron. Soc. **471**, 1012 (2017)

Busà, I., Aznar Cuadrado, R., Terranegra, L., Andretta, V., Gomez, M.T.: The Ca II infrared triplet as a stellar activity diagnostic. II. Test and calibration with high resolution observations. Astron. Astrophys. **466**, 1089 (2007)

Chabrier, G. Baraffe, I.: Structure and evolution of low-mass stars. Astron. Astrophys. **327**, 1039 (1997)

Christensen-Dalsgaard, J., Aguirre, V.S.: Ages for exoplanet host stars. arXiv:03125v1 (2018)

Cram, L.E., Mullan, D.J.: Formation of the H-alpha absorption line in the chromospheres of cool stars. Astrophys. J. **294**, 626 (1985)

Douglas, S.T., Agüeros, M.A., Covey, K.R., Bowsher, E.C., Bochanski, J.J., Cargile, P.A., Kraus, A., Law, N.M., Lemonias, J.J.: The factory and the beehive. II. Activity and rotation in Praesepe and the Hyades. Astrophys. J. **795**, 161 (2014)

France, K., Froning, C.S., Linsky, J.L., Roberge, A., Stocke, J.T., Tian, F., Bushinsky, R., Désert, J.-M., Mauas, P., Vietes, M., Walkowicz, L.: The ultraviolet radiation environment around M dwarf exoplanet host stars. Astrophys. J. **763**, 149 (2013)

France, K., Loyd, R.O.P., Youngblood, A., Brown, A., Schneider, P.C., Hawley, S.L., Froning, C.S., Linsky, J.L., Roberge, A., et al.: The MUSCLES Treasury Survey I: Motivation and overview. Astrophys. J. **820**, 89 (2016)

Garraffo, C., Drake, J.J., Cohen, O.: Magnetic complexity as an explanation for bimodal rotation populations among young stars. Astrophys. J. Lett. **807**, L6 (2015)

Garraffo, C., Drake, J.J., Dotter, A., Choi, J., Burke, D.J., Moschou, S.P., Alvarado-Gómez, J.D., Kashyap, V.L., Cohen, O.: The revolution revolution: Magnetic morphology driven spin-down. Astrophys. J. **862**, 90 (2018)

Gondoin, P.: Magnetic activity evolution on Sun-like stars. Astron. Astrophys. **616**, A154 (2018)

Güdel, M., Benz, A.O.: Title: X-ray/microwave relation of different types of active stars. Astrophys. J. Let. **405**, L63 (1993)

Güdel, M., Guinan, E.F., Skinner, S.L.: The x-Ray Sun in time: a study of the long-term evolution of coronae of solar-type stars. Astrophys. J. **483**, 947 (1997)

Guinan, E.F., Engle, S.G., Durbin, A.: Living with a red dwarf: rotation and X-ray and ultraviolet properties of the halo population Kapteyn's star. Astrophys. J. **821**, 81 (2016)

Hartmann, L., Baliunas, S.L., Noyes, R.W., Duncan, D.K.: A study of the dependence of Mg II emission on the rotational periods of main-sequence stars. Astrophys. J. **279**, 778 (1984)

Jackson, A.P., Davis, T.A., Wheatley, P.J.: The coronal X-ray-age relation and its implications for the evaporation of exoplanets. Mon. Not. R. Astron. Soc. **422**, 2024 (2012)

Jones, D.O., West, A.A.: A catalog of GALEX ultraviolet emission from spectroscopically confirmed M dwarfs. Astrophys. J. **817**, 1 (2016)

Kraft, R.P.: Studies of stellar rotation. V. The dependence of rotation on age among solar-type stars. Astrophys. J. **150**, 551 (1967)

Linsky, J.L.: Significance of the relative strengths of chromospheric and coronal emission in F–M stars. Astrophys. J. (submitted, 2019)

Linsky, J.L., France, K., Ayres, T.: Computing Intrinsic Ly-α fluxes of F5 V to M5 V stars. Astrophys. J. **766**, 69 (2013)

Lorenzo-Oliveira, D., Freitas, F.C., Meléndez, J., Bedell, M., Ramírez, I., Bean, J.L., Asplund, M., Spina, L., Dreizler, S., Alves-Brito, A., Casagrande, L.: The Solar twin planet search: The age - chromospheric activity relation. arXiv180608014L (2018)

Maggio, A., Sciortino, S., Vaiana, G.S., Majer, P., Bookbinder, J., Golub, L., Harnden, F.R.Jr., Rosner, R.: Einstein Observatory survey of X-ray emission from solar-type stars—The late F and G dwarf stars. Astrophys. J. **315**, 687 (1987)

McQuillan, A., Aigrain, S., Mazeh, T.: Measuring the rotation period distribution of field M dwarfs with Kepler. Mon. Not. R. Astron. Soc. **432**, 1203 (2013)

McQuillan, A., Mazeh, T., Aigrain, S.: Rotation periods of 34,030 Kepler main-sequence stars: The full autocorrelation sample. Astrophys. J. Suppl. Series **211**, 24 (2014)

Milligan, R.O., Kerr, G.S., Dennis, B.R., Hudson, H.S., Fletcher, L., Allred, J.C., Chamberlin, P.C., Ireland J., Mathioudakis, M., Keenan, F.P.: The radiated energy budget of chromospheric plasma in a major solar flare deduced from multi-wavelength observations. Astrophys. J. **793**, 70 (2014)

Newton, E.R., Irwin, J., Charbonneau, D., Berta-Thompson, Z.K., Dittmann, J.A., West, A.A.: The rotation and Galactic kinematics of id M dwarfs in the solar neighborhood. Astrophys. J. **821**, 93 (2016)

Newton, E.R., Irwin, J., Charbonneau, D., Berlind, P., Calkins, M.L., Mink, J.: The Hα emission of nearby M dwarfs and its relation to stellar rotation. Astrophys. J. **834**, 85 (2017)

Newton, E.R., Mondrik, N., Irwin, J., Winters, J.G., Charbonneau, D.: New rotation period measurements for M dwarfs in the southern hemisphere: an abundance of slowly rotating, fully convective stars. arXiv:1807.09365 (2018)

Noyes, R.W., Hartmann, L.W., Baliunas, S.L., Duncan, D.K., Vaughan, A.H.: Rotation, convection, and magnetic activity in lower main-sequence stars. Astrophys. J. **279**, 763 (1984)

Pace, G., Pasquini, L.: The age-activity-rotation relationship in solar-type stars. Astron. Astrophys. **426**, 1021 (2004)

Pallavicini, R., Golub, L., Rosner, R., Vaiana, G.S., Ayres, T., Linsky, J.L.: Relations among stellar X-ray emission observed from Einstein, stellar rotation and bolometric luminosity. Astrophys. J. **248**, 279 (1981)

Pevtsov, A.A., Fisher, G.H., Acton, L.W., Longcope, D.W., Johns-Krull, C.M., Kankelborg, C.C., Metcalf, T.R.: The relationship between X-Ray radiance and magnetic flux. Astrophys. J. **598**, 1387 (2003)

Pizzolato, N., Maggio, A., Micela, G., Sciortino, S., Ventura, P.: The stellar activity-rotation relationship revisited: Dependence of saturated and non-saturated X-ray emission regimes on stellar mass for late-type dwarfs. Astron. Astrophys. **397**, 147 (2003)

Redfield, S., Ayres, T.R., Linsky, J.L., Ake, T.B., Dupree, A.K., Robinson, R.D., Young, P.R.: A far ultraviolet spectroscopic explorer survey of coronal forbidden lines in late-type stars. Astrophys. J. **585**, 993 (2003)

Reiners, A., Mohanty, S.: Radius-dependent angular momentum evolution in low-mass stars. I. Astrophy. J. **746**, 43 (2012)

Reiners, A., Schüssler, M., Passegger, V.M.: Generalized investigation of the rotation-activity relation: favoring rotation period instead of Rossby number. Astrophys. J. **794**, 144 (2014)

Ribas, I., Guinan, E.F., Güdel, M., Audard, M.: Evolution of the solar activity over time and effects on planetary atmospheres. I. High-energy irradiances (1-1700 Å). Astrophys. J. **622**, 680 (2005)

Sanz-Forcada, J., Micela, G., Ribas, I., Pollock, A.M.T., Eiroa, C., Velasco, A., Solano, E., Garcia-Alvarez, D.: Estimation of the XUV radiation onto close planets and their evaporation. Astron. Astrophys. **532**, A6 (2011)

Schmidt, S.J., Hawley, S.L., West, A.A., Bochanski, J.J., Davenport, J.R.A., Ge, J., Schneider, D.P.:BOSS ultracool dwarfs. I. colors and magnetic activity of M and L dwarfs. Astron. J. **149**, 158 (2015)

Shkolnik, E.L., Barman, T.S.: HAZMAT. I. The evolution of far-UV and near-UV emission from early M stars. Astron. J. **148**, 64 (2014)

Shkolnik, E.L., Rolph, K.A., Peacock, S., Barman, T.S.: Predicting Ly and Mg II fluxes from K and M dwarfs using Galaxy Evolution Explorer ultraviolet photometry. Astrophys. J. Let. **796**, L20 (2014)

Simon, T., Herbig, G., Boesgaard, A.M.: The evolution of chromospheric activity and the spin-down of solar-type stars. Astrophys. J. **293**, 551 (1985)

Skumanich, A.: Time scales for Ca II emission decay, rotational braking, and Lithium depletion. Astrophys. J. **171**, 565 (1972)

Stelzer, B., Marino, A., Micela, G., López-Santiago, J., Liefke, C.: The UV and X-ray activity of the M dwarfs within 10 pc of the Sun. Mon. Not. R. Astron. Soc. **442**, 343 (2014)

Vaughan, A.H., Preston, G.W.: A survey of chromospheric Ca II H and K emission in field stars of the solar neighborhood. Publ. Astron. Soc. Pacific **92**, 385 (1980)

Vilhu, O., Walter, F.M.: Chromospheric-coronal activity at saturated levels. Astrophys. J. **321**, 958 (1987)

Walkowicz, L.M., Hawley, S.L.: Tracers of chromospheric structure. I. Observations of Ca II K and Hα in M dwarfs. Astron. J. **137**, 3297 (2009)

West, A.A., Hawley, S.L., Bochanski, J.J., Covey, K.R., Reid, I.N., Dhital, S., Hilton, E.J., Masuda, M.: Constraining the age-activity relation for cool stars: The Sloan Digital Sky Survey data release 5 low-mass star spectroscopic sample. Astron. J. **135**, 785 (2008)

West, A.A., Weisenburger, K.L., Irwin, J., Berta-Thompson, Z.K., Charbonneau, D., Dittmann, J., Pineda, J..: An activity-rotation relationship and kinematic analysis of nearby mid-to-late-type M dwarfs. Astrophys. J. **812**, 3 (2015)

Wood, B.E., Redfield, S., Linsky, J.L., Müller, H.-R., Zank, G.P.: Stellar Lyα lines in the Hubble Space Telescope archive: Intrinsic line fluxes and absorption from the heliosphere and astrospheres. Astrophys. J. Suppl. Series **159**, 118 (2005)

Wright, N.J., Drake, J.J., Mamajek, E.E., Henry, G.W.: The stellar-activity-rotation relationship and the evolution of stellar dynamos. Astrophys. J. **743**, 48 (2011)

Wright, N.J., Newton, E.R., Williams, P.K.G., Drake, J.J., Yadav, R.K.: The stellar rotation-activity relationship in fully convective M dwarfs. Mon. Not. R. Astron. Soc. **479**, 2351 (2018)

Youngblood, A., France, K., Loyd, R.O.P., Linsky, J.L., Redfield, S., Schneider, P.C., Wood, B.E., Brown, A., Froning, C., Miguel, Y., Rugheimer, S., Walkowicz, L.: The Muscles Treasury Survey II: Intrinsic Lyman Alpha and extreme ultraviolet spectra of K and M dwarfs with exoplanets. Astrophys. J. **824**, 101 (2016)

Youngblood, A., France, K., Loyd, R.O.P., Brown, A., Mason, J.P., Schneider, P.C., Tilley, M.A., Berta-Thompson, Z.K., Buccino, A., Froning, C.S., et al.: The MUSCLES Treasury Survey. IV. Scaling relations for ultraviolet, Ca II K, and energetic particle fluxes from M dwarfs. Astrophys. J. **843**, 31 (2017)

Chapter 10
Host Star Driven Exoplanet Mass Loss and Possible Surface Water

Whether an exoplanet retains its atmosphere and surface water depends on many factors. I will first describe the two main mass-loss mechanisms for exoplanet atmospheres. The main thermal process is hydrodynamical outflow driven by extreme ultraviolet and x-radiation from the host star that heat and inflate an exoplanet's exosphere leading to mass loss. Non-thermal mass loss occurs when ions and electrons in the host star's wind remove neutrals and ions in the exoplanet's upper atmosphere through charge exchange, ion pick-up, and other processes. Although there are simulations of thermal and non-thermal processes occurring in isolation, these processes can work together to enhance the mass-loss rate. There are, however, many unknowns or poorly known parameters including the exoplanet's initial water inventory, its orbital migration, the strength of its magnetic field, its outgassing rate, the history of the host star's UV, EUV and x-radiation, and the present and past flaring and coronal mass ejection (CME) rates. Thus for an exoplanet to be habitable today, it must survive a gauntlet of inadequately known challenges. Latter in the chapter, I describe what the term "habitable zone" is intended to be and include case studies of several exoplanets for which there are estimates of mass loss, surface liquid water, and habitability. Finally, I consider which types of host stars are most likely to support exoplanets that satisfy the wide variety of conditions that appear to be required for habitability.

10.1 Thermally Driven Mass Loss

The survival of an exoplanet's atmosphere depends upon both the XUV ($\lambda < 91.2$ nm) energy provided by the host star and the planet's gravitational potential. The XUV irradiation energizes mass loss by thermal processes, while the host star's wind is the source of non-thermal mass loss (see Sect. 10.2). Thermal mass-loss

© Springer Nature Switzerland AG 2019
J. Linsky, *Host Stars and their Effects on Exoplanet Atmospheres*,
Lecture Notes in Physics 955, https://doi.org/10.1007/978-3-030-11452-7_10

regimes are characterized by the value of the Jeans escape parameter,

$$\Lambda = GM_{pl}m_H/k_B T_{exo} R_{exo} = (v_{esc}/v_o)^2, \tag{10.1}$$

which is the ratio of the exoplanet's gravitational potential, GM_{pl}/R_{exo}, to the mean thermal energy of hydrogen atoms in the exoplanet's exosphere, $k_B T_{exo}/m_H$. The base of the exosphere (R_{exo}) is defined as the location where the density is so low that collisions cannot prevent energetic atoms and molecules from escaping. Here M_{pl} is the exoplanet's mass, k_B is the Boltzmann constant, and T_{exo} is the exosphere temperature. R_{exo} is the radial distance from the center of the exoplanet to the base of its exosphere, which can be significantly larger than the exoplanet's radius when the exosphere is extended as is often the case for hot Jupiters. The Jeans escape parameter can also be expressed as the square of the ratio of the speed of a hydrogen atom needed to escape the exoplanet's gravitational potential, v_{esc}, to the mean speed of hydrogen atoms at the temperature of the exosphere, v_0.

Figure 10.1 shows the importance of Λ for estimating the mass-loss rate. For a static atmosphere, the distribution of hydrogen atom speeds is given by the Maxwell–Boltzmann distribution, $n(v)dv \sim v^2 e^{-m_H v^2/2k_B T} dv$. Atoms and molecules escape an atmosphere when $v > v_{esc} = v_0\sqrt{\Lambda}$. As shown in Fig. 10.1, very few atoms have speeds many times larger than v_0. Therefore, for large values of the Jeans parameter Λ, there are very few atoms that can escape, but for small values of Λ nearly the entire Maxwell–Boltzmann distribution can escape. Since the probability of particles having speeds $5.5v_0$, corresponding to $\Lambda = 30$, is about

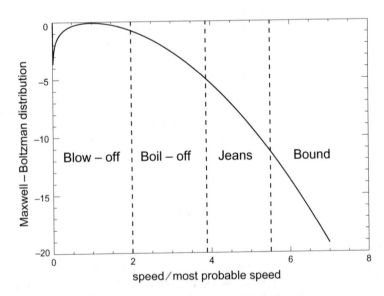

Fig. 10.1 The Maxwell–Boltzman distribution of particle speeds. The vertical lines show the approximate regions of rapid hydrodynamic mass loss (Blow-off), moderate hydrodynamic mass loss (Boil-off), very slow hydrostatic mass-loss (Jeans), and essentially no mass loss (Bound)

10^{-12}, there is essentially no mass loss and the atmosphere is bound for $\Lambda > 30$. At $v/v_0 = 3.8$, corresponding to $\Lambda = 15$, the probability of a particle having this speed is about 10^{-5} and slow mass loss is feasible. One could call the $\Lambda = 15$–30 interval the Jeans regime. The blow-off regime of rapid hydrodynamic mass loss occurs when the probability of escape exceeds 10^{-1}, corresponding to roughly $v/v_0 < 2$ or $\Lambda < 4$. The intermediate regime of a hydrodynamic atmosphere with moderate mass loss has been called "boil-off". Owen and Wu (2016) use this term to describe the loss of an extended hydrogen atmosphere by very young planets as they emerge from the stellar disk and before strong stellar XUV radiation energizes the "blow-off" regime. Fossati et al. (2017) described these different mass-loss regimes in more detail.

Erkaev et al. (2015) investigated the nature of hydrodynamic mass loss in the boil-off and blow-off regimes when the Jeans escape parameter Λ decreases to its critical value $\Lambda_{\mathrm{crit}} = 2.5$ in a hydrogen-rich exosphere. For values of Λ near 2.7, stationary hydrodynamical escape occurs with a transonic outflow. When $\Lambda < \Lambda_{\mathrm{crit}}$, the outflow velocity is supersonic even at the base of the thermosphere, resulting in extremely rapid mass loss. Then strong adiabatic cooling could increase Λ to exceed Λ_{crit}, thereby returning mass loss to a steady transonic outflow. Since Λ is proportional to $1/T_{\mathrm{exo}}$, the mass-loss rate depends on the strength of the incident XUV radiation and on the efficiency η with which the XUV flux is converted to heat and thus expansion compared to radiative losses.

To estimate the hydrodynamic mass-loss rate, consider the energy balance in the exoplanet's exosphere. Stellar XUV radiation is the input energy that first photoionizes hydrogen and helium atoms and then heats the atmosphere with the excess energy available after photoionization. Depending on the atmospheric density at the penetration level of the XUV radiation, this excess energy can either levitate the gas thereby driving mass loss or radiate the excess energy with no contribution to the mass loss. The efficiency factor η in the equation for the mass-lose rate,

$$L_{\mathrm{en}} = \eta \pi R_{\mathrm{pl}} R_{\mathrm{XUV}}^2 F_{\mathrm{XUV}} / a G M_{\mathrm{pl}} m_{\mathrm{H}}, \qquad (10.2)$$

is the ratio of the excess heat to the total input XUV flux. This rate will be larger when Roche lobe overflow is important. The term L_{en} introduced by Watson et al. (1981) is called the energy-limited escape rate. It's value is critical for estimating whether an exoplanet can retain its atmosphere, but there are a wide range of assumed and computed values of η in the literature. Since input XUV photon energy must be used to ionize neutral hydrogen (threshold 13.595 eV) and neutral helium (threshold 24.581 eV), η must be less than unity and depend upon the spectral energy distribution of the stellar XUV radiation. A harder XUV spectral energy distribution produced in a hotter host star corona will result in a larger value of η, because more energetic photons produce more energetic electrons during photoionizations and thus more heating. While the energy-limited mass-loss rate formula is commonly used, it can lead to errors in the mass-loss rates by factors or three to four for super-Earths and sub-Neptunes in strong radiation environments (cf. Erkaev et al. 2016).

Owen and Jackson (2012) showed that there are two regimes of thermally-driven flows: X-ray driven flows when the incident X-ray flux is high, and EUV driven flows when the X-ray flux is low because the star is old. X-ray driven flows are transonic because the strong X-ray absorption and heating occurs largely above the sonic point. Conversely, EUV driven flows occur when the X-ray and EUV fluxes are sufficiently weak that the flow is subsonic from its base where EUV photons ionize hydrogen. EUV driven flows are of two types: when the absorbed energy is converted mainly to heat and expansion (PdV work, where P is the pressure and dV is the change in volume) the flow is **energy limited**, but when the ionization is followed by recombination then the flow is **recombination limited** as the input energy is mostly radiated away resulting in weak mass loss. For example, the wind of the hot Jupiter HD 209458b studied by Murray-Clay et al. (2009) is EUV driven and energy limited, although non-thermal pick-up of protons contributes to the mass loss (Khodachenko et al. 2007). Also, mass loss has been detected in lines of C II (Linsky et al. 2010). Owen and Alvarez (2016) introduced a third term **photon limited** to describe thermal mass-loss flows from exoplanets with weak gravitational fields for which the outflow is only determined by the incoming flux of ionizing photons. Magnetic fields can suppress mass loss compared to the case of hydrodynamic outflows (e.g., Khodachenko et al. 2015).

An important consideration is the atmospheric density where the XUV photons are absorbed. This is where the exoplanet's gravitational potential becomes important. Exoplanets with higher gravities will have smaller atmospheric pressure height scales, $H = kT/\mu g$, where μ is the mean mass per particle. This means that the atmospheric density falls off more rapidly with height in exoplanets with higher gravities. As a consequence, XUV photons are absorbed at lower densities in high gravity exoplanets. Lower density means that for the same input energy the exosphere will be hotter, leading to much larger radiation by Lyman-α and molecular spectra, especially H_3^+. Thus high gravity exoplanets, such as the Earth and super-Earths should have smaller η values than bloated hot Jupiters (e.g., HD 189733b), warm Neptunes (e.g., GJ 436b), and mini-Neptunes (e.g., Gl 1214b). The detailed dependence on η and exoplanet gravity requires a calculation. For example, Bourrier et al. (2016) found that $\eta = 0.012 \pm 0.005$. Figure 10.2 shows the calculation of Salz et al. (2016) that identifies a critical value for the gravitational potential, $\Phi_G = GM_{pl}/R_{pl}$ erg g^{-1}. For exoplanet's with log $\Phi_G < 13.1$, the excess energy goes primarily into expansion (PdV work) rather than radiation, resulting in η values near 0.15. For planets with log $\Phi_G > 13.1$, η decreases very rapidly with increasing Φ_G as radiative cooling dominates the loss of this excess energy. Exoplanets with log $\Phi_G > 13.6$ are in the Jeans mass-loss regime, those with log Φ_G between 13.1 and 13.6 have hydrodynamic winds in the "boil-off" regime, and exoplanets with $\Phi_G < 13.1$ have hydrodynamic winds in the "blow-off" regime. The strength of the XUV flux impacting the exoplanet does not determine which mass-loss regime is appropriate, only the rate of hydrodynamic mass loss and thus the time scale required for an exoplanet to lose its atmosphere.

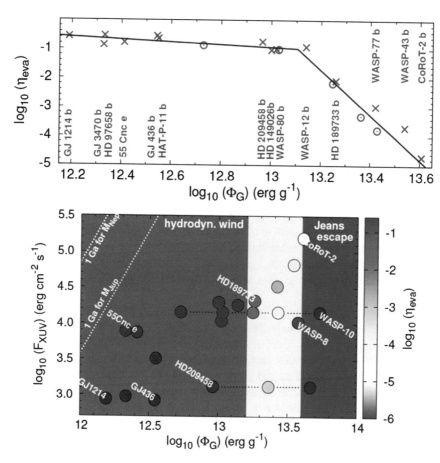

Fig. 10.2 *Upper figure:* Evaporation efficiencies as a function of the exoplanet's gravitational potential, Φ_G. Well characterized exoplanets are identified. *Lower figure:* Location of well characterized exoplanets in a plot of XUV flux vs gravitational potential. Evaporation efficiencies are color coded. To the upper left of the dotted lines are the parameter spaces where the entire atmosphere is lost in 10^9 years for a Jupiter-mass and Neptune-mass exoplanet. Figure from Salz et al. (2016). Reproduced with permission of ESO

Lammer et al. (2014) computed thermal mass loss rates for the hydrogen envelopes that protoplanets acquired from their protoplanetry disks. Their calculation of hydrodynamical mass loss is for planets with masses $(0.1–5)\,M_{\text{Earth}}$ located at 1 au from the young Sun. They assumed two values for η, 0.15 and 0.40, which they assumed bracket realistic values and the XUV emission rates from the young Sun following Claire et al. (2012). Planets with sub-Earth and Earth-like mass lose all of their initial hydrogen envelopes within 100 Myr, but super-Earths retain nearly all of their hydrogen atmospheres, for example the five super-Earth exoplanets of Kepler-11 (Kislyakova et al. 2014).

Using a 1-D hydrodynamical code, Kubyshkina et al. (2018) extended the previous analysis by computing mass-loss rates from K2-33, the youngest known exoplanet with an age <20 Myr. This exoplanet located about 0.04 au from its M-type host star receives an extremely high EUV flux estimated to be 2.4 × $10^5 \, \mathrm{erg \, cm^{-2} \, s^{-1}}$ about 10^5 times larger than the Earth receives from the quiet Sun. The mass-loss rate depends on the value of Λ Eq. (10.1), which depends on the exoplanet's gravitational potential and thus on its unknown mass. For assumed masses smaller than about 10 M_{Earth}, the exoplanet's weak gravity results in extremely high mass-loss rates in the boil-off regime. For assumed masses greater than about 10 M_{Earth} and thus higher gravitational potentials, stellar EUV heating balanced by adiabatic cooling results in more modest mass-loss rates and hotter exosphere gas temperatures. For $\Lambda \leq 5$, the case for assumed low masses of K2-33b, the computed hydrodynamic mass-loss rate can exceed the energy-limited rate Eq. (10.2) by orders of magnitude. The timescale for complete loss of its hydrogen-dominated atmosphere is about 20 Myr for a 7 M_{Earth} exoplanet and about 100 Myr for a 20 M_{Earth} exoplanet. This calculation shows that K2-33b and other close-in exoplanets will lose their initial atmosphere early in their evolution.

Bolmont et al. (2017) computed hydrogen and oxygen mass-loss rates from exoplanets with surface liquid water that are in orbit around low-mass (0.01–0.08 $M_{\odot}$) host stars. These mass-loss simulations were for 0.1–5.0 M_{Earth} exoplanets that receive EUV radiation scaled from two prescriptions for the X-ray luminosity. The simulations assume that the photolysis of H_2O by the host star's Lyman-α and other UV radiation is at least as fast as the escape of hydrogen atoms and ions and, therefore, does not constrain the mass-loss rate. The simulations include the decrease in total flux from the host star as it descends to the main sequence (or it decreases monotonically with time if the host star is a brown dwarf) causing the inner and outer edges of the habitable zone to move inward with time. They use the energy-limited escape formalism Eq. (10.2) with the fraction of the input EUV and X-ray energy that inflates the atmosphere and drives mass-loss, $\eta = 0.10$. Figure 10.3 shows the results of their simulations of the water inventory of exoplanets as a function of their mass, their distance from the host star, and the host star's mass. Exoplanets in the light blue region have lost less than two Earth oceans of water during their runaway greenhouse phase before the inner edge of the habitable zone reaches the exoplanet's orbit and the exoplanets then spend more then 500 Myr in the HZ. The dark blue region is for 1000 Myr in the HZ. Exoplanets lucky enough to be in these blue regions have a good chance of retaining water and being habitable. Planet e in the TRAPPIST-1 system is located in the light blue region.

An interesting way of characterizing exoplanet mass loss is by means of an "energy diagram" first introduced by Lecavelier des Etangs (2007) and extended with new data by Ehrenreich and Désert (2011). Energy level diagrams are plots of an exoplanet's gravitational potential, dE'_p, the energy needed per unit mass to escape the exoplanet's gravity, compared to the incident XUV flux from the

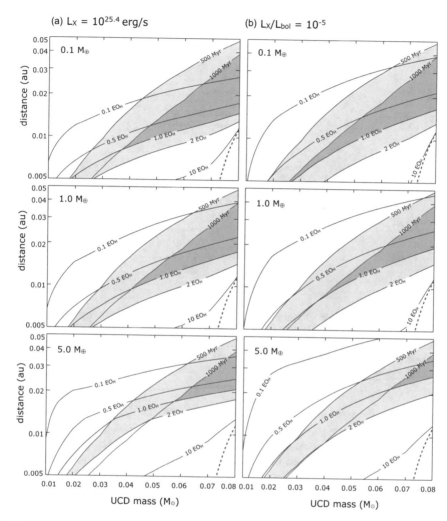

Fig. 10.3 Simulations of water loss from exoplanets as a function of planet mass (0.1–5.0) M_{Earth}, ultracool dwarf (UCD) host star mass (0.01–0.08) $M_{\odot}$, and distance of the exoplanet from the host star. The black solid lines are the total hydrogen loss as measured in Earth ocean equivalents over the lifetime of the host star. Exoplanets located in the light blue region have lost less than two Earth oceans of water before entering the inner edge of the habitable zone and are then located in the HZ for at least 500 Myr or 1000 Myr (dark blue zone). Below the dashed blue line, exoplanets are too close to the host star and never enter the HZ. Figure from Bolmont et al. (2017). Reproduced from MNRAS by permission of the Oxford University Press

host star, dE_{XUV}. For energy-limited thermal escape, these parameters are related by

$$-dE'_p = \eta \, dE_{XUV}, \qquad (10.3)$$

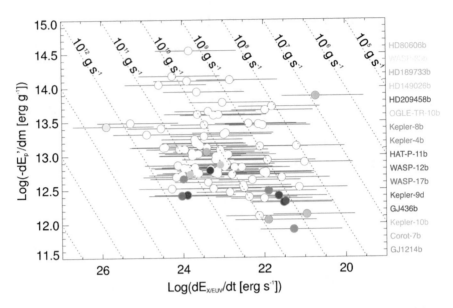

Fig. 10.4 An energy level diagram for exoplanet mass loss plotting the energy needed for a unit mass to escape the exoplanet's gravity vs. the XUV flux from the host star incident on the exoplanet. Diagonal dashed lines are mass-loss rates. Each exoplanet is plotted assuming a heating efficiency of $\eta = 0.15$ with horizontal error bars corresponding to the extreme values of $\eta = 0.01$ and 1.0. Figure from Ehrenreich and Désert (2011). Reproduced with permission of ESO

where η has the usual meaning of the fraction of the incident XUV flux that produces mass loss. Figure 10.4 shows the energy diagram proposed by Ehrenreich and Désert (2011) for exoplanets with well characterized parameters. In this diagram η is assumed to be 0.15 and the horizontal error bars correspond to the extreme values of $\eta = 0.01$ and 1.0. The exoplanet mass-loss rates are the diagonal lines. In principle, values of η can be inferred when the mass-loss rates and XUV fluxes are accurately known (see Sect. 6.2).

Is there evidence for the critical role that the exoplanet's gravitational potential plays concerning mass loss? The most direct evidence is the presence or absence of an extended hydrogen outer envelope observed in the hydrogen Lyman-α line when the exoplanet transits its host star. The first observation of the transit of a hot Jupiter (HD 209458b) by Vidal-Madjar et al. (2003) showed that the size of the exoplanet's atmosphere seen in the Lyman-α line is considerably larger than the size seen in the optical continuum. Transit observation of other hot Jupiters (e.g., HD 189777) and mini-Neptunes (e.g., Gl 1214b and Gl 876b) show that these exoplanets also have extended hydrogen envelopes indicative of hydrodynamic mass loss. See Ehrenreich et al. (2015) for a summary of these studies. Observations of comet-like Lyman-α absorption after an exoplanet transits its host star also indicates mass loss from these low gravity exoplanets. On the other hand, transits of high gravity Earth-like

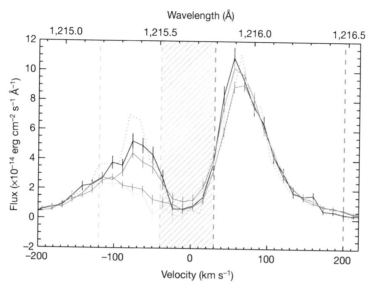

Fig. 10.5 Lyman-α line profiles observed during different phases of the transit of GJ 436b: out-of-transit (black), pre-transit (blue), in-transit (green), and post-transit (red). The hatched region is not observed because of strong interstellar absorption. The wavelength regions between the blue dashed vertical lines enclose the blue wing and the region between the red dashed lines enclose the red wing of the Lyman-α line. Figure from Ehrenreich et al. (2015)

and super-Earth exoplanets (e.g., Kepler 444b-f) show no evidence for extended hydrogen atmospheres and thus minimal mass loss rates.

The remarkable transits of the warm mini-Neptune GJ 436b observed in the Lyman-α line by Ehrenreich et al. (2015) and Lavie et al. (2017) show absorption beginning 2 h prior to ingress and extending to at least 10 h and perhaps as long as 25 h after egress. Despite the exoplanet barely grazing the host star's limb at mid-transit, the transit depth maximum is 56.3 $\pm$ 3.5%, indicating a very extensive envelope and tail of neutral hydrogen lost from the exoplanet's exosphere. Figure 10.5 shows the observed Lyman-α line profiles coadded over several transits. Compared to the out-of-transit profile, the blue wing (-120 to -40 km s^{-1}) shows absorption by neutral hydrogen moving towards the observer along the line of sight to the star during pre-transit, post-transit, and deepest for the in-transit time. In the spectra obtained during subsequent transits, Lavie et al. (2017) also detected absorption in the red wing ($+30$ to $+120$ km s^{-1}) of the Lyman-α line indicating hydrogen moving away from the observer. The huge neutral hydrogen cloud shown in Fig. 10.6 extends about 12 stellar radii in the plane of the sky and about 2.5 stellar radii in the perpendicular direction. The comet-like tail extending behind the exoplanet contains hydrogen atoms moving faster than the 26 km s^{-1} escape speed from the exoplanet consistent with mass loss. Using a 3-D numerical simulation of atmospheric escape to fit the observed Lyman-α line profiles, Ehrenreich et al. (2015) and Bourrier et al. (2016) estimated a mass-loss rate of 10^8 to 10^9 g s^{-1}

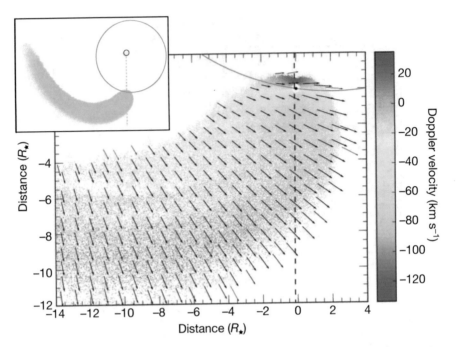

Fig. 10.6 *Upper left panel:* A polar view of the GJ 438 system showing the host star (small circle), the 2.64 day orbit of GJ 436b (green circle), and the neutral hydrogen tail trailing GJ 436b (blue). *Lower right panel:* An expanded view showing the exoplanet (small black dot), a portion of its orbit (green), and the hydrogen tail (colored according to its velocity relative to the exoplanet). Arrows show the flow direction of hydrogen atoms relative to the rest frame of the star. Figure from Ehrenreich et al. (2015)

corresponding to 1.6×10^{-18} to $1.6 \times 10^{-17}\, M_\odot \, \mathrm{yr}^{-1}$. The present mass-loss rate is far too small to deplete the atmosphere of GJ 436b, but far larger mass-loss rates when the host star was young and likely very active could have eroded much of the exoplanet's atmosphere.

XUV radiation includes all wavelengths shortward of 91.2 nm including the 10–91.2 nm region called the extreme ultraviolet (EUV) and X-rays at wavelengths shortward of 10 nm. Are these spectral regions equally important or does one region dominate? Figure 6.12 shows the solar EUV and X-ray spectrum. The EUV spectrum includes the hydrogen Lyman continuum (60–91.2 nm), He I continuum, (45–54 nm), chromospheric lines of He I (58.4 nm), He II (30.4 nm), several other emission lines formed in the chromosphere (T near 10,000 K) and transition region (T between 20,000 K and 1,000,000 K). There are also many coronal emission lines in this spectral region and coronal lines dominate the X-ray spectrum. Figure 10.7 shows the observed spectral energy distribution of the moderately active M1.5 dwarf star GJ 832 obtained from nearly simultaneous *HST* ultraviolet spectra and *Chandra* X-ray spectra as a part of the MUSCLES Treasury Survey. Inspection of the solar and M dwarf spectral energy distributions suggests that the EUV flux

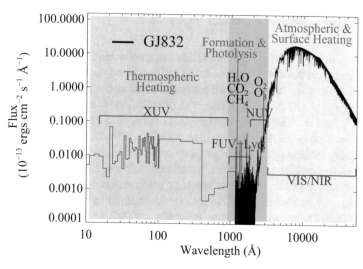

Fig. 10.7 The spectral energy distribution of the M1.5 V star GJ 832. The spectrum is a combination of near simultaneous *Chandra* X-ray spectra, *HST* UV spectra, ground-based optical and infrared spectra, and reconstructed Lyman-α and extreme ultraviolet flux. Figure from France et al. (2016). Reproduced by permission of the AAS

is much larger than the X-ray flux. The observed EUV/X-ray flux ratio for the 6 moderately active M dwarfs and 4 K stars observed in the MUSCLES Treasury Survey (Loyd et al. 2016; Youngblood et al. 2017) shows that the EUV flux dominates by roughly a factor of 10. Since the photoionization cross-section of hydrogen decreases rapidly with decreasing wavelength, the contribution of X-rays to the photoionization and thus mass-loss from exoplanet atmospheres is very small compared to the EUV flux (cf. Bourrier et al. 2016). For more active stars with spectral energy distributions skewed toward X-rays, the contribution of X-rays to the photoionization of hydrogen will be more important than for the less active stars. King et al. (2018), for example, find that the flux ratio F_{EUV}/F_{x-ray} decreases monotonically with increasing F_{x-ray} from about 10 to nearly 4 for G-M dwarf stars.

There are other sources of heating in additional to XUV radiation for close-in exoplanets. For example, Cohen et al. (2018) estimate an upper limit to the Ohmic dissipation of stellar winds in the ionospheres of the TRAPPIST-1 exoplanets e, f, and g, which lie within 0.045 au of their M8 V host star. They estimate that the ionospheric heating rate could be as large as 0.5–1 W m^{-2}, which is about 1% of the total stellar irradiance and 5–15 times the EUV irradiance. This heating source, if efficiently transmitted to the exoplanet's upper atmosphere, could speed up mass loss and perhaps explain the expanded size of hot Jupiters like HD 209458b and HD 189733b.

10.2   Non-thermally Driven Mass Loss

Ionized stellar winds with their embedded magnetic fields can impact exoplanet atmospheres, chemically interact with neutrals and ions, and extract these particles as non-thermal mass loss. There are a number of processes involved as summarized by Lammer et al. (2003) and by Dong et al. (2018). Atomic and molecular hydrogen, either acquired from the protoplanetary disk or produced by photolysis of H_2O, can be photoionized by EUV radiation, charge exchanged with stellar wind protons, or collisionally ionized by stellar wind electrons. The resulting H^+ and H_2^+ ions can then be picked up by the stellar wind's magnetic field and thereby expelled from the exoplanet's atmosphere. These are important mechanisms for water loss from Mars.

The loss of heavy atoms and molecules by thermal hydrodynamic outflows is difficult from Earth-mass and lower mass planets such as Mars. Instead there are several efficient non-thermal mass-loss processes involving stellar winds. Dissociative recombination reactions can produce hot atoms. For example,

$$O_2^+ + e \rightarrow O^* + O^* \tag{10.4}$$

can produce energetic neutral oxygen atoms with enough energy to escape an exoplanet's gravitational field. The magnetic field in the stellar wind can also pick-up ions including O^+, O_2^+ and CO_2^+.

Finally, collisions of stellar wind protons with neutral atoms and molecules can give these particles enough energy to escape the exoplanet's gravitational field. This process is called sputtering. The relative importance of these different non-thermal mass-loss processes for Mars as a function of look-back time will be discussed later in this chapter.

Computing the non-thermal mass-loss rates from an exoplanet's atmosphere requires estimating the range of penetration depths of the stellar wind into an exoplanet's atmosphere and the chemical composition and density at these depths. If an exoplanet has a dipolar magnetic field like the Earth, but unlike present day Venus and Mars, then the penetration depth will depend on magnetic latitude with deeper penetration near the magnetic poles. Near the magnetic equator, the effect of the exoplanet's magnetic field is to restrict the stellar wind flow to larger heights where the atmospheric densities are lower and local mass-loss rates lower. Near the magnetic poles, however, mass-loss rates will increase due to the funneling of the stellar wind into a smaller area and the ambipolar diffusion of ions into the upper atmosphere. The combination of shielding near the magnetic equator and amplification near the poles predicts that the total mass-loss rates will increase with the exoplanet's magnetic field strength at least initially, although the orientation of this field with respect to the stellar wind inflow direction is important (e.g., Egan et al. 2018).

One must also compute or estimate the stellar wind dynamic and magnetic pressures, guess the exoplanet's magnetic field strength as there are no measurements except for the solar system planets, and know how these variables change over the lifetime of the planet. To obtain some insight concerning the stand-off distance of exoplanetary magnetic fields against the ram pressure of stellar winds, See et al. (2014) computed this parameter for hypothetical Earths located in the centers of the habitable zones of solar-like stars with $M = (0.8–1.4)\,M_\odot$. Using a variety of stellar wind models, they found that these exoplanets with Earth-like magnetic fields have magnetospheric stand-off distances $\geq 7\,R_{Earth}$. This presents a favorable picture for Earth-like planets of roughly solar-mass stars maintaining substantial atmospheres with liquid surface water.

However, the strength of stellar winds can be orders of magnitude larger than the relatively benign simulations previously described for more active stars and planets located much closer to their host stars. As described in the following case studies for Proxima b and TRAPPIST-1, exoplanets in close-in orbits of M dwarfs are exposed to stellar wind dynamic pressures 10^3 or more times larger than the Earth encounters from the present day Sun. For the extreme case of V374 Peg, a very active M4 V star with a rotational period of 0.44 days, Vidotto et al. (2011) computed a wind dynamic pressure 10^5 times larger than the solar wind seen at Earth. As a result, strong stellar winds can penetrate close to the surfaces of close-in exoplanets surrounding their magnetic polar caps to produce rapid mass loss and desiccation. Even for the more benign stellar wind that Mars has faced after it lost its magnetic field about 4.1 Gyr ago, the planet has lost all of its surface water.

At the base of an exoplanet's exosphere, upwardly propagating photoelectrons can outrun the more slowly moving heavy ions, leading to an ambipolar electric field that accelerates the upflowing ions (Airapetian et al. 2017). This process partially cancels an exoplanet's gravitational field leading to enhanced mass loss. Also, Joule resistive heating occurs where electric fields produced by interacting stellar wind and planetary magnetic fields enhance collisions of ions with neutrals (cf. Cohen et al. 2014). This process can heat gas in an exosphere, enhancing both thermal and non-thermal mass-loss rates. Joule heating is especially important for close-in exoplanets that are impacted by strong stellar winds, magnetic reconnection events, and CMEs.

Another type of nonthermal heating and mass loss is produced by the input of electrons accelerated where the magnetic fields of the star and a close-in planet interact and recombine. These electrons flow along field lines to penetrate to column densities 10^{23} to $10^{25}\,m^{-2}$ near the planet's magnetic poles with energies comparable to the EUV radiation from the star (Lanza 2013). The effect is to add to the enhance the heating of the planet's outer atmosphere that drives mass loss.

10.3   Habitable Zone: What Does It Mean?

The term "habitable zone" is often misinterpreted. As used in the astrobiological literature, the term does not mean that an exoplanet is inhabited or could be inhabited by life forms. The usual definition of "habitable zone" is the range of distances from a given host star where liquid water is stable on the surface of an exoplanet. The possibility of life forms existing without liquid water has been discussed extensively, but I will here follow the usual definition of what habitable zone (HZ) signifies. The location of the HZ depends on the luminosity of the host star and to a lesser extent its spectral energy distribution, together with the albedo, atmospheric composition, gravity, internal heat sources, synchronous or non-synchronous rotation, and other aspects of the exoplanet. For a detailed review of HZ calculations see Kaltenegger (2017) and sources cited therein. The following is a short overview.

At present, the Earth receives an irradiance of $S_0 = 1370\,\text{W}\,\text{m}^{-1}$ from the Sun. Figure 10.8 shows the change in the Sun's luminosity, effective temperature, and radius with time. If the Earth had no atmosphere and thus no greenhouse effect but its present albedo of 30%, its equilibrium temperature would be $T_{eq} = 255\,\text{K}$ or $-18\,^\circ\text{C}$. The Earth would then be in a snowball phase and uninhabitable. Instead, $T_{eq} = 288\,\text{K}$, a comfortable $+15\,^\circ\text{C}$, because H_2O, CO_2, and other atmospheric gasses provide a greenhouse increase in surface temperature of 33 K. Thus it is essential to assume an atmospheric composition and pressure when determining the location of the HZ. To a lesser extent, the host star's spectral energy distribution plays a role in this calculation because liquid water and ice absorb near-infrared radiation more efficiently than optical radiation. As a result, M stars with their spectral energy distribution skewed to the near-infrared (see Fig. 10.9) provide proportionally more absorbed energy to an exoplanet than warmer stars.

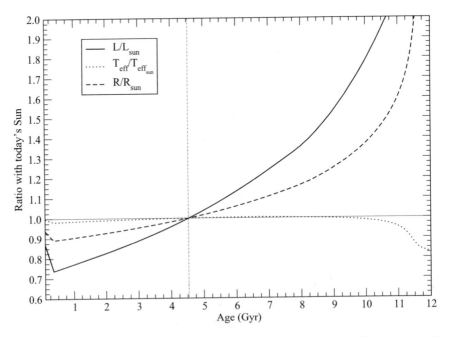

Fig. 10.8 The evolution of the luminosity (solid line), effective temperature (dotted line), and radius (dashed line) of the Sun from the zero-age main sequence to the start of its red giant phase. Units are relative to the present day Sun (vertical dotted line). See Kim et al. (2002) and Yi et al. (2003) for the solar models. Figure from Ribas (2009). Reproduced with permission of ESO

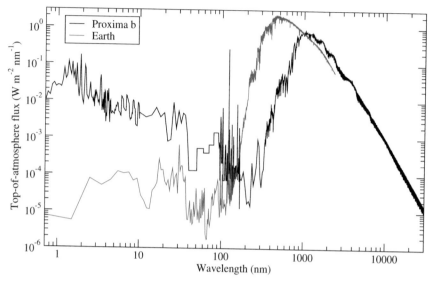

Fig. 10.9 Comparison of the spectral energy distributions at the top of the atmospheres of Proxima b (black) and the Earth (red). Flux units are $W\,m^{-2}\,nm^{-1}$ at 1 au for the Earth and 0.0485 au for Proxima b. Figure from Ribas et al. (2017)

Kasting et al. (1993) introduced the classical definition of the habitable zone or CHZ. As modified by Kopparapu et al. (2013, 2014), the CHZ assumes a geologically active planet with a rocky surface and an atmosphere containing N_2, H_2O, and CO_2. The inner edge of the CHZ is determined by the insolation level from the host star that forces the exoplanet's surface temperature to exceed the triple point of water leading to a runaway greenhouse and rapid loss of water by photolysis and hydrogen escape. The outer edge is fixed by the maximum greenhouse effect from CO_2. Kopparapu et al. computed that the outer limits for the present day Sun are 0.95 au and 1.67 au, responsively. According to this narrowly defined CHZ, Venus at 0.723 au sees $1.9S_0$ and is, therefore, inside of the CHZ, Earth is barely inside of the CHZ, and Mars at 1.523 au sees $0.4S_0$ and is also inside of the CHZ. Mars, however, does not have liquid water on its surface, because with its very thin atmosphere and a very small greenhouse effect, the CHZ model is inappropriate. The CHZ model does not encompass a range of possible cloud covers, orbital eccentricities, or exoplanet gravities.

A much broader approach to defining the habitable zone is based on the possibility of water on the surfaces of Venus up to 1 Gyr ago (recent Venus or RV) and Mars 3.8 Gy ago (early Mars or EM). Venus then received $S = 1.76S_0$ from the young Sun that was 8% fainter than today, and the early Mars received $S = 0.32S_0$, but it had a thick atmosphere with a significant greenhouse effect. These two criteria set the outer limits of the RVEM habitable zone of today's Sun as 0.75 au to 1.77 au. Figure 10.10 shows the outer limits of the CHZ and RVEM as a function of stellar effective temperature.

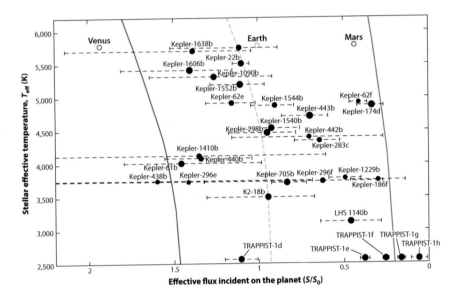

Fig. 10.10 Detected exoplanets with $R < 2R_E$ and effective flux relative to the Earth as a function of stellar effective temperature. Symbol size indicates planetary radius. Solid red and blue lines indicate the inner and outer edges of the RVEM (recent Venus to early Mars) habitable zones today. The dashed line is the inner edge of the classical habitable zone (CHZ). Figure from Kaltenegger (2017)

Figure 10.8 shows the evolution of the solar luminosity, radius, and effective temperature relative to the present day Sun. Stars with different masses show similar evolution after reaching the zero-age main sequence but with longer timescales for lower mass stars. The 30% lower luminosity of the young Sun should have caused the Earth's surface to be completely frozen. That this did not happen is referred to as the "faint young Sun paradox" (Sagan and Mullen 1972).

Habitability on a planet's surface also requires that the atmosphere contains sufficient ozone to block the penetration of UV light where it can sterilize all life forms (O'Malley-James and Kaltenegger 2017).

10.4 Can Exoplanets Retain Their Atmospheres in Their Space Climate Environment?

As described in this chapter, the ability of an exoplanet to maintain water on its surface requires both an atmosphere resistant to the thermal and nonthermal erosion processes over the exoplanet's lifetime and the proper amount of optical and near-IR insolation from its host star. The two requirements are coupled because greenhouse heating and the exoplanet's atmospheric and surface albedo control the surface thermal environment. The following case studies of exoplanets of two M dwarfs and warmer stars show that these two requirements must be satisfied for an exoplanet to be truly habitable.

10.4.1   Case Study: Proxima B

To understand Proxima Cen b's atmosphere, it will also be important to thoroughly characterize its host star.—Meadows et al. (2018)

In an astounding discovery, Anglada-Escudé et al. (2016) announced that Proxima Centauri (M5.5 V), the nearest star to the Sun at only 1.295 pc, is the host star for a super-Earth ($M > 1.3\,M_{Earth}$). With a semi-major axis of only 0.0485 au, this exoplanet is located near the center of the hoist star's CHZ defined by the possible presence of liquid water (Kasting et al. 1993; Kopparapu et al. 2016). The planet's equilibrium temperature without including possible greenhouse heating is 220–240 K, somewhat cooler than the globally averaged surface temperature of the Earth (288 K) but sufficiently Earth-like conditions to raise the question of whether this closest exoplanet, which could be resolvable with new instrumentation, is indeed habitable. Figure 10.9 compares the spectral energy distribution (SED) received by Proxima b compared to that received by the Earth, showing the far stronger EUV and X-ray emission and weaker optical and near-UV emission received by Proxima b (Ribas et al. 2017). A similar SED obtained by the *HST* MUSCLES Treasury Survey team (France et al. 2016) is available at https://archive.stsci.edu/prepds/muscles.

Several authors have explored the question of whether liquid water is possible on the surface of Proxima b. For example, Boutle et al. (2017) used the Met Office Unified Model that was developed to simulate global circulation on the Earth to simulate the atmosphere of Proxima b assuming an Earth-like atmospheric composition and either a tidally-locked or a 3:2 spin-orbit resonance like Mercury. For the tidally-locked case, the mean surface temperature in the day side hot spot is well above freezing (290 K) but the night side is very cold. For the 3:2 spin-orbit case, warm temperatures near the equator always permit liquid water when the eccentricity is above about 0.1 and surface temperatures increase with increasing eccentricity.

Meadows et al. (2018) computed a wide range of possible model atmospheres for Proxima b that range from a hot Venus-like high pressure and dry CO_2 dominated atmosphere to a cold desiccated CO_2 dominated Mars-like atmosphere to a more Earth-like N_2 and O_2 dominated atmosphere that is potentially habitable. They described scenarios in which Proxima b could evolve into each of these atmosphere models, but there are major uncertainties in deciding which type of model is the more likely. Important uncertainties include whether the planet had migrated from a more distant environment with significant H_2 in its atmosphere and whether its present orbit is circular and, therefore, tidally locked or has a significant eccentricity, in which case its revolution is asynchronous with its orbital period and tidal heating could important. However, the most important criterion for determining whether Proxima b retains its atmosphere and thus its surface water is the host star's radiation, wind, and high-energy proton emissions. These drivers control the environment of Proxima b today and the much harsher environment when its host star was young, more luminous, and more active. Since low-mass M dwarfs are thought to have high luminosity for $\geq 10^8$ years before they reach the main sequence, Proxima b could have lost all or most of its initial atmosphere during a thermal runaway phase when its host star was too bright for Proxima b to be in the habitable zone at that time.

Since Proxima b is 20 times closer to its host star than the Earth is to the Sun, atmospheric erosion by the stellar wind is a major concern. Garraffo et al. (2016) investigated this question by simulating the host star's wind using the BATS-R-US MHD code and the star's magnetic field obtained from a Zeeman–Doppler image of the more rapidly rotating late-M star GJ 51. Depending on the presently unknown orientation and eccentricity of Procyon b's orbit, the wind dynamic pressure is >2000 times larger than is present at the Earth and highly variable as Proxima b crosses the stellar current sheet twice each 11.2 day orbit (see Fig. 10.11). If the exoplanet has an Earth-like magnetic field, then the powerful stellar wind dynamic pressure compresses the planet's magnetosphere to be very close to the planet's surface twice a day. The effect of this deep penetration of the ionized magnetic stellar wind into the planet's mostly neutral upper atmosphere is to enhance mass loss by sputtering, charge-exchange, ion pick-up, and other processes described in Sect. 10.2. A detailed calculation is needed to estimate the time scale for these atmospheric stripping processes and the extension and density of the exoplanet's upper atmosphere due to heating by the star's EUV radiation and Joule heating (Cohen et al. 2014).

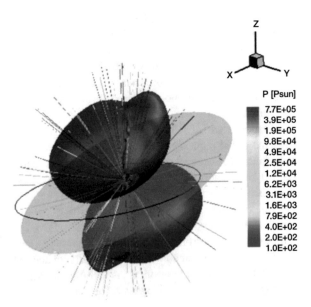

Fig. 10.11 Model of the magnetosphere and wind of Proxima Centauri for a mean magnetic field strength of 600 G. The Alfvén surface is blue and the green plane is the current sheet. The thin black line is the orbit of Proxima b, which crosses the current sheet twice per orbit. Color coding represents the wind dynamical pressure divided by the solar value at 1 au. Selected magnetic field lines are grey. Figure from Garraffo et al. (2016). Reproduced by permission of the AAS

P [Psun]

7.7E+05
3.9E+05
1.9E+05
9.8E+04
4.9E+04
2.5E+04
1.2E+04
6.2E+03
3.1E+03
1.6E+03
7.9E+02
4.0E+02
2.0E+02
1.0E+02

Dong et al. (2017) then estimated the escape rate of different species from Proxima b's atmosphere assuming that it is dominated by CO_2 like Mars and Venus. They simulated the host star's wind properties using the BATS-R-US MHD code and computed escape rates for the O^+, O_2^+, and CO_2^+ ions produced by photolysis, impact ionization, and charge-exchange reactions. They considered the cases of Proxima b having no magnetic field or a field with a dipole moment one-third that of the Earth. For both the magnetic and non-magnetic cases, the ion escape rates are in the range 2.0×10^{26} to 3.7×10^{27} ions per second, one to two orders of magnitude larger than for the solar system terrestrial planets. Depending on assumptions concerning the stellar wind and the exoplanet's magnetic field and atmospheric chemical composition, Dong et al. (2017) concluded that Proxima b should lose its atmosphere in 0.1–1.0×10^8 years and most likely it no longer has an atmosphere.

Although many critical parameters about Proxima b's present and past are unknown, Ribas et al. (2016) explored whether the exoplanet could retain liquid water on its surface assuming that the exoplanet initially had a substantial water inventory, that it has migrated from outside the snow line of its host star or has always been near its present orbit, and that it now is either locked into a synchronous or a 3:2 spin-orbit configuration. During the first 100–200 Myr after formation, the much brighter host star forced Proxima b into a runaway phase in which surface water evaporated into a steam atmosphere. At that time, hydrodynamic mass loss of water together with hydrogen from photolysis of water could have removed up to the equivalent of an Earth's ocean of water, according to the simulations of Bolmont et al. (2017), leaving an oxygen rich atmosphere. As the host star cooled and Proxima b entered into a classical insolation habitable zone, surface liquid water became feasible. However, the EUV radiation from the host star is about 30 times that seen by the Earth today and likely larger at earlier times. Over the roughly 4.6 Gyr since Proxima b has been in the habitable zone, this EUV radiation could have powered the hydrodynamical loss of many Earth oceans of water. Ribas et al. (2016) estimated that the stellar wind seen by Proxima b is 4–80 times larger than is seen by the Earth today. The rate of stellar wind driven mass loss depends on (a) the location of the magnetopause, which depends on the wind ram pressure and the planet's magnetic field, and (b) the extension of the planet's exosphere, which responds to the EUV heating rate. Since the magnetopause could be as close to the planet as $1.5 R_{\text{planet}}$ and the base of the exosphere could extend further out, there are scenarios in which the stellar wind by itself could remove all of the planet's remaining water or have minimal effect. Even if Proxima b has a sufficiently large magnetic field to push its magnetopause well beyond the base of its exosphere, the penetration of the stellar wind into the planet's atmosphere at the magnetic poles could lead to significant loss of ions through pick-up and ambipolar diffusion and eventually the entire atmosphere (Garcia-Sage et al. 2017; Airapetian et al. 2017). Thus there are plausible scenarios in which Proxima b could be either be a desiccated life-less exoplanet or the nearest truly habitable planet.

10.4.2　Case Study: The TRAPPIST-1 Exoplanets

The M8 V star TRAPPIST-1 located at a distance of 12 pc is the host for seven exoplanets that can best studied in great detail by transmission spectroscopy. Gillon et al. (2016) discovered the first three planets (b, c, and d) with semi-major axes 0.011, 0.0152, and 0.0214 au. Subsequent ground based photometry and Spitzer satellite monitoring led Gillon et al. (2017) to identify four additional transiting planets (e, f, g, and h) with semi-major axes 0.0282, 0.0371, 0.0451, and 0.063 au. Transit durations measure planetary radii between 0.8 and 1.1 times R_{Earth}. Transit timing differences due to Laplace resonances among adjacent triplets of planets (cf. Luger et al. 2017) permitted Gillon et al. (2017) to determine that their masses lie between 0.4 and 1.4 times M_{Earth} and that their densities are between 0.6 and 1.2 times that of the Earth. Thus all seven planets appear to be rocky with Earth-like compositions. Also, planets d and e could be in the RVEM habitable zone with insolation values $1.143 S_0$ and $0.662 S_0$, respectively. As a result of this rich find, there have been many studies to assess whether any of these exoplanets could be habitable.

To determine which if any of the TRAPPIST-1 exoplanets could now have liquid surface water over at least a portion of its surface, Wolf (2017, 2018) used a modification of the 3D global climate model developed for the National Center for Atmospheric Research to calculate temperatures across the exoplanet surfaces. He assumed that the exoplanets are in synchronous rotation, which is consistent with their low eccentricity (Gillon et al. 2017). He also assumed that the atmospheres consist only of N_2, H_2O, and CO_2 with a range of partial pressures. He found that the host star's radiation, which is peaked in the near-IR where water and ice are more absorbant than in the optical, would quickly lead to thermal runaway and surface desiccation of the inner planets (b, c, and d). The outer planets (f, g, and h) would be completely frozen, except if planet f has an atmosphere with at least 2 bars of CO_2. This leaves only planet e as possibly habitable. At least 20% of its surface would have liquid water for an atmosphere with 1 bar of N_2 and at least 10^{-4} partial pressure of CO_2. This calculation assumes that planet e initially had a large inventory of H_2O, because it may have migrated from beyond the snow line. However the model ignores the high irradiation from the young host star at planet e's present orbit and the many ways in which EUV radiation and stellar winds can deplete a planet's atmosphere. There are several estimates of the EUV radiation received by the TRAPPIST-1 exoplanets and time scales for water loss. Starting from the X-ray luminosity, $L_x = (3.8–7.9) \times 10^{26}\,\mathrm{erg\,s^{-1}}$ observed with *XMM-Newton*, Wheatley et al. (2017) used scaling relations (Chadney et al. 2015) to infer the EUV irradiation levels and the mass-loss rates for each planet assuming that the heating efficiency in the planetary exospheres $\eta = 0.10$ (see Eq. (10.2)). Although this analysis has many uncertainties, it predicts that planets b and c will be completely desiccated and planet d may retain significant water. Taking a different approach, Bourrier et al. (2017) reconstructed the Lyman-α line luminosity from STIS spectra and used the relation between Lyman-α and the EUV emission (cf.

Linsky et al. 2014) to infer the EUV irradiation of each planet. Assuming the present value of the EUV flux and $\eta = 0.01$ rather than 0.10, they also concluded that planets b and c would have lost their water in 1–3 Gyr.

The hydrodynamical mass-loss calculations by Bolmont et al. (2017) also show that planets b and c are almost certainly desiccated worlds because they are always located inside of the habitable zone as the host star cools and they lose the water equivalent to 3–13 Earth's oceans over the lifetime of the star. According to Bolmont et al. (2017), planet d at a distance of 0.0214 au and irradiance level 1.143 that of the Earth (Gillon et al. 2017) would have lost 1.3–4.0 Earth oceans (depending on the input EUV flux) if it rotates synchronously before entering the inner edge of the habitable zone. If planet d is instead in a 3:2 spin-orbit resonance, it would have lost more water and never entered the habitable zone. Thus planet d under favorable circumstances could have retained sufficient water to be habitable. Planet e at 0.0281 au with 0.662 times insolation of the Earth is the more likely planet to be habitable as near-infrared radiation of the host star has a higher absorption by H_2O and CO_2 than optical radiation emitted by warmer host stars. The planets lying further from TRAPPIST-1 are likely to be snowballs (Wolf 2017) and uninhabitable.

The effect of very strong stellar wind dynamic pressure on the TRAPPIST-1 planets can be very detrimental to their habitability. To test this, Garraffo et al. (2017) computed the stellar wind and magnetic field structure of the host star using the BATS-R-US 3D MHD code and two values for the star's mean magnetic field. The total magnetic pressure (magnetic and dynamic pressure) and the wind dynamic pressure alone shown in Fig. 10.12 are 10^3–10^5 times larger than seen by the Earth at 1 au and vary by more than an order of magnitude as a planet crosses the current sheet near the stellar equatorial plane. As a result, the standoff distance of the magnetopause is not located far from the planet like the Earth (about $10 R_{Earth}$) but rather close to the planetary surface as shown in the figure. Garraffo et al. (2017) showed that the planets spend much of their time in the sub-Alfvénic region of the stellar wind, where the extreme wind pressure opens up a large polar cup allowing the wind to penetrate all of the way down to the surface of planet f and the interior planets leading to very rapid erosion of their atmospheres. All of the TRAPPIST-1 planets are, therefore, unlikely to be habitable.

10.4.3   Case Study: Solar-Like Stars

Since the habitable zones of G and K stars lie much farther from their host stars than for the M stars, exoplanets of warmer stars will not suffer the harsh radiation and wind environment that M dwarf exoplanets must endure. Figure 10.10 shows a recent inventory of warm host star exoplanets observed by *Kepler* during transits (Kaltenegger 2017). These exoplanets likely have rocky surfaces as their radii are less than twice that of the Earth. They are located inside of the CHZ or RVEM habitable zones but are typically located at a distance of several hundred parsecs. The search for Earth-mass and super-Earth mass exoplanets in the habitable zones

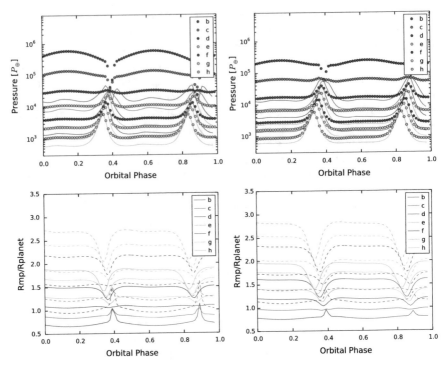

Fig. 10.12 *Top panels:* Simulations of the total magnetic pressure (filled circles) and wind dynamic pressure (lines) for the TRAPPIST-1 planets with the mean magnetic field of 600 G (left) and 300 G (right) for the host star. Units of pressure are relative to the solar wind at Earth. *Bottom panels:* Magnetopause standoff distance in planetary radii for an assumed 0.5 G planetary magnetic field (dashed lines) and 0.1 G (solid lines). $R_{mp}/R_{planet} = 1$ corresponds to the magnetopause located at the planet's surface. Figure from Garraffo et al. (2017). Reproduced by permission of the AAS

of warm stars by exploiting the *Kepler* and *K2* data sets continues but with very limited success so far. For example, Dressing et al. (2017) searched the first seven *K2* campaigns and identified only one super-Earth in the RVEM habitable zone (EPIC 206209135) but the host star is an M2 dwarf. Two other Neptune-mass planets were also identified inside of the RVEM habitable zone. Another Earth-size exoplanet in its habitable zone is Kepler-186f but it's host star is M1 V with $T_{eff} = 3788 \pm 54$ K (Quintana et al. 2014). The prime objective of the new TESS mission is detect and study exoplanets in the habitable zones of nearby stars, especially warmer stars.

Table 10.1 lists exoplanets located inside of or near the CHZs of nearby warm stars studied by radial velocity techniques. Three solar system planets are included for comparison. The table lists the host star luminosities, effective temperatures, and distances plus the super-Earth masses and distances from the host star. Also listed are the inner and outer boundaries of the CHZ computed by Kopparapu et al. (2013, 2014). HD 20794d studied by Feng et al. (2017) and HD 85512b studied

Table 10.1 Possibly habitable planets of nearby warm stars

Host star	T_{eff} (K)	$L/L_\odot$	d (pc)	Habitable zone (au)	Planet	a (au)	M/M_E
Sun	5777	1.00	–	0.95–1.67	Venus	0.723	0.815
					Earth	1.00	1.00
					Mars	1.523	0.107
Kepler-452	5757	1.21	–	0.95–1.68	b	1.046	–
HD 20794	5401	0.656	6.06	0.76–1.34	d	0.35 ± 0.006	≥ 4.8
τ Cet	5344	0.52	3.7	0.70–1.32	e	0.538 ± 0.006	≥ 3.93
					f	$1.334^{+0.017}_{-0.044}$	≥ 3.93
HD 40307	4956	0.23	13.0	0.54–0.97	g	0.60 ± 0.034	≥ 7.1
HD 85512	4715	0.126	11.15	0.47–0.85	b	0.26 ± 0.005	≥ 3.6

by Pepe et al. (2011) lie inside of their CHZs, and HD 40307g is definitely inside of its CHZ (Tuomi et al. 2013). τ Cet has two interesting super-Earth exoplanets, planet e lies close to the inner boundary and planet f lies close to the outer boundary. Although its distance is not known, Kepler-452 is now the best example of an Earth-size presumably rocky exoplanet in its habitable zone (Jenkins et al. 2015). In a detailed study of its habitability and possible biosignatures, Silva et al. (2017) found that Kepler-452b is slightly warmer than the Earth but should be in the CHZ if the CO_2 partial pressure is less than 0.04 bars.

10.4.4 Case Study: Mars

There is an extensive literature concerning the origin and evolution of Venus, Earth, and Mars addressing such questions as the origin of their very different atmospheres and biospheres given the same host star whose spectral energy distribution and wind decrease by large factors over time. Lammer et al. (2018) have recently published an extensive review of this fascinating topic.

Mars provides is an interesting case study of the interaction of the wind and radiation from a G2 V star with a sub-Earth mass planet from the time of their formations to the present. This study has many fewer unknowns and uncertainties than comparable studies of exoplanets, because the present atmospheric composition of Mars is known, there is strong evidence that Mars had a thick atmosphere and surface water at least episodically in the past (cf. Wordsworth 2016), the present solar spectral energy distribution and wind properties are known, and reasonable estimates of their past properties have been inferred from younger solar-type stars as described in Chaps. 7–9.

Dong et al. (2018) presented a recent assessment of the present and past mass-loss rates of different atoms and ions from the Martian atmosphere. They used state-of-the-art 3D global models for the ionosphere, exosphere, and magnetosphere that include radiative transfer, chemical reactions, and kinematics. These models

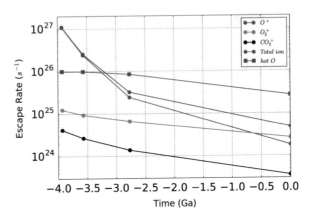

Fig. 10.13 Dissociative recombination rates in the atmosphere of Mars as a function of look back time. Figure from Dong et al. (2018). Reproduced by permission of the AAS

include as input the decrease in the solar EUV flux and wind as the Sun has evolved. Figure 10.13 shows the escape rates for different species. At present, the largest escape rate is by hot oxygen atoms produced by dissociative recombination reactions following ionization of molecular oxygen. Before 3.0 Gyr ago, the largest escape rate was by O^+. At all times, the O_2^+ and CO_2^+ escape rates involving pick-up by the solar wind particles are important but not dominant. This relative change in the dominant escape process is probably due to the importance of ion pick-up by the young Sun's strong magnetized solar wind that escaping hot oxygen atoms do not need. The computed loss of surface H_2O by evaporation, photolysis, and subsequent escape of hydrogen over the last 4 Gyr is equivalent to a water ocean covering all of Mars to a depth of 2.6 m. Most of the water and atmosphere loss occurred near the beginning of this period when the solar EUV flux was 100 times larger than today and the solar wind was also much stronger. This conclusion is supported by observations of the $^{38}Ar/^{36}Ar$ ratio in the upper atmosphere of Mars obtained with the *Mars Atmosphere and Volatile Evolution Mission (MAVEN)* (Jakosky et al. 2017). *MAVEN* observations also showed that the ionization rates of O^+ and O_2^+ increase with the incidence EUV flux (Dubinin et al. 2017). Since the Dong et al. models do not include sputtering processes, which may be very important, the total escape rate may be larger and water loss occurred at an even earlier time.

Observations of the Martian atmosphere and input solar wind and EUV radiation over the course of a year (Jakosky et al. 2018) are consistent with the calculations of Dong et al. (2018), but the data quantify the uncertainties in the present mass-loss rates and the far larger uncertainties in the mass-loss rates 3.5–4.2 billion years ago when Mars likely had a thick atmosphere, magnetic field, and surface water. Jakosky et al. (2018) found that at present hydrogen is lost to space by Jeans escape and that oxygen is lost primarily as hot oxygen atoms produced by dissociative recombination of O_2^+ with ambient electrons. They estimated that this process dominated the oxygen mass-loss rate until about 3.5 billion years ago when O^+ pick-up ions may have dominated, but there are many uncertainties in this speculation. The importance of charge exchange processes in the Martian atmosphere is demonstrated by *MAVEN* observations of auroral emission produced by energetic neutral hydrogen atoms resulting from change exchange between solar wind protons and hydrogen in the Martian atmosphere (Deighan et al. 2018).

10.5 Which Star Types Are Best for Hosting a Habitable Planet?

There is an unresolved discussion as to whether the stars most likely to host habitable exoplanets are M dwarfs or the warmer G and K stars. In order to better understand the issue, I list the pros (P) and cons (C) of these two options. The arguments in favor of M stars rather than their more massive and warmer stars are:

P1-Star density: M dwarfs are the most abundant stars with about 75% of the stellar population in the Galaxy. The relative population of K and G stars decreases rapidly with increasing mass. Both types of stars have rich planetary systems, for example, TRAPPIST-1 (M8 V) has seven known planets and 55 CnC (G8 V) has five known exoplanets.

P2-Nearest exoplanets: Since M dwarfs are the most common stars, their exoplanets will likely be closest to the Sun. The nearest star, Proxima Centauri, harbors the nearest exoplanet. Proximity is a great benefit for observations of transits and direct imaging.

P3-Transits: Since M stars are faint, their habitable zones lie close to the star. Exoplanets close to a star have a higher probability of being located close to the line of sight to the star and thus to have observable transits. Since M stars are smaller and fainter than more massive stars, transiting planets cover a larger fraction of the stellar surface with a larger contrast between the emission or absorption from the plane compared to the star.

P4-Detection: Exoplanets in the close-in habitable zones of M stars produce relatively large orbital radial velocity variations in their low mass host stars. This facilitates exoplanet detection by radial velocity monitoring. The large orbital radial velocity amplitude of these exoplanets also favors the separation of their emission in the infrared from that of their cool host star.

There are also arguments in favor of warmer stars being the better hosts of habitable exoplanets:

C1-Irradiation: The habitable zone exoplanets of G and K stars are located much further away from their host star than for M stars. For example, Proxima b is located at 0.0485 au near the center of its habitable zone. If the X-ray, EUV, and UV emission from the host star are the same as the Sun, then the X-ray to UV irradiation level at Proxima b would be 425 times larger. The dynamic pressure of the stellar wind would also be larger by this factor.

C2-Activity time scale: Young rapidly rotating stars are active with many strong flares, superflares and CMEs that can produce rapid mass loss by thermal and nonthermal processes. This high level of activity can persist for 10–100 Myr for G-type stars until the decrease in their rotation reduces their energetic radiation from saturation to far lower levels. For M stars the corresponding decrease in high-energy radiation takes place over several billion years, especially for the lowest mass M dwarfs. The result is that exoplanets of M dwarfs lose mass and water from their atmospheres for much longer times than exoplanets of more massive stars.

C3-Existence proof: The only known habitable planet that is indeed inhabited is the Earth which has a G2 V host star.

A number of authors have considered whether M dwarfs or more massive stars are more likely to host habitable exoplanets using the arguments outlined above, but there are additional considerations. Cuntz and Guinan (2016) proposed an Habitable-Planetary-Real-Estate Parameter (HabPREP) to identify which star types have a higher probability of hosting habitable planets. The HabPREP is the product of the area in the climatological habitable zone (CLI-HZ), which decreases rapidly to cooler stars, and the relative population of stars of a given effective temperature, which increases rapidly to cooler stars. Cuntz and Guinan (2016) summarize the different ways of defining the CLI-HZ and follow the Kopparapu et al. (2013, 2014) model of the CHZ for which the inner edge is determined from a runaway greenhouse (or moist greenhouse) and the outer edge is determined by the possibility of liquid surface water with maximum greenhouse effect (cf., Haqq-Misra et al. 2016). These CLI-HZ limits are 0.97–0.99 au and 1.70 au, respectively, for the present day Sun. For stars, these limits change with the total stellar flux and to a lesser extent planetary albedo, cloudiness, mass, and atmospheric chemical composition (cf., Kasting et al. 2014). Also changes in the star's luminosity over time will move the CLI-HZ boundaries in which case an exoplanet could be located too close to the star when its luminosity is high during in its premain-sequence phase, then too far from the fainter star during its early main-sequence phase, and finally too close to the star as it brightens during its late main-sequence and giant phases (see Fig. 10.8).

The CLI-HZ and IMF stellar population as a function of effective temperature are shown in Fig. 10.14. According to the HabPREP, the stars that are most likely to have habitable exoplanets are stars with effective temperatures 4900–5300 K, corresponding to spectral types K2 V to G8 V, although the HabPREP does not decline rapidly to cooler or hotter stars. The HabPREP also rises for the coolest stars $T_{eff} < 3500$ K, corresponding to spectral types cooler than M2 V. The HabPREP is a useful tool for searching for habitable planets, but it does not include parameters (e.g., C1 and C2) that actually determine habitability. As described in the case studies of Proxima b and the TRAPPIST-1 planets, the extremely high radiation levels and stellar wind dynamic pressures that close-in exoplanets face make it unlikely that exoplanets in the CLI-HZ's of late-M dwarfs and probably also early-M dwarfs will be habitable.

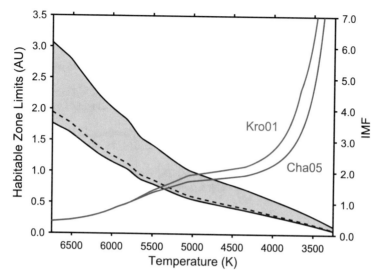

Fig. 10.14 Solid black lines are the outer and inner limits of the climatological habitable zone as a function of stellar effective temperature. The dashed line describes the location of total irradiation similar to the Earth. The relative population of stars in the initial mass function (IMF) is shown by the red line (Kroupa 2001) or the blue line (Chabrier et al. 2005). Figure from Cuntz and Guinan (2016). Reproduced by permission of the AAS

Lingam and Loeb (2017, 2018) pursued a different approach to the probability of life forms on the exoplanets. They developed a model for the timescale of the evolution of biodiversity on Earth and compared it to the time scales for stellar evolution and exoplanets retaining their atmospheres given the erosion by stellar winds and thermally driven hydrodynamic outflows. Their model predicts that the most likely exoplanets to have a diverse biosphere are those with host star masses in the range 0.38–1.0 $M_\odot$ with a peak near 0.55 $M_\odot$. This mass range corresponds to spectral types M2 V to G2 V with a peak near K7 V. Late-M dwarfs are definitely excluded due to the rapid loss of their atmospheres.

Finally, See et al. (2014) suggest that the host star mass range (0.6–0.8) $M_\odot$ may be the most favorable for exoplanet habitability on the basis that Earth-like planets in the center of their habitable zone should have the largest magnetospheres for protection against host star winds and their nonthermal mass-loss processes.

The evidence accumulated so far indicates that G and K dwarfs are prime candidates for having habitable exoplanets. Future studies in the next few years with the *TESS* and *JWST* satellites and large ground-based observatories may answer this question.

References

Airapetian, V.S., Glocer, A., Khazanov, G.V., Loyd, R.O.P., France, K., Sojka, J., Danchi, W.C., Liemohn, M.W.: How hospitable are space weather affected habitable zones? The role of ion escape. Astrophys. J. Lett. **836**, L3 (2017)

Anglada-Escudé, G., Amado, P.J., Barnes, J., Berdinas, Z.M., Butler, R.P., Coleman, Gavin A.L., de La Cueva, I., Dreizler, S., Endl, M., et al.: A terrestrial planet candidate in a temperate orbit around Proxima Centauri. Nature **536**, 437 (2016)

Bolmont, E., Selsis, F., Owen, J.E., Ribas, I., Raymond, S.N., Leconte, J., Gillon, M.: Water loss from terrestrial planets orbiting ultracool dwarfs: implications for the planets of TRAPPIST-1. Mon. Not. R. Astron. Soc. **464**, 3728 (2017)

Bourrier, V., Lecavelier des Etangs, A., Ehrenreich, D., Tanaka, Y.A., Vidotto, A.A.: An evaporating planet in the wind: stellar wind interactions with the radiatively braked exosphere of GJ 436 b. Astron. Astrophys. **591**, A121 (2016)

Bourrier, V., Ehrenreich, D., Wheatley, P.J., Bolmont, E., Gillon, M., de Wit, J., Burgasser, A.J., Jehin, E., Queloz, D., Triaud, A.H.M.J.: Reconnaissance of the TRAPPIST-1 exoplanet system in the Lyman-α line. Astron. Astrophys. **599**, L3 (2017)

Boutle, I.A., Mayne, N.J., Drummond, B., Manners, J., Goyal, J., Hugo L.F., Acreman, D.M., Earnshaw, P.D.: Exploring the climate of Proxima b with the Met Office Unified Model. Astron. Astrophys. **601**, 120 (2017)

Chabrier, G., Baraffe, I., Allard, F., Hauschildt, P.H.: Review on low-mass stars and brown dwarfs. arXiv:astro-ph/0509798 (2005)

Chadney, J.M., Galand, M., Unruh, Y.C., Koskinen, T.T., Sanz-Forcada, J.: XUV-driven mass loss from extrasolar giant planets orbiting active stars. Icarus **250**, 357 (2015)

Claire, M.W., Sheets, J., Cohen, M., Ribas, I., Meadows, V.S., Catling, D.C.: The evolution of solar flux from 0.1 nm to 160 μm: quantitative estimates for planetary studies. Astrophys. J. **757**, 95 (2012)

Cohen, O., Drake, J.J., Glocer, A., Garraffo, C., Poppenhaeger, K., Bell, J.M., Ridley, A.J., Gombosi, T.I.: Magnetospheric structure and atmospheric Joule heating of habitable planets orbiting M-dwarf stars. Astrophys. J. **790**, 57 (2014)

Cohen, O., Glocer, A., Garraffa, C., Drake, J.J., Bell, J.M.: Energy dissipation in the upper atmospheres of TRAPPIST-1 planets. Astrophys. J. Lett. **856**, L11 (2018)

Cuntz, M., Guinan, E.F.: About exobiology: the case for dwarf K stars. Astrophys. J. **827**, 79 (2016)

Deighan, J., Jain, S.K., Chaffin, M.S., Fang, X., Halekas, J.S., Clarke, J.T., Schneider, N.M., Stewart, A.I.F., Chaufray, J.-Y., Evans, J. S., et al.: Discovery of a proton aurora at Mars. Nat. Astron. **2**, 802 (2018)

Dong, C., Lingam, M., Ma, Y., Cohen, O.: Is Proxima Centauri b habitable? A study of atmospheric loss. Astrophys. J. Lett. **837**, L26 (2017)

Dong, C., Lee, Y., Ma, Y., Lingam, M., Bougher, S., Luhmann, J., Curry, S., Toth, G., Nagy, A.: Modeling Martian atmospheric losses over time: implications for exoplanetary climate evolution and habitability. Astrophys. J. Lett. **859**, L14 (2018)

Dressing, C.D., Vanderburg, A., Schlieder, J.E., Crossfield, I.J.M., Knutson, H.A., Newton, E.R., Ciardi, D.R., Fulton, B.J., Gonzales, E.J., et al.: Characterizing K2 candidate planetary systems orbiting low-mass stars. II. Planetary systems observed during campaigns 1–7. Astron. J. **154**, 207 (2017)

Dubinin, E., Fraenz, M., Pätzold, M., McFadden, J., Mahaffy, P.R., Eparvier, F., Halekas, J.S., Connerney, J.E.P., Brain, D., et al.: Effects of solar irradiance on the upper ionosphere and oxygen ion escape at Mars: MAVEN observations. J. Geophys. Res. Space Phys. **122**, 7142 (2017)

Egan, H., Ma, Y., Dong, C., Modolo, R., Jarvinen, R., Bougher, S., Halekas, J., Brain, D., Mcfadden, J., Connerney, J., Mitchell, D., Jakosky, B.: Comparison of global Martian plasma models in the context of MAVEN observations. J. Geophys. Res. Space Phys. **123**, 3714 (2018)

Ehrenreich, D., Désert, J.-M.: Mass-loss rates for transiting exoplanets. Astron. Astrophys. **529**, A136 (2011)

Ehrenreich, D., Bourrier, V., Wheatley, P.J., Lecavelier des Etangs, A., Hébrard, G., Udry, S., Bonfils, X., Delfosse, X., Désert, J.-M., Sing, D.K., Vidal-Madjar, A.: A giant comet-like cloud of hydrogen escaping the warm Neptune-mass exoplanet GJ 436b. Nature **522**, 459 (2015)

Erkaev, N.V., Lammer, H., Odert, P., Kulikov, Y.N., Kislyakova, K.G.: Extreme hydrodynamic atmospheric loss near the critical thermal escape regime. Mon. Not. R. Astron. Soc. **448**, 1916 (2015)

Erkaev, N.V., Lammer, H., Odert, P., Kislyakova, K.G., Johnstone, C.P., Güdel, M., Khodachenko, M.L.: EUV-driven mass-loss of protoplanetary cores with hydrogen-dominated atmospheres: the influences of ionization and orbital distance. Mon. Not. R. Astron. Soc. **460**, 1300 (2016)

Feng, F., Tuomi, M., Jones, H.R.A., Barnes, J., Anglada-Escudé, G., Vogt, S.S., Butler, R.P.: Color difference makes a difference: four planet candidates around τ Ceti. Astron. J. **154**, 135 (2017)

Fossati, L., Erkaev, N.V., Lammer, H., Cubillos, P.E., Odert, P., Juvan, I., Kislyakova, K.G., Lendl, M., Kubyshkina, D., Bauer, S.J.: Aeronomical constraints to the minimum mass and maximum radius of hot low-mass planets. Astron. Astrophys. **598**, 90 (2017)

France, K., Loyd, R.O.P., Youngblood, A., Brown, A., Schneider, P.C., Hawley, S.L., Froning, C.S., Linsky, J.L., Roberge, A., et al.: The muscles treasury survey I: motivation and overview. Astrophys. J. **820**, 89 (2016)

Garraffo, C., Drake, J.J., Cohen, O.: The space weather of Proxima Centauri b. Astrophys. J. Lett. **833**, L4 (2016)

Garraffo, C., Drake, J.J., Cohen, O., Alvarado-Gómez, J.D., Moschou, S.P.: The threatening magnetic and plasma environment of the TRAPPIST-1 planets. Astrophys. J. Lett. **843**, L33 (2017)

Garcia-Sage, Glocer, A., Drake, J.J., Gronoff, G., Cohen, O.: On the magnetic protection of the atmosphere of Proxima Centauri b. Astrophys. J. Lett. **844**, L13 (2017)

Gillon, M., Jehin, E., Lederer, S.M., Delrez, L., de Wit, J., Burdanov, A., Van Grootel, V., Burgasser, A.J., Triaud, A.H.M.J., et al.: Temperate Earth-sized planets transiting a nearby ultracool dwarf star. Nature **533**, 221 (2016)

Gillon, M., Triaud, A.H.M.J., Demory, B.-O., Jehin, E., Agol, E., Deck, K.M., Lederer, S.M., de Wit, J., Burdanov, A., Ingalls, J.G., et al.: Seven temperate terrestrial planets around the nearby ultracool dwarf star TRAPPIST-1. Nature **542**, 456 (2017)

Haqq-Misra, J., Kopparapu, R.Kumar, Batalha, N.E., Harman, C.E., Kasting, J.F.: Limit cycles can reduce the width of the habitable zone. Astrophys. J. **827**, 120 (2016)

Jenkins, J.M., Twicken, J.D., Batalha, N.M., Caldwell, D.A., Cochran, W.D., Endl, M., Latham, D.W., Esquerdo, G.A.: Discovery and validation of Kepler-452b: a 1.6 R_{Earth} super Earth exoplanet in the habitable zone of a G2 star. Astron. J. **150**, 56 (2015)

Jakosky, B.M., Slipski, M., Benna, M., Mahaffy, P., Elrod, M., Yelle, R., Stone, S., Alsaeed, N.: Mars' atmospheric history derived from upper-atmosphere measurements of $^{38}Ar/^{36}Ar$. Science **355**, 1408 (2017)

Jakosky, B.M., Brain, D., Chaffin, M., Curry, S., Deighan, J., Grebowsky, J., Halekas, J., Leblanc, F., Lillis, R., Luhmann, J.G., et al.: Loss of the Martian atmosphere to space: present-day loss rates determined from MAVEN observations and integrated loss through time. Icarus **315**, 146 (2018)

Kaltenegger, L.: How to characterize habitable worlds and signs of life. Ann. Rev. Astron. Astrophys. **55**, 433 (2017)

Kasting, J.F., Whitmire, D.P., Reynolds, R.T.: Habitable zones around main sequence stars. Icarus **101**, 108 (1993)

Kasting, J.F., Kopparapu, R., Ramirez, R.M., Harman, C.E.: Remote life-detection criteria, habitable zone boundaries, and the frequency of Earth-like planets around M and late K stars. Proc. Natl. Acad. Sci. **111**, 12641 (2014)

Khodachenko, M.L., Lammer, H., Lichtenegger, H.I.M., Langmayr, D., Erkaev, N.V., Grießmeier, J.-M., Leitner, M., Penz, T., Biernat, H.K., et al.: Mass loss of "Hot Jupiters" implications for

CoRoT discoveries. Part I: the importance of magnetospheric protection of a planet against ion loss caused by coronal mass ejections. Plan. Space Sci. **55**, 631 (2007)

Khodachenko, M.L., Shaikhislamov, I.F., Lammer, H., Prokopov, P.A.: Atmosphere expansion and mass loss of close-orbit giant exoplanets heated by stellar XUV. II. Effects of planetary magnetic field; structuring of inner magnetosphere. Astrophys. J. **813**, 50 (2015)

Kim, Y.-C., Demarque, P., Yi, S.K., Alexander, D.R.: The Y^2 isochrones for α-element enhanced mixtures. Astrophys. J. Suppl. **143**, 499 (2002)

King, G.W., Wheatley, P.J., Salz, M., Bourrier, V., Czesla, S., Ehrenreich, D., Kirk, J., Lecavelier des Etangs, A., Louden, T., Schmitt, J., Schneider, P.C.: The XUV environments of exoplanets from Jupiter-size to super-Earth. Mon. Not. R. Astron. Soc. **478**, 1193 (2018)

Kislyakova, K.G., Johnstone, C.P., Odert, P., Erkaev, N.V., Lammer, H., Lüftinger, T., Holmström, M., Khodachenko, M.L., Güdel, M.: Stellar wind interaction and pick-up ion escape of the Kepler-11 "super-Earths". Astron. Astrophys. **562**, A116 (2014)

Kopparapu, R.K., Ramirez, R., Kasting, J.F., Eymet, V., Robinson, T.D., Mahadevan, S., Terrien, R.C., Domagal-Goldman, S., Meadows, V., Deshpande, R.: Habitable zones around main-sequence stars: new estimates. Astrophys. J. **765**, 131 (2013)

Kopparapu, R.K., Ramirez, R.M., SchottelKotte, J., Kasting, J.F., Domagal-Goldman, S., Eymet, V.: Habitable zones around main-sequence stars: dependence on planetary mass. Astrophys. J. Lett. **787**, L29 (2014)

Kopparapu, R., Wolf, E.T., Haqq-Misra, J., Yang, J., Kasting, J.F., Meadows, V., Terrien, R., Mahadevan, S.: The inner edge of the habitable zone for synchronously rotating planets around low-mass stars using general circulation models. Astrophys. J. **819**, 84 (2016)

Kroupa, P.: On the variation of the initial mass function. Mon. Not. R. Astrtron. Soc. **322**, 231 (2001)

Kubyshkina, D., Lendl, M., Fossati, L., Cubillos, P.E. Lammer, H., Erkaev, N.V., Johnstone, C.P.: Young planets under extreme UV irradiation. I. Upper atmosphere modelling of the young exoplanet K2-33b. Astron. Astrophys. **612**, 25 (2018)

Lammer, H., Lichtenegger, H.I.M., Kolb, C., Ribas, I., Guinan, E.F., Abart, R., Bauer, S.J.: Loss of water from Mars: implications for the oxidation of the soil. Icarus, **165**, 9 (2003)

Lammer, H., Stökl, A., Erkaev, N.V., Dorfi, E.A., Odert, P., Güdel, M., Kulikov, Y. N., Kislyakova, K.G., Leitzinger, M.: Origin and loss of nebula-captured hydrogen envelopes from 'sub'- to 'super-Earths' in the habitable zone of Sun-like stars. Mon. Not. R. Astron. Soc. **439**, 3225 (2014)

Lammer, H., Zerkle, A.L., Gebauer, S., Tosi, N., Noack, L., Scherf, M., Pilat-Lohinger, E., Güdel, M., Grenfell, J.L., Godolt, M., Nikolaou, A.: Origin and evolution of the atmospheres of early Venus, Earth and Mars. Astron. Astrophys. Rev. **26**, 2 (2018)

Lanza, A.F.: Star-planet magnetic interaction and evaporation of planetary atmospheres. Astron. Astrophys. **557**, A31 (2013)

Lavie, B., Ehrenreich, D., Bourrier, V., Lecavelier des Etangs, A., Vidal-Madjar, A., Delfosse, X., Gracia Berna, A., Heng, K., Thomas, N., Udry, S., Wheatley, P.J.: The long egress of GJ 436b's giant exosphere. Astron. Astrophys. **605**, L7 (2017)

Lecavelier des Etangs, A.: A diagram to determine the evaporation status of extrasolar planets. Astron. Astrophys. **461**, 1185 (2007)

Lingam, M., Loeb, A.: Reduced diversity of life around Proxima Centauri and TRAPPIST-1. Astrophys. J. Lett. **846**, L21 (2017)

Lingam, M., Loeb, A.: Physical constraints on the likelihood of life on exoplanets. Int. J. Astrobiology **17**, 116 (2018)

Linsky, J.L., Yang, H., France, K., Froning, C.S., Green, J.C., Stocke, J.T., Osterman, S.N.: Observations of mass loss from the transiting exoplanet HD 209458b. Astrophys. J. **717**, 1291 (2010)

Linsky, J.L., Fontenla, J., France, K.: The intrinsic extreme ultraviolet fluxes of F5 V to M5 V stars. Astrophys. J. **780**, 61 (2014)

Loyd, R.O.P., France, K., Youngblood, A., Schneider, C., Brown, A., Hu, R., Linsky, J., Froning, C.S., Redfield, S., Rugheimer, S., Tian, F.: The MUSCLES treasury survey III: X-ray to infrared spectra of 11 M and K stars hosting planets. Astrophys. J. **824**, 102 (2016)

Luger, R., Sestovic, M., Kruse, E., Grimm, S.L., Demory, B.-O., Agol, E., Bolmont, E., Fabrycky, D., Fernandes, C.S., et al.: A seven-planet resonant chain in TRAPPIST-1. Nat. Astron. **1**, 129 (2017)

Meadows, V.S., Arney, G.N., Schwieterman, E.W, Lustig-Yaeger, J., Lincowski, A.P., Robinson, T., Domagal-Goldman, S.D., Deitrick, R., Barnes, R.K., et al.: The habitability of Proxima Centauri b: environmental states and observational discriminants. Astrobiology **18**, 133 (2018)

Murray-Clay, R.A., Chiang, E.I., Murray, N.: Atmospheric escape from hot Jupiters. Astrophys. J. **693**, 23 (2009)

O'Malley-James, J.T., Kaltenegger, L.: UV surface habitability of the TRAPPIST-1 system. Mon. Not. R. Astron. Soc. Lett. **469**, L260 (2017)

Owen, J.E., Alvarez, M.A.: UV driven evaporation of close-in planets: Energy-limited, recombination-limited, and photon-limited Flows. Astrophys. J. **816**, 340 (2016)

Owen, J.E., Jackson, A.P.: Planetary evaporation by UV and X-ray radiation: basic hydrodynamics. Mon. Not. R. Astron. Soc. **425**, 2531 (2012)

Owen, J.E., Wu, Y.: Atmospheres of low-mass planets: the "boil-off". Astrophys. J. **817**, 1070 (2016)

Pepe, F., Lovis, C., Ségransan, D., Benz, W., Bouchy, F., Dumusque, X., Mayor, M., Queloz, D., Santos, N.C., Udry, S.: The HARPS search for Earth-like planets in the habitable zone. I. Very low-mass planets around HD 20794, HD 85512, and HD 192310. Astron. Astrophys. **534**, A58 (2011)

Quintana, E.V., Barclay, T., Raymond, S.N., Rowe, J.F., Bolmont, E., Caldwell, D.A., Howell, S.B., Kane, S.R., et al.: An Earth-sized planet in the habitable zone of a cool star. Science **344**, 277 (2014)

Ribas, I.: The Sun and stars as the primary energy input in planetary atmospheres. In: Kosovichev, A.G., Andrei, A.H., Rozelot, J.-P. (eds.) Solar and Stellar Variability: Impact on Earth and Planets. Proceedings IAU Symposium No. 264 (2009)

Ribas, I., Bolmont, E., Selsis, F., Reiners, A., Leconte, J., Raymond, S.N., Engle, S.G., Guinan, E.F., Morin, J.: The habitability of Proxima Centauri b. I. Irradiation, rotation and volatile inventory from formation to the present. Astron. Astrophys. **596**, A111 (2016)

Ribas, I., Gregg, M.D., Boyajian, T.S., Bolmont, E.: The full spectral radiative properties of Proxima Centauri. Astron. Astrophys. **603**, 58 (2017)

Sagan, C., Mullen, G.: Earth and Mars: evolution of atmospheres and surface temperatures. Science **177**, 52 (1972)

Salz, M., Schneider, P.C., Czesla, S., Schmitt, J.H.M.M.: Energy-limited escape revised. The transition from strong planetary winds to stable thermospheres. Astron. Astrophys. **585**, L2 (2016)

See, V., Jardine, M., Vidotto, A.A., Petit, P., Marsden, S.C., Jeffers, S.V., do Nascimento, J.D.: The effects of stellar winds on the magnetospheres and potential habitability of exoplanets. Astron. Astrophys. **570**, A99 (2014)

Silva, L., Vladilo, G., Murante, G., Provenzale, A.: Quantitative estimates of the surface habitability of Kepler-452b. Mon. Not. R. Astron. Soc. **470**, 2270 (2017)

Tuomi, M., Anglada-Escudé, G., Gerlach, E., Jones, H.R.A., Reiners, A., Rivera, E.J., Vogt, S.S., Butler, R.P.: Habitable-zone super-Earth candidate in a six-planet system around the K2.5V star HD 40307. Astron. Astrophys. **549**, A48 (2013)

Vidotto, A.A., Jardine, M., Opher, M., Donati, J.F., Gombosi, T.I.: Powerful winds from low-mass stars: V374 Peg. Mon. Not. R. Astron. Soc. **412**, 351 (2011)

Vidal-Madjar, A., Lecavelier des Etangs, A., Désert, J.-M., Ballester, G.E., Ferlet, R., Hébrard, G., Mayor, M.: An extended upper atmosphere around the extrasolar planet HD 209458b. Nature **422**, 143 (2003)

Watson, A.J., Donahue, T.M., Walker, J.C.G.: The dynamics of a rapidly escaping atmosphere - Applications to the evolution of Earth and Venus. Icarus **48**, 150 (1981)

Wheatley, P.J., Louden, T., Bourrier, V., Ehrenreich, D., Gillon, M.: Strong XUV irradiation of the Earth-sized exoplanets orbiting the ultracool dwarf TRAPPIST-1. Mon. Not. R. Astron. Soc. Lett. **465**, L74 (2017)

Wolf, E.T.: Assessing the habitability of the TRAPPIST-1 system Using a 3D climate model. Astrophys. J. Lett. **839**, L1 (2017)

Wolf, E.T.: Erratum: "Assessing the habitability of the TRAPPIST-1 system using a 3D climate model". Astrophys. J. Lett. **855**, L14 (2018)

Wordsworth, R.D.: The climate of early Mars. Ann. Rev. Earth Planet. Sci. **44**, 381 (2016)

Yi, S.K., Kim, Y.-C., Demarque, P.: The Y^2 stellar evolutionary track. Astrophys. J. Suppl. **144**, 259 (2003)

Youngblood, A., France, K., Loyd, R.O.P., Brown, A., Mason, J.P., Schneider, P.C., Tilley, M.A., Berta-Thompson, Z.K., Buccino, A., Froning, C.S., et al.: The MUSCLES tresury survey IV. Scaling relations for ultraviolet Ca II, and energetic particle fluxes from M dwarfs. Astrophys. J. **843**, 31 (2017)

Chapter 11
Host Star Driven Photochemistry in Exoplanet Atmospheres

> Most of these false-positive mechanisms are dependent on properties of the host star and are often strongest for planets orbiting M dwarfs. (Meadows 2017)

Photochemical reactions often dominate over collision-based equilibrium chemistry at atmospheric pressures less than 1 mbar, corresponding to about 0.1% of the Earth's surface pressure and heights above 60 km for Earth-like rocky planets. These reactions occur when photons, usually at ultraviolet wavelengths, dissociate or ionize molecules leading to subsequent reactions that produce abundances of important species that are very different from equilibrium values in the lower atmosphere. I describe in this chapter the important photochemical reactions driven by the host star's UV radiation and atmospheric models that include photochemistry. The resulting abundances of important molecules depend on the spectral energy distribution of the host star, but O_3 is an extreme case with bearing on the detection of biosignatures in exoplanet atmosphere. As the study of exoplanets moves from discovery to atmospheric characterization and the search for habitability, disequilibrium chemical models that include photochemical and mixing processes are required to correctly interpret absorption and emission line spectra. In particular, photochemical haze produced by far-UV driven dissociation of CH_4 is the likely cause of the observed featureless near-IR spectra of many sub-Neptune exoplanets (e.g., Morley et al. 2015).

11.1 Photochemistry of Important Molecules

The rate of photochemical reactions is the integral of the incident photon flux $f(\lambda)$ times the photodissociation cross section $\sigma(\lambda)$ (for all pathways) from the threshold wavelength of the transition λ_{thr},

$$R = \int_{\lambda thr}^{\infty} \sigma(\lambda) f(\lambda) d\lambda. \tag{11.1}$$

© Springer Nature Switzerland AG 2019
J. Linsky, *Host Stars and their Effects on Exoplanet Atmospheres*,
Lecture Notes in Physics 955, https://doi.org/10.1007/978-3-030-11452-7_11

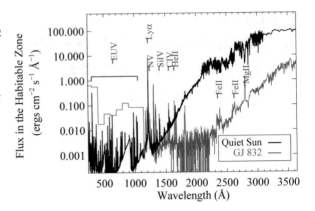

Fig. 11.1 Comparing the spectra of the Sun and GJ 832 (M1.5 V) as seen from their respective habitable zones. The figure is adapted from France et al. (2012). Reproduced by permission of the AAS

Figure 11.1 shows that the spectral energy distributions (SEDs) of stars can be qualitatively different even when viewed from distances with the same bolometric insolation. In this figure the flux from the M1.5 V star GJ 832 at a distance of 0.16 au is compared to the flux from the Sun at a distance of 1 au. The near-UV spectrum (160–350 nm) is a factor of 1000 times weaker from GJ 832 than from the Sun because this portion of the M dwarf's spectrum is formed in the cool upper photosphere where the emission at wavelengths much shorter than the peak of an equivalent black body is very weak. This short wavelength emission is similar to the exponential portion of the blackbody function. While the emission from stellar photospheres is not a blackbody at the star's effective temperature, the effect is similar. By contrast, the far-UV spectrum (120–160 nm) is formed in the chromosphere (see Chap. 4) where the thermal structures of the M dwarf and Sun are similar producing comparable emission line spectra. At shorter wavelengths, the emission in the extreme-UV (EUV, 10–91 nm) and X-ray (< 10 nm) regions is much stronger for the M dwarf than the Sun as the M dwarf has a hotter corona. The very different SEDs of active M dwarfs compared to a relatively inactive G2 dwarf has important effects on the photochemistry.

Figure 11.2 shows the photodissociation cross sections of eight important molecules in exoplanet atmospheres (Loyd et al. 2016 and references therein). They all have large cross sections in the far-UV. O_2, H_2O, and N_2O also have significant cross sections in the near-UV, and O_3 (not shown) has its most important cross sections in the near-UV and the blue region of the optical spectrum. With its large dissociation energy of 11.14 eV, CO is dissociated by photons at wavelengths shorter than 111.3 nm. The bottom panel of the figure shows the SEDs in photon rather than energy units normalized so that the bolometric insolation is the same. The three plots in the bottom panel show the relative SEDs for an active K2 V star (ϵ Eri), a relatively low activity M5 V dwarf (GJ 581), and the low activity Sun. The molecules with large cross sections in the near-UV have low photoexcitation rates from M stars. The important example of O_3 will be discussed in Sect. 11.3.

Note that the brightest spectral feature in the far-UV for all three star types is the Lyman-α line at 121.5 nm. The Lyman-α line is especially important for

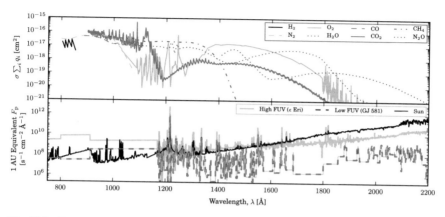

Fig. 11.2 *Upper panel:* Photodissociation cross sections for eight molecules. *Lower panel:* Spectral energy distributions of an active K2 V star (ϵ Eri), a low activity M5 V star (GJ 581), and the Sun (G2 V). The y axis is in photons normalized such that the bolometric energy flux from each star is the same as the Earth's insolation. The figure is from Loyd et al. (2016). Reproduced by permission of the AAS

M dwarfs because of their weak photospheric emission. The wavelength of the Lyman-α line corresponds to high photodissociation cross sections for CH_4 and H_2O. The importance of Lyman-α as a photodissociation driver is illustrated in Fig. 11.3 in which the cumulative dissociation fraction is plotted for eight molecules as a function of wavelength. The sharp rise in the cumulative dissociation fraction at 121.5 nm for H_2O and CH_4 shows that Lyman-α is the primary cause of photoionization for these molecules. H_2, N_2, CO, and CO_2 are primarily dissociated by shorter wavelength radiation, while O_2 and N_2O are primarily dissociated by radiation longward of 140 nm including the emission lines of Si IV and C IV. For each molecule, the three line types are for irradiation from ϵ Eri (faint line), GJ 581 (dashed line), and the Sun (dark line). The wavelength regions where photodissociation occurs is nearly same for the three stars, except for O_2 and N_2O, where the M star's weak near-UV radiation forces the photodissociation to occur at shorter wavelengths by chromospheric radiation.

11.2 Photochemical Atmospheres

A critical issue when computing chemical models and resulting molecular mixing ratios in exoplanet atmospheres is whether chemical equilibrium is a realistic assumption or whether disequilibrium processes, including photodissociation, bulk vertical flows, diffusion, and chemical quenching must be taken into account. Chemical quenching occurs when upflows bring gas from a region where chemical reactions dominate into a region of lower density where the slow chemical reaction rates result in molecular mixing ratios that are "frozen" into their previous values.

Fig. 11.3 Cumulative photdissociation spectra for eight molecules. Each curve shows the fraction of the dissociation of the molecule from photons with wavelengths from the photodissociation threshold to the present wavelength. The grey curves are for the spectral energy distribution of ϵ Eri, the dashed curves are for the SED of GJ 581, and the solid line curves are for the SED of the Sun. The figure is from Loyd et al. (2016). Reproduced by permission of the AAS

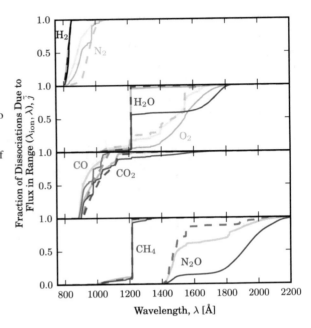

Whether collisional reaction rates are faster or slower than the disequilibrium rates determines whether or not a gas is in chemical equilibrium. Timescales for chemical equilibrium reactions are faster at the high densities in the lower atmosphere and are faster at the high atmospheric temperatures of close-in exoplanets. Conversely, disequilibrium timescales dominate over chemical reaction rate time scales at low pressures and the low temperatures of more distant exoplanets. Thus, photochemistry is important deeper in the atmospheres of cooler exoplanets as chemical reaction rates that support chemical equilibrium decrease to lower temperatures. Accurate UV irradiances are especially important for lower temperature exoplanets where photochemistry strongly effects the determination of molecular abundances from the analysis of infrared absorption spectra.

In their evaluation of chemical equilibrium processes, Madhusudhan et al. (2016) identified three regions in the atmospheres of giant exoplanets: chemical equilibrium in the lower layers, transport-induced quenching in the middle layers, and photochemistry in the upper layers typically at pressures below 1 mbar. The locations of the boundaries between these regions depends upon the chemical composition of the atmosphere, the UV insolation, the molecules available for photodisociation, the "limiting step" in chemical reactions that govern the onset of quenching, and the poorly known vertical mixing and diffusion rates. The chemical composition of the lower atmospheric layers of the hydrogen-rich atmospheres of Jupiters and sub-Neptunes are dominated by H_2, H_2O, CH_4, and NH_3, while the lower layers of hydrogen-poor low-mass exoplanets (Earth-like and super-Earths) are dominated by CO_2 and N_2 with very little H_2O (cf. Madhusudhan et al. 2016). The low amount of H_2 and He in the atmospheres of rocky planets means that the proportion of metals (e.g., C, N, O, and heavier elements) to hydrogen is far higher than in the Sun. These high metallicity exoplanet atmospheres are referred to as "metal rich". The sensitivity to metal richness and C/O ratio for chemical equilibrium models has been explored by Moses et al. (2013a), Hu and Seager (2014), and others. The chemical composition in higher disequilibrium layers of exoplanet atmospheres differs from that in the chemical equilibrium lower layers because of the various disequilibrium processes and hydrodynamic mass loss for highly irradiated exoplanets.

The first generation of photochemical models assumed either the extremely weak UV emission from stellar photospheres without chromospheres or the UV emission measured from G and K stars but very few M stars by the *IUE* satellite. In particular, these one-dimensional (1D) models used the observed flux of the important Lyman-α line, which is severely attenuated by interstellar absorption (see Sect. 6.1) and must be reconstructed to provide the realistic irradiance of an exoplanet's atmosphere. An example of these disequilibrium chemical models is the study by Kopparapu et al. (2012) of WASP-12b, a hot-Jupiter in a short period orbit around its G0 V host star. Using a one-dimensional photochemical code with heritage back to Kasting (1982), they computed the mixing ratios of 31 chemical species in an atmosphere extending up to pressures of 10^{-8} bars. Photodissociation of H_2O by UV radiation dominates the H_2O photochemistry, producing OH which reacts with H_2 to increase the abundance of atomic H relative to H_2. They explained the lack of observable H_2O features in infrared spectra and abundance CH_4 in WASP-12b as resulting from photolysis of H_2 and a carbon rich atmosphere.

There are other photochemistry models of exoplanet atmospheres based on UV irradiances available prior to *HST*. Miguel and Kaltenegger (2014) used the code developed by Kopparapu et al. (2014) to model the atmospheres of sub-Neptunes and hot-Jupiters. For these oxygen-rich atmospheres, H_2O and CH_4 are most sensitive to UV irradiance, but the effect of photodissociation occurs at high levels in the atmosphere (above 10^{-5} bars) because the assumed Lyman-α flux was much smaller than its reconstructed level and the UV irradiance from M dwarfs was only that from their photospheres. If the unknown rate of vertical mixing is weak, then the effect of photodissociation occurs deeper in the atmosphere.

The cumulative dissociation fractions shown in Fig. 11.3 assume that the host star's radiation is not shielded by absorption from overlying molecules and atoms in an exoplanet's atmosphere. To properly include the many photochemical reactions, shielding, diffusion, and flows requires disequilibrium chemical models. Figure 11.4 shows the volume mixing ratios computed for models of the Neptune-mass exoplanet GJ 436b orbiting its M3.5 V host star at a distance of 0.0287 au. Even at this close distance, this exoplanet has only a modest equilibrium temperature of about 640 K, much lower than many hot Jupiters.

Miguel et al. (2015) computed a one-dimensional model for the atmosphere of GJ 436b including photodissociation driven by the UV spectrum of its low activity host star GJ 436 observed by *HST*. Since photodissociation occurs high in the atmosphere where densities are low, recombination can be ignored. Their models assume that the exoplanet is irradiated by the observed UV flux including the

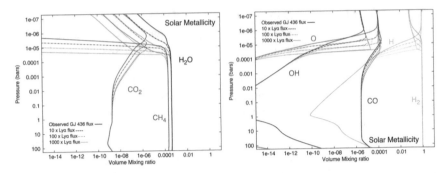

Fig. 11.4 Volume mixing ratios for different molecules in the atmosphere of the sub-Neptune GJ 436b. The mixing ratios are computed including photodissociation by the observed UV flux from the host star with the Lyman-α flux equal to the observed value and 10, 100, and 1000 times the observed value. The figure is from Miguel et al. (2015). Reproduced from MNRAS by permission of the Oxford University Press

reconstructed Lyman-α flux (France et al. 2013). To demonstrate the role played by Lyman-α photoionization, Miguel et al. (2015) computed models with the observed Lyman-α flux multiplied by factors of 1, 10, 100, and 1000. Enhanced Lyman-α fluxes are typical of very active M dwarfs and flares (e.g., Loyd et al. 2018). They also computed models with the Lyman-α flux multiplied by factors of 0.1, 0.01, and 0.001 to simulate possible absorption of the host star's Lyman-α emission by atomic hydrogen in the exospheres of some close-in exoplanets. Figure 11.4 shows that for an exoplanet atmosphere with solar metalicity, Lyman-α photodissociation severely destroys H_2O and CH_4 above a pressure level of 0.1 mbar, producing orders of magnitude increases in the mixing ratios of the daughter products O, OH, and H. H_2O absorbes most of the Lyman-α radiation shielding CH_4, which is instead dissociated mostly by FUV radiation near 130 nm. These models show which molecules are photodissociated and where in the atmosphere this occurs for an exoplanet with solar composition and a modest range of temperatures. Further development is needed to quantify the importance of vertical mixing processes such as diffusion and atmospheric flow patterns and the effect of the uncertain atmospheric thermal structure on the mixing rates. Of critical importance is the timescale needed to mix the very different molecular populations of the upper and lower atmosphere, because vertical mixing processes can alter the molecular mixing ratios in the lower atmosphere.

Rugheimer et al. (2015) extended the previous study to compute photochemical models for Earth-like exoplanets in orbit around stars with spectral types between M0 V and M9 V. They included UV irradiation from active M dwarfs scaled from the observed spectrum of AD Leo, irradiation from inactive stars without chromospheres, and irradiation from the observed MUSCLES stars (France et al. 2016). Their coupled 1D radiative-convective code (EXO-Prime) calculated the atmospheric thermal structure, molecular mixing ratios, and emergent optical and infrared spectra of 55 molecular species. Figure 11.5 shows the molecular mixing

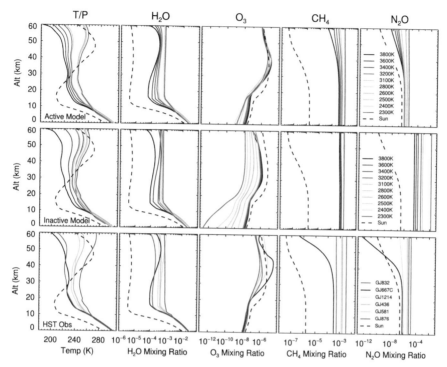

Fig. 11.5 Mixing ratios for different molecules computed by Rugheimer et al. (2015), including photodissociation rates based on observed UV spectra with reconstructed Lyman-α fluxes. The models are for the atmospheres of Earth-like exoplanets in orbit around M0 V ($T_{\text{eff}} = 3800$ K) to M9 V (2300 K) stars. *Left column:* Atmospheric temperatures from the 0 km (surface) to 60 km (about 0.1 mbar pressure) for the M dwarfs (color coded) and Sun (dashed line). *Top row:* Mixing ratios assuming UV insolation from active star models scaled from AD Leo. *Second row:* Mixing ratios assuming UV insolation from inactive models without chromospheres. *Bottom row:* Mixing ratios assuming the UV insolation measured for the MUSCLES stars. The figure is from Rugheimer et al. (2015). Reproduced by permission of the AAS

ratios from the exoplanet surface (0 km) at 1 bar pressure to 60 km (about 0.1 mbar), assuming that the chemical abundances are similar to the Earth's atmosphere and that the exoplanets receive the same bolometric flux as the Earth. Although the largest effect of UV radiation occurs higher in the atmosphere than 0.1 mbar, which is higher than these models extend, UV-driven photodissociation becomes more important as the UV flux increases from the inactive to active stars and from the M9 V to M0 V stars. As shown in Fig. 11.5, the effect of increasing UV irradiation is to decrease the mixing ratios of H_2O, CH_4, and N_2O. However, increasing UV irradiation increases the mixing ratio of O_3 as will be described in the next section. The authors computed the optical and infrared emission spectra at secondary eclipse for these models to determine the spectral features that future spacecraft could detect as biosignatures in the atmospheres of terrestrial exoplanets. Their analysis indicated that the detection of O_2 and O_3 together with N_2O and the reducing gas

CH$_4$ should be a feasible biosignature for early-M dwarfs than for late-M dwarfs. Transmission spectra obtained during transits are most sensitive to the products of photochemistry high in an exoplanet's atmosphere.

The use of UV fluxes from *HST*, including reconstructed Lyman-α fluxes, or observed solar fluxes for solar-like host stars are essential input for detailed atmospheric models of individual stars and their synthetic spectra. Some examples of photochemical models are: (1) the super-Earth and sub-Neptune exoplanets of the host stars 55 Cnc (G8 V), HD 97658 (K1 V), and GJ 1214 (M4.5 V) by Hu and Seager (2014); (2) models for hot Neptunes like GJ 436b and hot Jupiters (Moses et al. 2013a,b; Morley et al. 2017), and (3) a model of the warm Neptune-like GJ 3470b (Venot et al. 2014).

The extent to which the atmospheric properties of an exoplanet are different between their day and night sides has been explored in a set of pseudo-2D calculations by Agúndez et al. (2014) for the hot Jupiters HD 209458b and HD 189733b. These calculations of molecular abundance distributions include chemical disequilibrium produced by photochemistry on the day side, vertical mixing by eddy diffusion, and horizontal transport produced by a strong superrotating jet primarily at equatorial latitudes predicted by global circulation models. At pressures below 1 mbar, the zonal winds homogenize the chemical abundances to their subsolar point equilibrium composition. Higher in the atmosphere, the interplay between horizontal winds, vertical mixing, and chemical processes produces large longitudinal variations in the abundances of some molecules (e.g., CH$_4$, CO$_2$, NH$_3$, and HCN) but only small longitudinal variations in other molecules (e.g., CO, H$_2$O, and N$_2$). Abundance calculations that take into account longitudinal variations in photochemical rates and temperature are essential for the analysis of spectra obtained during and outside of transits.

11.3 Is Oxygen a Reliable Biosignature?

The determination of reliable biosignatures and their detection in exoplanet spectra is a key driver of exoplanet research. Kaltenegger (2017) provides a summary of the molecules that have been proposed as biosignatures. Since the oxygen abundance in the terrestrial atmosphere is produced by plant photosynthesis, the question arises whether oxygen either atomic or molecular can be a reliable biosignature, in other words whether the detection of oxygen in an exoplanet's atmosphere is a sufficient condition for the presence of bioforms. Also, the absorption of near-UV radiation by O$_2$ and O$_3$ (atmospheric sunscreens) is essential for shielding biological material on planetary surfaces from this harmful radiation, although life under water would not be affected (Rugheimer et al. 2015). For these reasons it is essential to understand whether a host star can control the oxygen photochemistry in an exoplanet's atmosphere.

Meadows (2017) provided a comprehensive summary of our present understanding of the chemistry and photochemistry of oxygen in exoplanet atmospheres including the roles that cold traps, surface outgasing and chemical reactions with oceans and surface crust can play in determining the atmospheric abundance of oxygen. This paper describes three scenarios that can produce significant abiotic oxygen in an exoplanet's atmosphere that are false positive biosignatures for the photosynthetic production of oxygen. One mechanism is the photolysis of H_2O and subsequent loss of H in an atmosphere that does not have a cold trap for H_2O because the atmosphere is too warm or the abundance of N_2 is too low. In this case, H_2O from oceans can rise to the heights where photolysis occurs. A second mechanism can occur when the star is very bright during its pre-main sequence evolution so that a planet is then inside of the habitable zone leading to rapid photolysis of H_2O and oxygen production before the star settles down to its main sequence life and the planet is located inside the HZ. After the Sun's pre-main sequence phase during which the Earth lost it initial gas envelope, there is geological evidence (e.g., Marty et al. 2013; Som et al. 2016) that the secondary atmosphere had less than half of the present column density 2.7 billion years ago and that outgasing and other processes produced the present N_2-rich atmosphere.

The abundances of O, O_2, and O_3 in an exoplanet's atmosphere are a "poster child" for the importance of the host star's spectral energy distribution. Figure 11.6 illustrates an abiotic path for producing significant amounts of neutral and molecular oxygen in an exoplanet's atmosphere that is the third false positive biosignature discussed by Meadows (2017). Stellar extreme-UV and far-UV radiation, in particular the Lyman-α emission line, photodissociates CO_2, H_2O, and other O-bearing molecules to produce atomic oxygen, which then forms O_2 by three-body

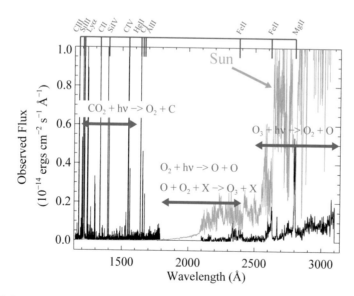

Fig. 11.6 Photodissociation of CO_2 and H_2O and formation of neutral oxygen, O_2, and O_3 in exoplanet atmospheres. The black spectrum is for GJ 832, and the yellow spectrum is for the Sun as observed from their respective habitable zones. Figure courtesy of K. France

recombination and subsequently O_3. The four reactions in this process is called the "Chapman mechanism". O_2 and O_3 have photdissociation cross sections that peak in the near-UV and the short-wavelength portion of the optical. Thus far-UV irradiation creates atomic oxygen and near-UV radiation dissociates the resulting O_2 and O_3 molecules to produce atomic oxygen, which can then be sequestered in other molecules or surface rocks. The figure shows that a host star with a relatively strong far-UV spectrum but weak near-UV spectrum can produce exoplanet atmospheres with abundant O_2 and O_3 in their atmospheres. M stars fit this description as shown in Fig. 11.1. Tian et al. (2014) computed a detailed model atmosphere for a hypothetical Earth-mass planet with the observed UV irradiation from its M5 V host star that demonstrates the abiotic buildup O_2 and O_3 in its upper atmosphere. Thus the exoplanets of M and likely also K stars could have abiotic oxygen-rich atmospheres while exoplanets of F and G stars like the Sun will not. This is shown in the exoplanet atmosphere models computed by Harman et al. (2015) that include outgasing and other interactions with the exoplanet's rocky surface and ocean. They find that the buildup of O_2 is controlled principally by the host star's NUV flux. In a photochemistry model for the F2 V star σ Boo including reactions with reduced radicals like NH_3, Domagal-Goldman et al. (2014) found that the abiotic column density of O_2 is a factor of 3000 below that of the present day Earth, but the O_3 column density is only a factor of 6 smaller and could be detectable.

The upflow of H_2O into the photoionizing upper layers of a terrestrial planet's atmosphere requires passage through a potential cold trap near the top of the tropopause, but Wordsworth and Pierrehumbert (2014) argue that the cold trap is ineffective when the abundance of noncondensing gasses such as N_2, Ar, and O_2 is low. Thus the buildup of O_2 will be self-limiting by the increasing effective cold trap restricting the upflow of H_2O. Another abiotic path for oxygen buildup in an exoplanet's atmosphere is the UV photodissociation of H_2O followed by hydrodynamical escape of H. The resulting O_2 builds up in the atmosphere and can oxydize surface rocks as is the case of present day Mars. The buildup of O_2 by this path will also be self-limiting.

11.4   What Are Reliable Biosignatures?

Lovelock (1965) and Lederberg (1965) proposed that a good biosignature would be the observation of a thermodynamic disequilibrium atmosphere characterized by the presence of both oxidizing molecules (e.g., O_2 and O_3) and reducing molecules (e.g., H_2 and CH_4). Sagan et al. (1993a,b) used this argument to infer the presence of biological sources on the Earth for the O_2, CH_4 and N_2O observed by the Galileo spacecraft when it obtained optical spectra during an Earth flyby. This idea has been tested with models by a number of authors beginning with Selsis et al. (2002). See the recent reviews of this topic by Meadows et al. (2018) and Schwieterman et al. (2018) that also discuss possible false positive biosignatures. The simultaneous presence of the reducing-oxydizing pair CH_4 and CO_2 could be one useful biosignature (Catling et al. 2018). Domagal-Goldman et al. (2014) investigated the detectability of reliable biosignatures including O_2 and O_3 as the oxidizing molecules. They computed photochemical models of lifeless Earth-like exoplanets subject to irradiation by F3 V (σ Boo) to M4 V (GJ 876) host stars including the Sun. In their atmosphere models they included CO_2, H_2O, and N_2, the production of abiotic oxygen from photodissociation of CO_2 and H_2O, and the destruction of O_3 by near-UV irradiation and chemical reactions with CH_4 and H_2 that emerge from volcanos and ocean floor vents. Because the abundance of atmospheric oxygen decreases with increasing venting of reducing species and stellar near-UV irradiation, the presence of oxygen in an atmosphere containing reducing species is a strong indicator of a continuous biological source of the oxygen. A robust biosignature would include observations of the Hartley bands (200–300 nm) and 9.6 μm feature of O_3 and, if feasible, the 0.76 μm line of O_2, together with the 1.7 and 3.3 μm bands of CH_4. This biosignature search requires both UV and near-IR spectroscopy and high S/N to separate the biosignature absorption features from spectral blends. Detection of O_2 in an exoplanet's spectrum will be difficult due to blending with terrestrial O_2 absorption, but detection of the O_3 and CH_4 features should be feasible by spectrometers on future spacecraft.

Seager et al. (2013) proposed classifying biosignature gases into three categories according to the physical processes responsible for their production. Their objective was to identify robust single molecule biosignatures available in a wide variety of exoplanet atmospheres. In their classification scheme, Type I biosignature gases are by-products of microbial metabolic reactions that capture energy from environmental chemical reactions. An important example is CH_4 but other examples include H_2, CO_2, N_2O, NO_2, and H_2S. False positives are a serious problem for identifying these gases as biosignatures because these molecules can be formed abiotically from available atoms and molecules under favorable conditions and are released by volcanos and vents. In an H_2-rich reducing atmosphere, NH_3 would be a useful biosignature especially for inactive host stars as EUV radiation creates OH radicals that react with biosignature molecules (e.g., Seager 2013). Type II biosignature gasses are biproducts of metabolic reactions that require input energy, for example the capture of carbon into sugars by photosynthesis. False positives are unlikely for these complex molecules. Type III biosynthesis gasses are the more complex molecules including CH_3Cl, sulfur-based molecules, and organic carbon compounds produced by specific organisms that require input energy for their production. By virtue of their complexity, these biosignatures are unlikely to have false positives, but their abundances in an exoplanet atmosphere and may be below detection.

As the first stage in a search for organic molecules in exoplanet atmospheres, Seager et al. (2016) compiled a list of 622 organic molecules that are produced by terrestrial life forms and contain atoms of C, N, O, P, S, and up to six H atoms. These molecules are volatile at standard temperature and pressure (STP) on Earth. Whether any of these molecules can be detected in exoplanet spectra is unclear as their absorption wavelengths and cross-sections are often poorly known, their abundances in the terrestrial atmosphere are very low, they may be easily destroyed by chemical and photochemical reactions on short timescales, and the strength of their absorption features in exoplanet spectra are likely to be very small. Nevertheless, there will be searches for organic molecules in infrared spectra by JWST and other future observatories. Some potentially interesting organic molecules previously suggested as biosignatures include methyl chloride (CH_3Cl) with absorption bands at 6.5–7.5 μm, 9.3–10.3 μm, and 13–14.8 μm (cf. Segura et al. 2005) and dimethyl sulfide (CH_3SCH_3) with absorption bands just shortward of 7 μm (cf. Domagal-Goldman et al. 2011). Both molecules are present in the terrestrial atmosphere.

11.5   Can Featureless Absorption Spectra Be Explained by High Altitude Photochemical Haze?

Both "clouds" and "haze" can obscure the emergent spectrum formed lower in an exoplanet's atmosphere, but they are different phenomena having different origins and different spectral signatures. Morley et al. (2013) and Madhusudhan et al. (2016) describe a cloud as consisting of liquid or solid condensates formed as a result of a gas cooling below its saturation vapor pressure. For example, Powell et al. (2018) computed the condensation of $MgSiO_3$ and TiO_2 for clouds in hot Jupiters. Haze, on the other hand, consists of aerosol particles that are formed by photochemistry either directly or after cooling from the gas phase.

Figure 11.7 shows the HST/WFC3 grism spectrum of the super-Earth GJ 1214b observed during transits. Here I will call all exoplanets with masses between Earth and Neptune as sub-Neptunes. Berta et al. (2012) called attention to the flat 1.1–1.7 μm spectrum without a trace of the expected absorption feature due to H_2O or other molecules. The figure shows that the observed spectrum is inconsistent with solar composition models, even models with enhanced metals and H_2O. They argued that the flat spectrum could be explained by either a high molecular weight atmosphere (for example more than 50% H_2O) or by optically thick clouds or

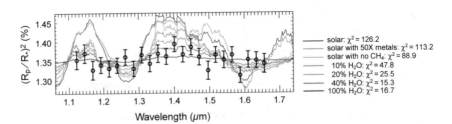

Fig. 11.7 Comparison of observed HST WFC3 spectra of GJ 1214b during transits (black circles with error bars) with cloud and haze-free atmospheric models. The color-coded spectra are for models with solar and 50× solar abundances and with different amounts of H_2O. The χ^2 values for the fits of the models to the data are included on the right. The figure is from Berta et al. (2012). Reproduced by permission of the AAS

hydrocarbon haze extending above a pressure level of 10 mbars. Featureless flat spectra of GJ 1214b had been previously observed at 0.78–1.0 μm (Bean et al. 2010) and in the infrared by Désert et al. (2011) and by Crossfield et al. (2011). Subsequently, Kreidberg et al. (2014) obtained a very precise 1.1–1.7 μm spectrum of GJ 1214b based on the summation of 12 transits observed by *HST*/WFC3. They argued that this featureless spectrum is inconsistent with a high mean molecular weight H_2O atmosphere, leaving only a grey optically thick opacity source located high in the atmosphere as the viable explanation. If the opacity source is ZnS or KCl clouds, then the clouds must be lofted from their formation at much deeper layers in the atmosphere. Hydrocarbon haze naturally formed high in the atmosphere by photochemistry could explain the flat spectrum without a lofting mechanism.

In their analysis of the 1.1–1.7 μm spectrum of the sub-Neptune exoplanet WASP-107b observed during transit, Kreidberg et al. (2018) found that the H_2O absorption feature is much weaker than expected from a cloud-free atmosphere. They found that the observed spectrum can be fit by absorption high in the atmosphere ($P = 0.01–3$ mbar) that could be explained either by warm clouds or by a hydrocarbon haze with particles roughly 0.3 μm in size. It is no surprise that haze particles exist high in the atmosphere where stellar UV radiation photodissociates CH_4 to begin the process of haze particle formation (Arney et al. 2016, 2017).

Crossfield and Kreidberg (2017) searched for the reason why some sub-Neptune exoplanets have high-level absorption and other do not. With a sample of seven sub-Neptune-like exoplanets with equilibrium temperatures between 500 and 1000 K and transit spectra observed by *HST*/WFC, they found that the four lower-temperature sub-Neptunes (e.g., GJ 1214b and GJ 436b) have haze absorption and the two higher-temperature sub-Neptunes have clear atmospheres. The separation between these two groups occurs at equilibrium temperatures between 800 and 900 K. However, Morley et al. (2017) found that very high metalicity (1000 times solar) could also produce a nearly featureless 1.1–1.7 μm spectrum for GJ 436b, but such high metallicities may not be realistic for Neptune-mass exoplanets with their gaseous envelopes.

Morley et al. (2015) computed equilibrium and disequilibrium exoplanet atmospheric models for a range of metallicity and equilibrium temperatures to understand the conditions that can form clouds and photochemical haze and possibly explain the featureless near-IR transmission spectra seen in GJ 1214b and other warm exoplanets with masses between those of the Earth and Neptune. They found that warm temperature clouds containing condensed KCl and ZnS particles could explain the featureless spectra for very high metallicities and inefficient sedimentation, that is the velocity of downfalling particles is much smaller than convective velocities. High altitude photochemical haze can also explain the featureless spectra when the conversion efficiency of CH_4 into polycyclic hydrocarbon (PAH) particles smaller than 1 μm, often referred to as "soot", exceeds about 10%. The upper atmosphere haze observed in Saturn's moon Titan consisting of PAHs or other hydrocarbon molecules could be produced by the same photochemical process (Sagan et al. 1993b; López-Puertas et al. 2013). Photodissociation of CH_4 to produce soot precursors (e.g., C_2H_2, C_2H_6, HCN, etc.) requires CH_4 to initiate the conversion process, but at temperatures above about 1000 K, chemical reactions force CO rather than CH_4 to be the primary carrier of C atoms. This sets an upper limit of about 1000 K for a sub-Neptune exoplanet to have high altitude haze. A further constraint is that the UV irradiation not be so large as to produce a thermal inversion in the exopalnet's outer atmosphere that would raise the gas temperature above this limit. Thermal inversions are now being observed, for example in high-resolution spectra of the 0.62–0.88 μm TiO lines of the hot Jupiter WASP-33b (Nugroho et al. 2017). Morley et al. (2015) suggested that future near-IR spectra covering a wide wavelength range should be able to distinguish clouds from haze in the thermal emission and reflected light emission from these exoplanets.

The analysis of transit spectra of GJ 1214b in the optical region (450–926 nm) by Rackham et al. (2017) provided new insights concerning upper atmosphere haze. They found that the best fit to both the optical and near-IR spectra was with a thick photochemical haze model irradiated by the host star with a heterogenous temperature distribution. The best fit parameters for the exoplanet's haze are aerosol particles with a mean size of 0.1 μm and 10% efficiency for the photochemical conversion of CH_4 into haze particles. The best fit for the host star's atmosphere includes 3.2% coverage by regions that are 350 K hotter than the mean temperature. Such regions are analogous to solar faculae.

There is now strong evidence that photochemical haze is the source of the obscuring opacity in the spectra of sub-Neptunes cooler than about 800 K. The reasoning behind this conclusion involves the following steps: (1) The haze particles are likely PAHs and other large hydrocarbon molecules whose formation is initiated by the photodissociation of CH_4. (2) The haze must be located in the upper atmospheres of these exoplanets at pressures lower than 1 mbar to obscure the H_2O spectrum formed in the lower atmosphere. (3) The far-UV radiation from the host star penetrates to this atmospheric level and can photodissociate CH_4 and large hydrocarbon molecules. Thus the host star's far-UV irradiation is essential for explaining the featureless spectra of these exoplanets. Also, the host star's near-UV irradiation can photodissociate the sulfur molecules HS and S_2 high in the atmospheres of hot Jupiters leading to complex sulfur chemistry and local heating (Zahnle et al. 2009).

Arney et al. (2016, 2017) computed photochemically driven reaction networks for the formation of organic hazes in the atmospheres of exoplanets located in the habitable zones of F, G, K , and M stars. Far-UV stellar radiation, particularly in the 140–160 nm band initiates photochemical reactions leading to the formation of C_4H_2, C_5H_4 and more complex organic particles and fractals, but strong far-UV illumination can photoionize CO_2 in an exoplanet's atmosphere producing oxygen atoms and radicals that react with and remove haze precursors. They concluded that habitable zone exoplanets around F stars and active M stars have far-UV radiation too strong for the development of hazes, but less active G-M star can have hazy exoplanet's. Thus the strength of a hoststar's far-UV emission is critical for creating and destroying hazes. They also compute reflectance, thermal radiation, and transit transmission spectra for exoplanets with and without hazes as a guide to future observations.

In summary, host star driven photochemistry plays essential roles in exoplanet atmospheres including surface desiccation resulting from the photodissociation of atmospheric H_2O and subsequent loss of hydrogen, abiotic formation of oxygen-rich atmospheres, initiation of reactions leading to photochemical haze, and the formation of ions that provide pathways for chemical reactions. The host star's spectral energy distribution, especially in the ultraviolet must be known to accurately model these processes. Finally, the identification of useful biosignatures and the analysis of their observations often requires including photochemical processes.

References

Agúndez, M., Parmentier, V., Venot, O., Hersant, F., Selsis, F.: Pseudo 2D chemical model of hot-Jupiter atmospheres: application to HD 209458b and HD 189733b. Astron. Astrophys. **564**, A73 (2014)

Arney, G., Domagal-Goldman, S.D., Meadows, V.S. Wolf, E.T., Schwieterman, E., Charnay, B., Claire, M., Hébrard, E., Trainer, M.G.: The pale orange dot: the spectrum and habitability of hazy archean Earth. Astrobiology **16**, 873 (2016)

Arney, G.N., Meadows, V.S., Domagal-Goldman, S.D., Deming, D., Robinson, T.D., Tovar, G., Wolf, E.T., Schwieterman, E.: Pale orange dots: the impact of organic haze on the habitability and detectability of Earthlike exoplanets. Astrophys. J. **836**, 49 (2017)

Bean, J.L., Miller-Ricci Kempton, E., Homeier, D.: A ground-based transmission spectrum of the super-Earth exoplanet GJ 1214b. Nature **468**, 669 (2010)

Berta, Z.K., Charbonneau, D., Désert, J.-M., Miller-Ricci Kempton, E., McCullough, P.R., Burke, C.J., Fortney, J.J., Irwin, J., Nutzman, P., Homeier, D.: The flat transmission spectrum of the super-Earth GJ 1214b from Wide Field Camera 3 on the Hubble Space Telescope. Astrophys. J. **747**, 35 (2012)

Catling, D.C., Krissansen-Totton, J., Kiang, N.Y., Crisp, D., Robinson, T.D., DasSarma, S., Rushby, A.J., Del Genio, A., Bains, W., Domagal-Goldman, S.: Exoplanet biosignatures: a framework for their assessment. Astrobiology **18**, 709 (2018)

Crossfield, I.J.M., Kreidberg, L.: Trends in atmospheric properties of Neptune-size exoplanets. Astron. J. **154**, 261 (2017)

Crossfield, I.J.M., Barman, T., Hansen, B.M.S.: High-resolution, differential, near-infrared transmission spectroscopy of GJ 1214b. Astrophys. J. **736**, 132 (2011)

Désert, J.-M., Bean, J., Miller-Ricci Kempton, E., Berta, Z.K., Charbonneau, D., Irwin, J., Fortney, J., Burke, C.J., Nutzman, P.: Observational evidence for a metal-rich atmosphere on the super-Earth GJ1214b. Astrophys. J. Lett. **731**, L40 (2011)

Domagal-Goldman, S.D., Meadows, V.S., Claire, M.W., Kasting, J.F.: Using biogenic sulfur gases as remotely detectable biosignatures on anoxic planets. Astrobiology **11**, 419 (2011)

Domagal-Goldman, S.D., Segura, A., Claire, M.W., Robinson, T.D., Meadows, V.S.: Abiotic ozone and oxygen in atmospheres similar to prebiotic Earth. Astrophys. J. **792**, 90 (2014)

France, K., Linsky, J.L., Tian, F., Froning, C.S., Roberge, A.: Time-resolved ultraviolet sectroscopy of the M-dwarf GJ 876 exoplanetary system. Astrophys. J. Lett. **750**, L32 (2012)

France, K., Froning, C.S., Linsky, J.L., Roberge, A., Stocke, J.T., Tian, F., Bushinsky, R., Désert, J.-M., Mauas, P., Vietes, M., Walkowicz, L.: The ultraviolet radiation environment around M dwarf exoplanet host stars. Astrophys. J. **763**, 149 (2013)

France, K., Loyd, R.O.P., Youngblood, A., Brown, A., Schneider, P.C., Hawley, S.L., Froning, C.S., Linsky, J.L., Roberge, A., et al.: The muscles treasury survey I: motivation and overview. Astrophys. J. **820**, 89 (2016)

Harman, C.E., Schwieterman, E.W., Schottelkotte, J.C., Kasting, J.F.: Abiotic O_2 levels on planets around F, G, K, and M stars: possible false positives for life? Astrophys. J. **812**, 137 (2015)

Hu, R., Seager, S.: Photochemistry in terrestrial exoplanet atmospheres. III. Photochemistry and thermochemistry in thick atmospheres on super Earths and mini Neptunes. Astrophys. J. **784**, 63 (2014)

Kaltenegger, L.: How to characterize habitable worlds and signs of life. Ann. Rev . Astron. Astrophys. **55**, 433 (2017)

Kasting, J.F.: Stability of ammonia in the primitive terrestrial atmosphere. J. Geophys. Res. **87**, 3091 (1982)

Kopparapu, R., Kasting, J.F., Zahnle, K.J.: A photochemical model for the carbon-rich planet WASP-12b. Astrophys. J. **745**, 77 (2012)

Kopparapu, R.K., Ramirez, R.M., SchottelKotte, J., Kasting, J.F., Domagal-Goldman, S., Eymet, V.: Habitable zones around main-sequence stars: dependence on planetary mass. Astrophys. J. **787**, 29 (2014)

Kreidberg, L., Bean, J.L., Désert, J.-M., Benneke, B., Deming, D., Stevenson, K.B., Seager, S., Berta-Thompson, Z., Seifahrt, A., Homeier, D.: Clouds in the atmosphere of the super-Earth exoplanet GJ1214b. Nature **505**, 69 (2014)

Kreidberg, L., Line, M.R., Thorngren, D., Morley, C.V., Stevenson, K.B.: Water, high-altitude condensates, and possible methane depletion in the atmosphere of the warm super-Neptune WASP-107b. Astrophys. J. **858**, 6 (2018)

Lederberg, J.: Signs of life: criterion-system of exobiology. Nature **207**, 9 (1965)

López-Puertas, M., Dinelli, B.M., Adriani, A., Funke, B., García-Comas, M., Moriconi, M.L., D'Aversa, E., Boersma, C., Allamandola, L.J.: Large abundances of polycyclic aromatic hydrocarbons in Titan's upper atmosphere. Astrophys. J. **770**, 132 (2013)

Lovelock, J.E.: A physical basis for life detection experiments. Nature **207**, 568 (1965)

Loyd, R.O.P., France, K., Youngblood, A., Schneider, M.C., Brown, A., Hu, R., Linsky, J.L., Froning, C.S., Redfield, S., Rugheimer, S., Tian, F.: The MUSCLES Treasury Survey III. X-ray to infrared spectra of 11 M and K stars hosting planets. Astrophys. J. **824**, 102 (2016)

Loyd, R.O.P., France, K., Youngblood, A., Schneider, C. Brown, A., Hu, R., Segura, A., Linsky, J., Redfield, S., Tian, F., Rugheimer, S., Miguel, T., Froning, C.S.: The MUSCLES Treasury Survey. V. FUV flares on active and inactive M dwarfs. Astrophys. J. **867**, 71 (2018)

Madhusudhan, N., Agúndez, M., Moses, J.I., Hu, Y.: Exoplanetary atmospheres-chemistry, formation conditions, and habitability. Space Sci. Rev. **205**, 285 (2016)

Marty, B., Zimmermann, L., Pujol, M., Burgess, R., Philippot, P.: Nitrogen isotopic composition and density of the Archean atmosphere. Science **342**, 101 (2013)

Meadows, V.S.: Reflections on O_2 as a biosignature in exoplanetary atmospheres. Astrobiology **17**, 1022 (2017)

Meadows, V.S., Reinhard, C.T., Arney, G.N., Parenteau, M.N., Schwieterman, E.W., Domagal-Goldman, S.D., Lincowski, A.P., Stapelfeldt, K.R., Rauer, H., et al.: Exoplanet biosignatures: understanding oxygen as a biosignature in the context of its environment. Astrobiology **18**, 630 (2018)

Miguel, Y., Kaltenegger, L.: Exploring atmospheres of hot mini-Neptunes and extrasolar giant planets orbiting different stars with application to HD 97658b, WASP-12b, CoRoT-2b, XO-1b, and HD 189733b. Astrophys. J. **780**, 166 (2014)

Miguel, Y., Kaltenegger, L., Linsky, J.L., Rugheimer, S.: The effect of Lyman α radiation on mini-Neptune atmospheres around M stars: application to GJ 436b. Mon. Not. R. Astron. Soc. **446**, 345 (2015)

Morley, C.V., Fortney, J.J., Kempton, E.M.-R., Marley, M.S., Visscher, C., Zahnle, K.: Quantitatively assessing the role of clouds in the transmission spectrum of GJ 1214b. Astrophys. J. **775**, 33 (2013)

Morley, C.V., Fortney, J.J., Marley, M.S., Zahnle, K., Line, M., Kempton, E., Lewis, N., Cahoy, K.: Thermal emission and reflected light spectra of super Earths with flat transmission spectra. Astrophys. J. **815**, 110 (2015)

Morley, C.V., Knutson, H., Line, M., Fortney, J.J., Thorngren, D., Marley, M.S., Teal, D., Lupu, R.: Forward and inverse modeling of the emission and transmission spectrum of GJ 436b: investigating metal enrichment, tidal heating, and clouds. Astron. J. **153**, 86 (2017)

Moses, J.I., Line, M.R., Visscher, C., Richardson, M.R., Nettelmann, N., Fortney, J.J., Barman, T.S., Stevenson, K.B., Madhusudhan, N.: Compositional diversity in the atmospheres of hot Neptunes, with application to GJ 436b. Astrophys. J. **777**, 34 (2013a)

Moses, J.I., Madhusudhan, N., Visscher, C., Freedman, R.S.: Chemical consequences of the C/O ratio on hot Jupiters: examples from WASP-12b, CoRoT-2b, XO-1b, and HD 189733b. Astrophys. J. **763**, 25 (2013b)

Nugroho, S.K., Kawahara, H., Masuda, K., Hirano, T., Kotani, T., Tajitsu, A.: High-resolution spectroscopic detection of TiO and a stratosphere in the day-side of WASP-33b. Astron. J. **154**, 221 (2017)

Powell, D., Zhang, X., Gao, P., Parmentier, V.: Formation of silicate and titanium clouds on hot Jupiters. Astrophys. J. **860**, 18 (2018)

Rackham, B., Espinoza, N., Apai, D., López-Morales, M., Jordán, A., Osip, D.J., Lewis, N.K., Rodler, F., Fraine, J.D., Morley, C.V., Fortney, J.J.: Access I. An optical transmission spectrum of GJ 1214b reveals a heterogeneous stellar photosphere. Astrophys. J. **834**, 151 (2017)

Rugheimer, S., Kaltenegger, L., Segura, A., Linsky, J., Mohanty, S.: Effect of UV radiation on the spectral fingerprints of Earth-like planets orbiting M dwarfs. Astrophys. J. **809**, 57 (2015)

Sagan, C., Thompson, W.R., Carlson, R., Gurnett, D., Hord, C.: A search for life on Earth from the Galileo spacecraft. Nature **365**, 715 (1993a)

Sagan, C., Khare, B.N., Thompson, W.R., McDonald, G.D., Wing, M.R., Bada, J.L., Vo-Dinh, T., Arakawa, E.T.:Polycyclic aromatic hydrocarbons in the atmospheres of Titan and Jupiter. Astrophys. J. **414**, 399 (1993b)

Schwieterman, E.W., Kiang, N.Y., Parenteau, M.N., Harman, C.E., DasSarma, S., Fisher, T.M., Arney, G.N., Hartnett, H.E., Reinhard, C.T., et al.: Exoplanet biosignatures: a review of remotely detectable signs of life. Astrobiology **18**, 663 (2018)

Seager, S.: Exoplanet habitability. Science **340**, 577 (2013)

Seager, S., Bains, W., Hu, R.: A biomass-based model to estimate the plausibility of exoplanet biosignature gases. Astrophys. J. **775**, 104 (2013)

Seager, S., Bains, W., Petkowski, J.J.: Toward a list of molecules as potential biosignature gases for the search for life on exoplanets and applications to terrestrial biochemistry. Astrobiology **16**, 465 (2016)

Segura, A., Kasting, J.F., Meadows, V., Cohen, M., Scalo, J., Crisp, D., Butler, R.A.H., Tinetti, G.: Biosignatures from Earth-Like planets around M dwarfs. Astrobiology **5**, 706 (2005)

Selsis, F., Despois, D., Parisot, J.-P.: Signature of life on exoplanets: can Darwin produce false positive detections? Astron. Astrophys. **388**, 985 (2002)

Som, S.M., Buick, R., Hagadorn, J.W., Blake, T.S., Perreault, J.M., Harnmeijer, J.P., Catling, D.C.: Earth's airpressure 2.7 billion years ago constrained to less than half of modern levels. Nat. Geosci. **9**, 448 (2016)

Tian, F., France, K., Linsky, J.L., Mauas, P.J.D., Vieytes, M.C.: High stellar FUV/NUV ratio and oxygen contents in the atmospheres of potentially habitable planets. Earth Planet. Sci. Lett. **385**, 22 (2014)

Venot, O., Agúndez, M., Selsis, F., Tessenyi, M., Iro, N.: The atmospheric chemistry of the warm Neptune GJ 3470b: influence of metallicity and temperature on the CH4/CO ratio. Astron. Astrophys. **562**, 51 (2014)

Wordsworth, R., Pierrehumbert, R.: Abiotic oxygen-dominated atmospheres on terrestrial habitable zone planets. Astrophys. J. **785**, L20 (2014)

Zahnle, K., Marley, M.S., Freedman, R.S., Lodders, K., Fortney, J.J.: Atmospheric sulfur photochemistry on hot Jupiters. Astrophys. J. **701**, 20 (2009)

Chapter 12
Space Weather: The Effects of Host Star Flares on Exoplanets

If the energetic proton fluxes and coronal mass ejection energies scale with radiated flare energy, the impact on the atmosphere and magnetosphere of any hypothetical terrestrial planet would be catastrophic.—Osten et al. (2016)

Exoplanets have intimate relationships with their host stars as the Earth does with its host star. While the Earth's environment produced by the Sun is usually benign, exoplanets located close to their host stars, especially active M dwarfs, must suffer through powerful flares, stellar winds, CMEs, and very high energy radiation. The environment in which exoplanets must live is now called "stellar space weather" in analogy with the extensively monitored "space weather" that is the environment of the Earth. In this chapter I describe flares and superflares on the Sun and stars and how repeated flares destroy O_3 in the atmospheres of exoplanets possibly leading to the sterilization of their surfaces and loss of habitability.

12.1 Important Characteristics of Flares

Flares seen on the Sun and stars exhibit an enormous range of phenomena for which there is a rich literature. Benz (2017) provides a comprehensive review of solar flare observations and physical processes. It is generally assumed that the physical processes responsible for solar and stellar flares differ only in scale, but this assumption has not been tested in detail. In this chapter, I will describe those aspects of flares that are important sources of space weather as they change the environment of exoplanets. The reviews by Haisch et al. (1991) and Osten (2016) provide a more complete description of flare phenomena and their likely physical explanations. The following is a short summary of the phenomena that characterize flares:

- The entire electromagnetic spectrum from X-rays to radio wavelengths brightens during flares with the largest relative enhancements compared to preflare

© Springer Nature Switzerland AG 2019
J. Linsky, *Host Stars and their Effects on Exoplanet Atmospheres*,
Lecture Notes in Physics 955, https://doi.org/10.1007/978-3-030-11452-7_12

conditions at the shortest wavelengths where the emitting plasma is hottest. The relative flux enhancements at X-ray and ultraviolet wavelengths are very large because the preflare emission is generally faint at these wavelengths, but a large component of the emitted energy is often in the optical region (white light flares) where the contrast will be small for G stars but can be very large for M dwarfs with their much fainter continua. Flare emission at cm to km wavelengths by gyrosynchrotron or coherent processes (plasma emission and electron cyclotron maser emission) can also be far brighter than quiescent emission.

- Flares generally show two time scales. There is an initial very rapid increase in flux at X-ray, UV, and optical wavelengths (impulsive phase) followed by an exponential decrease (gradual phase) that lasts from minutes (X-ray, UV and optical) to hours or even days (gyrosynchrotron radio emission). This time development is shown in UV emission lines observed during a flare on the M5 V star GJ 876 (France et al. 2016; Loyd et al. 2018) and in X-ray emission from Proxima Cen (Haisch et al. 1983) and from a variety of stars (Pye et al. 2015). Soft X-ray emission often peaks about 5 min after the initial prompt emission.

- Rapid magnetic recombination produces a very hot (10^7–10^8 K) or perhaps nonthermal plasma. This plasma produces X-ray emission that may be very hard, for example the 50–100 keV X-rays seen by the *Swift* satellite during extreme flares on CC Eri and DG CVn (Karmakar et al. 2017; Osten et al. 2016). Young rapidly rotating stars have the hottest flares. γ-rays have been detected from solar flares and must surely be produced in stellar flares but have not yet been detected. High-energy protons ($E > 10$ MeV) are often emitted during strong solar flares. Energetic proton events on host stars have not yet been detected but likely play an important role in exoplanet habitability as they lead to the destruction of atmospheric ozone as described later in this chapter

- The flare energy source is almost certainly the rapid conversion of magnetic energy that both heats the plasma and accelerated particles. During solar flares the initial non-thermal hard X-ray emission is a diagnostic of the energy distribution of the accelerated particles, but this non-thermal emission has not yet been unambiguously detected during stellar flares. Many solar and stellar flare models (e.g., Kowalski et al. 2015) include beams of relativistic electrons and protons that penetrate into the upper photosphere to produce the $T \approx 10{,}000$ K plasma that emits continua and emission lines of hydrogen at optical and near-UV wavelengths. Young stars with their rapid rotation and strong magnetic fields have far more energetic and more numerous flares than older stars like the Sun with their weaker magnetic fields.

- During the impulsive stages of solar flares, spatially coincident downflowing chromospheric gas and upflowing coronal gas are observed with the coronal gas upflow laging the downflow by 60–75 s (Graham and Cauzzi 2015). They used the *Interface Region Imaging Spectrograph (IRIS)* to observe individual flaring pixels at 0.5 arcsec resolution that show downflows as large as 40 km s^{-1} in the Mg II emission lines and upflows as large as 300 km s^{-1} in the Fe XXI 135.4 nm emission line formed at 10 MK. This dynamic scenario is in qualitative agreement with MHD flare models (e.g., Kowalski and Allred 2018 and references therein) in which nonthermal electron beams produce the explosive heating responsible for the flows. UV emission lines observed during stellar flares, for example the giant flare on the young (50 Myr) solar-mass star EK Dra (Ayres 2015), show that 10^5 K material is flowing down toward the chromosphere at about 10 km s^{-1}. Upflows of coronal gas during stellar flares have not yet been detected, but searches for this flare diagnostic are underway.

- As described in Sect. 8.2.5, some strong solar flares eject matter into space, events called coronal mass ejections (CMEs). The kinetic energy in solar CMEs is comparable to the bolometric radiation of the flare. CMEs contain cool gas (10^4 K) but shock fronts can accelerate energetic protons. There are several programs underway to search for stellar CMEs (e.g., Crosley and Osten 2018) either by searching for coherent radio emission or emission line Doppler shifts. In a numerical study of a large-scale dipolar magnetic fields, Alvarado-Gómez et al. (2018) found that a coronal field of 75 G can suppress CMEs with kinetic energies less than about 3×10^{32} erg, corresponding to *GOES*-class X35 flares, and reduce the speed of CMEs with all kinetic energies. Thus for active stars with strong magnetic fields there may be far fewer CMEs than predicted by solar analogy and less resulting risk for exoplanet habitability.

12.2 Flares and Superflares on the Sun

Flares on the Sun have been studied for many years—first at optical wavelengths and then at radio wavelengths and at UV, X-ray, and γ-ray energies with many satellite instruments. The standard classification for solar flares is in terms of the peak X-ray flux in the 0.1–0.8 nm band measured by the *Geostationary Operational Environmental Satellites (GOES)* at 1 au. Strong flares on the Sun with X-ray flux measured at the Earth of 0.1 erg cm^{-2} s^{-1}, corresponding to a luminosity of 2.8×10^{26} erg s^{-1}, are called X-class flares. These occur on average about once every 500 h (Youngblood et al. 2017). The *GOES* classification scheme for stronger flares is linear in X-ray flux with X10 flares having luminosities 10 times larger than X-class flares.

Figure 12.1 shows a recent very bright solar flare observed at a wavelength of 13.1 nm by the Atmospheric Imaging Assembly (AIA) on the *Solar Dynamics Observatory (SDO)*. This X9.3 flare observed on 2017 September 6 had a peak X-ray luminosity of 2.6×10^{27} erg s^{-1}. It produced high-frequency radio blackouts, bright aurorae, and high flux of γ-rays, but the Earth was saved from more severe effects because the CME and high-energy protons associated with the flare missed the Earth. The 2003 Nov 4 flare (about X45) is the largest in the NOAA records (cf Tsurutani et al. 2003). The most powerful recorded solar flare is the so-called Carrington flare of 1859 September 1 observed by Carrington (1859) as a rare white light flare. This flare sent electrical signals along telegraph lines shocking telegraph operators and produced bright aurorae observed as far south as Mexico and Columbia (Cárdenas et al. 2016). Cliver and Dietrich (2013) estimated that the

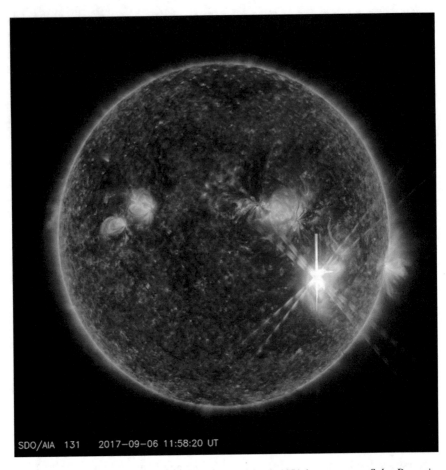

SDO/AIA 131   2017-09-06 11:58:20 UT

Fig. 12.1 The 2017 September 6 solar flare observed by the AIA instrument on *Solar Dynamics observatory (SDO)* satellite. The image is at wavelength 13.1 nm with emission from lines of Fe VIII, Fe XX, and Fe XXIII. Figure courtesy of NSO/GSFC/SDO

Carrington flare would be classified as X45 today, that the total radiative energy was $\sim 5 \times 10^{32}$ erg, and the total bolometric energy including the kinetic energy of the CME was $\sim 2 \times 10^{33}$ erg. A strong CME and high-energy protons associated with a solar X9.3 flare hitting the Earth's ionosphere would severely impact today's electrical and telecommunications infrastructure. The Great Storm of 1967, May 23–30 with its 76 flares (up to X6-class), intense white light flares, flux of greater than 10 MeV protons, largest recorded geomagnetic storm, and extreme disruption of radio communications demonstrated the many effects that large flares can have on Earth and stimulated the development of space weather as a scientific discipline (Knipp et al. 2016; Knipp 2011). Space weather is now monitored continuously because very large flares can also disrupt electrical equipment including the electric power grid (cf. Schrijver 2015).

Fig. 12.2 The Hertzsprung-Russell diagram of stellar flares seen in the *Kepler* data. Large (green) dots are for stars with at least one detected flare, and small (red) dots are for stars with no detected flares in 3 years of continuous photometry. Figure from Balona (2015). Reproduced from MNRAS by permission of the Oxford University Press

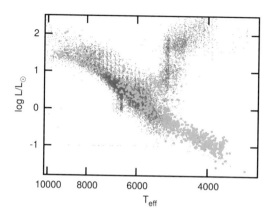

The total radiative energy in a solar flare integrated over wavelength and flare duration (typically a few thousand seconds) extends up to 10^{32} ergs for the Carrington flare. The power-law distribution in Fig. 12.5 predicts that a Carrington-like flare could occur on the Sun once in 500 years. Arctic and Antarctic ice core samples of ^{10}Be, ^{14}C, and ^{36}Cl provide evidence for very high >100 MeV proton fluxes occurred in AD 774/775 and AD 993/994 (Mekhaldi et al. 2015). They argued that these data are consistent with very energetic solar flares that likely exceeded, perhaps by a large factor, the energy in the Carrington event because ice cores for 1859 do not show enhancements of ^{10}Be. Enhancements of ^{14}C in Japanese cedar trees for the years AD 774/775 and AD 993/994 provide further evidence for very energetic solar flares (Miyake et al. 2013). Consistent with the flare data, measurements of high helium and neon abundances in the oldest meteorites indicate that the young Sun was a copious source of very energetic protons (Kööp et al. 2018).

12.3 Flares and Superflares on Stars

On 1948 September 25, Joy and Humason (1949) noticed that the secondary star in the L726-8 system was much brighter than usual with very bright continuum and emission lines in the hydrogen Balmer series, He I, and He II. This flare-up in emission now called a flare on the star now called UV Ceti was the first of many programs to study what types of stars flare and the physical properties of the flaring plasma. This work is reviewed, for example, by Haisch et al. (1991) and Gershberg (2005).

Prior to the era of long term continuous monitoring of many stars by the Kepler spacecraft, it was generally presumed that flares are a common phenomenon on K and M stars but less common and less energetic on stars with larger effective temperatures and decreasing convective zone depths. The *Kepler* spacecraft changed that picture. In an analysis of *Kepler* data, Balona (2015) found that the incidence of flares seen in broad-band optical light (white light flares) on A and F stars is only about a factor of 4 less frequent than on K and M stars. The Hertzprung-Russel (H-R) diagram comparing stars observed by *Kepler* with detected and no detected flares (see Fig. 12.2) shows that even A-type stars ($T_{eff} \approx 10,000$ K) flare. Balona (2015) argued that the flares seen on A-type stars are not from low mass binary companions, because the average flux level seen in flares on A-type stars is about 100 times larger than the average seen on M stars. Flares on A-type stars were unexpected, because these stars have very shallow convective zones that are not thought to be conducive to the dynamo generation of magnetic fields. About half of the *Kepler* stars with $T_{eff} = 8300$–12,000 K, however, show evidence of rotational modulation due to starspots (Balona 2017). Balona (2015) also found that stars that flare rotate faster and have larger starspots, roughly proportional to the stellar radius, than stars that are not observed to flare. Comparing *Kepler* flare rates and energies of K and M stars, Walkowicz et al. (2011) found that M dwarfs flare more frequently for shorter durations and with higher energy compared to their quiescent flux level than K dwarfs. Therefore, exoplanets orbiting M stars will see a larger cumulative effect of flares compared to quiescent emission than will exoplanets of K-type and probably also warmer stars. This difference implies that when all else is equal exoplanets of K-type stars will more likely be habitable than exoplanets of M-type stars.

Flares are detected as enhanced emission across the electromagnetic spectrum, but are most easily seen against the faint background of stellar X-ray, UV, and radio emission. In addition to their enhanced flux, UV flare emission lines often show red-and blue-shifted emission components extending to ± 100 km s^{-1} and beyond, indicating upflows and downflows of explosively heated gas. Examples are emission lines of C II, Si III, and C IV observed by *HST* during flares on the MUSCLES M dwarfs (Loyd et al. 2018) and on AD Leo (Hawley et al. 2003), and by emission lines of C III and O VI observed by *FUSE* during flares on AU Mic (Redfield et al. 2002). These rapid flows of explosively heated gas are consistent with radiative hydrodynamic simulations of flaring plasmas heated by non-thermal electron beams (e.g., Allred et al. 2006). The low spectral resolution of X-ray emission lines makes it difficult to study coronal kinematics during flares, but searches for such data are underway and will be facilitated by the next generation of higher resolution X-ray spectrometers.

Loyd et al. (2018) compared the cumulative flare frequency distribution of active and inactive M dwarfs and found that the distributions are identical when normalized by the stellar quiescent flux. This is one of many observations that indicate that the physical processes responsible for flares are the same independent on stellar activity level or star type.

Flares on active stars, especially M dwarfs, can be orders of magnitude more luminous than even the most luminous solar flares. Table 12.1 compares the brightest solar flares to the brightest detected stellar flares, however, the comparison is approximate because the solar and stellar data have different wavelength bands. A giant flare on the well-known flare star AD Leo observed in the ultraviolet by *IUE* and in the optical had a peak X-ray flux corresponding to a solar X18,000 flare (Hawley and Pettersen 1991). Osten (private communication) estimated the integrated radiative emission of this flare at 10^{34} ergs. An even more powerful X-ray flare on CC Eri was observed by the *Swift* satellite at energies 0.3–50 keV (Karmakar et al. 2017). This flare would be classified X570,000 on the *GOES* scale, about 1000 times more luminous than the Carrington flare on the Sun. Since a hypothetical exoplanet in the center of the habitable zone of CC Eri would be about four times closer to its host star than the Earth, the peak X-ray flux seen by the exoplanet would be 25,000 times larger than the Carrington flare at the Earth. Instruments on the *Swift* satellite detected two even more extreme flares on the young (30 Myr) M3 Ve+M3 Ve binary DG CVn (see Fig. 12.3) with peak temperature (likely thermal) $T \approx 290$ MK (Osten et al. 2016). The second of the two flares would be classified on the *GOES* scale as X60,000,000 with total X-ray radiative emission in the 0.01–100 keV band $\approx 10^{36}$ erg and the corresponding optical and near-IR emission $\approx 2 \times 10^{35}$ erg. The radiation component of superflares, therefore, can be far more serious for exoplanets than the relatively tame solar flare radiation is for the Earth.

To assess exoplanet habitability, it is important to characterize the distribution of flare energies for different types of stars with different levels of activity. With its very precise broad-band (400–850 nm) monitoring of about 160,000 stars, the *Kepler* satellite has provided an enormous data base to answer this question. Davenport

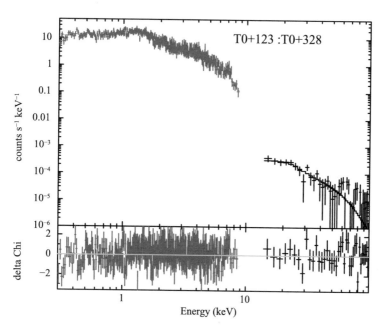

Fig. 12.3 X-ray spectrum of the 2014 April 23 superflare on DG CVn observed by the *Swift* satellite. The 0.3–10 keV emission detected by the XRT instrument is plotted in red and the 15–100 keV emission detected by the BAT instrument is plotted in black. The solid black line is the thermal model fit to the BAT data. Figure from Osten et al. (2016). Reproduced by permission of the AAS

Table 12.1 Comparison of powerful flares on the sun and stars

Star	Date	Peak luminosity ($\mathrm{erg\,s^{-1}}$)	GOES flare class (1.5–8 keV)
Solar flare	9/06/2017	2.6×10^{27}	X9.3
Carrington flare	9/01/1859	1.3×10^{28}	X45
AD Leo (M3 V)	4/12/1985	5×10^{30}	X18,000
CC Eri (K7 V)	10/16/2012	1.6×10^{32}	X570,000
DG CVn (M3 + M3)	4/23/2014	1.7×10^{34}	X60,000,000

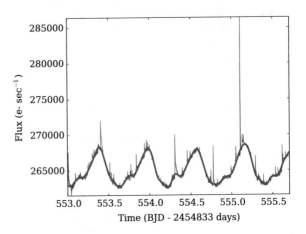

Fig. 12.4 *Kepler* observations with 1 s cadence of the active M dwarf GJ 1243 as a function of time in days. The blue solid line is the nonflare emission of the star and the red solid lines are flare emissions. Figure from Davenport (2016). Reproduced by permission of the AAS

(2016) searched the entire available *Kepler* data base (Data Release 24) for flares. After establishing criteria for measuring flares above variable stellar and other noise sources, he identified 851,168 flare events from 4041 flare stars. *Kepler* observations of the active M dwarf GJ 1243 (Fig. 12.4) shows one very bright flare and many less energetic flares superimposed on the nonflare stellar background. The rotation of active regions and starspots onto the visible surface of the star explains the periodic nonflare emission. The average flare energy for this large sample is about 10^{35} ergs with some flares perhaps as energetic as 10^{39} ergs. The radiative emission of all flares from a star divided by the star's bolometric emission has a similar dependence on the Rossby number as other activity indicators (X-ray, UV, Ca II K, H-α) with a negative power law slope with increasing Rossby number and possibly saturation at high rotation rates and low Rossby numbers $R_0 < 04$.

In their study of 500 days of *Kepler* data, Shibayama et al. (2013) identified 1547 superflares (flares with radiative emission greater than 10^{33} ergs) on 279 G-type dwarf stars. The flare radiative emissions fit a power-law distribution, $dN/dE \sim E^{-\alpha}$ with a power-law index $\alpha \sim 1.8$. Figure 12.5 shows that the power-law distribution for solar flare emissions from the lowest radiative emission nanoflares to the Carrington flare is consistent with the power law for the more slowly rotating ($P_{\mathrm{rot}} > 10\,$days) and thus less active Sun-like dwarfs observed by *Kepler*. The flare occurrence rate for these stars suggests that the Sun could have a 10^{34} erg flare,

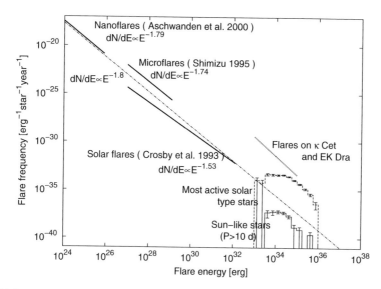

Fig. 12.5 Distribution of flare radiative emissions for the Sun and the solar-like stars observed by *Kepler*. Figure from Shibayama et al. (2013). Reproduced by permission of the AAS

100 times the Carrington flare radiative emission, once in 800–5000 years (Shibata et al. 2013). On the other hand, the faster-rotating more active G-type stars such as κ Cet and EK Dra have superflares 1000 more often than the less active solar-type stars. The similar power-law index over 14 orders of magnitude in radiative emission suggests a common physical origin but vastly different scales for these flares. For the active stars, superflares with luminosities of 10^{34} ergs occur once in 100 days and even more energetic superflares are likely. The most active G-type star (KIC10422252) even exhibited a superflare occurrence rate greater than once in 10 days. Shibayama et al. (2013) argued that superflares occur in starspots about ten times larger than the largest observed sunspot.

The All-Sky Automated Survey for Supernovae (ASAS-SN) provides a different approach to studying energetic flares on M dwarfs. Since this survey identifies as transient sources only stars that brighten more then one magnitude in the visual band (centered at 550 nm), it is biased to cooler stars for which the flux contrast of flares is larger for stars with faint quiescent emission. Schmidt et al. (2018) identified flares from 47 dwarfs, mostly spectral types M4–M8 but including one K5 star and one L0 star. The lower limits to the V-band flare energies are log E_V = 32–35, which classifies nearly all of these transients as superflares. It is interesting that these superflaring stars are only slightly younger than M dwarfs in the solar neighborhood.

12.4   Space Weather and Habitability

A useful way of developing insight into the many possible effects of stellar space weather on the atmospheres of exoplanets is to consider specific host stars and their observed or hypothetical exoplanets incorporating the available stellar and exoplanet properties. Segura et al. (2010) simulated the effects of the massive 1985 April 12 flare on AD Leo observed at UV and optical wavelengths (see Fig. 12.6) by Hawley and Pettersen (1991). This X18,000 superflare on the *GOES* scale had a total energy of 10^{34} ergs. Using models that incorporate radiative transfer, convection, and photochemistry, they computed the effects of this flare on a hypothetical exoplanet with an Earth-like oxygen-rich atmosphere located at 0.16 au, the distance from the M3 V star with same input flux as the Earth receives from the Sun. They also included the fluence of protons with energies greater than 10 MeV that is 200 times larger than what is estimated for the Carrington flare. This estimated proton flux was based on the correlation of energetic proton flux with X-ray flux during large solar flares (Belov et al. 2005). In an analysis of 28 years of solar activity, they found that all solar flares more energetic than X5 have associated >10 MeV proton events (cf. Dierckxsens et al. 2015). Flare UV radiation reduces the ozone column density by

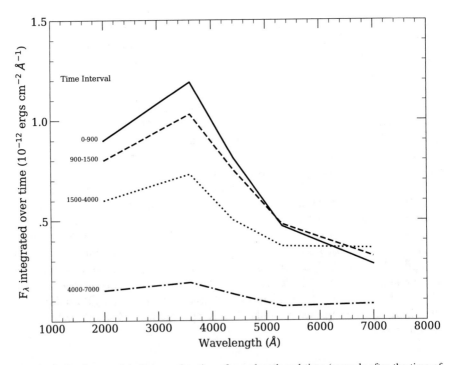

Fig. 12.6 Continuum emission as a function of wavelength and time (seconds after the time of flare onset) during the great flare of 1985 April 12 on AD Leo. Figure from Hawley and Pettersen (1991). Reproduced by permission of the AAS

only 1% at flare peak with recovery in about 10^5 s. However, the inclusion energetic protons is far more significant as they dissociate N_2 into excited N atoms that react with O_2 to produce nitrogen oxides (NO and NO_2) in the stratosphere (Tilley et al. 2019) that over time diffuse into the lower atmosphere and destroy ozone. The ozone column density is reduced by 94% 2 years after the flare and does not fully recover until 50 years later. Despite this long term major decrease in the ozone column density, the estimated decrease in UV dose rate for DNA damage is only 14% at peak ozone depletion.

This simulation for the effects of a superflare on AD Leo assumes that the flare was an isolated event, but the *Kepler* data shows that superflares with 10^{34} ergs on active M dwarfs occur on average once per 100 days and that less energetic flares occur far more frequently. Since these superflares occur much more rapidly than the effects of the previous superflare have time to subside, repeated superflares can enhance the decrease in ozone column density and other effects. Tilley et al. (2019) simulated multiple flare events for the M4 V flare star GJ 1243 with a hypothetical Earth-like planet located in the habitable zone at 0.16 au from the star. They assumed that the planet has an atmosphere with 21% oxygen, no haze to absorb UV photons, and no magnetic field to deflect high energy protons associated with large flares. Based on the distribution of flare radiative emissions ($10^{30.5}$ erg) with time observed by with the *Kepler* spacecraft, the long term effect of flare radiation alone is to decrease the ozone column density by only 37% with minimal decrease in the UV radiation incident at the exoplanet's surface.

However, the inclusion of energetic proton events accompanying the flare radiation is far more important (see Fig. 12.7) even taking into account the 0.08–0.25 fraction of time that protons events impact the planet. Tilley et al. (2019) found that essentially complete destruction of the ozone column density occurs over the course of 500 Myr of flaring by the observed flare-energy distribution or over 8 years by AD Leo size flares (10^{34} erg) occurring once each month. Both scenarios predict that the transmitted stellar UVA and UVB radiation will sterilize the planet's surface. The existence of a planetary magnetic field and atmospheric haze could mitigate the ozone loss, but by an amount that needs to be tested. Since GJ 1243 is a moderately active M dwarf, exoplanets of more active stars or those with closer orbits like Proxima Cen b (0.0485 au) could suffer more severe effects.

Tabatab-Vakili et al. (2016) also simulated the effects of high-energy protons on the destruction of ozone in Earth-like planets in the habitable zones on M dwarfs taking into account the full 16 MeV–0.5 TeV energy range of protons measured during the very energetic September 30, 1989 solar particle event and recent photochemical reaction rates. Contrary to previous studies (e.g., Segura et al. 2010; Grenfell et al. 2012) that predicted complete destruction of O_3 from repeated flaring, Tabatab-Vakili et al. (2016) found that during energetic proton events the inclusion of reactions involving HO_x molecules results in significantly less destruction of O_3, the decrease of atmospheric CH_4, a powerful greenhouse gas, and the formation of HNO_3, which may be observable in absorption near 11.2 μm. They concluded that even for the case of an active M dwarf host star

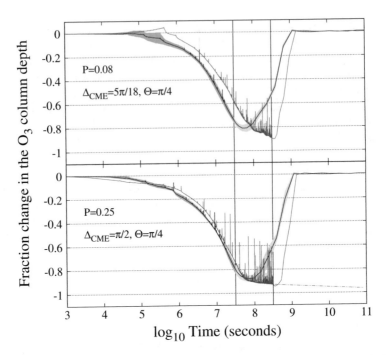

Fig. 12.7 Simulated depletion of ozone on an Earth-like planet located at 0.16 au from the M4 V host star GJ 1243. The simulations are for the distribution of flare energies observed by *Kepler* including both radiation and the fluence of high-energy protons assuming that the fraction of the time that the proton event impacts the planet is 0.08 (top) or 0.25 (bottom). The calculation is for 1 year duration (black line with error shading), 10 years (solid red line), and estimated long term (dashed red line). Figure from Tilley et al. (2019)

that produces repeated energetic proton events, biological material at the surface of an unmagnetized exoplanet would be partially shielded from destructive UV radiation by the remaining atmospheric O_3. Further study is needed to determine the conditions necessary for the survival of biological material and thus habitability on the surfaces of exoplanets near M dwarf stars.

References

Allred, J.C., Hawley, S.L., Abbett, W.P., Carlsson, M.: Radiative hydrodynamic models of optical and ultraviolet emission from M dwarf flares. Astrophys. J. **644**, 484 (2006)

Alvarado-Gómez, J.D., Drake, J.J., Cohen, O., Moschou, S.P., Garraffo, C.: Suppression of coronal mass ejections in active stars by an overlying large-scale magnetic field: a numerical study. Astrophys. J. **862**, 93 (2018)

Ayres, T.R.: The flare-ona of EK draconis. Astron. J. **150**, 7 (2015)

Balona, L.A.: Flare stars across the H-R diagram. Mon. Not. R. Astron. Soc. **447**, 2714 (2015)

Balona, L.A.: Starspots on A stars. Mon. Not. R. Astron. Soc. **467**, 1830 (2017)

Belov, A., Garcia, H., Kurt, V., Mavromichalaki, H., Gerontidou, M.: Proton enhancements and their relation to the X-ray flares during the three last solar cycles. Solar Phys. **229**, 135 (2005)

Benz, A.O.: Flare observations. Living Rev. Sol. Phys. **14**, 2 (2017)

Cárdenas, F.M., Sánchez, S.C., Domínguez, S.V.: The grand aurorae borealis seen in Columbia in 1859. Adv. Space Res. **57**, 257 (2016)

Carrington, R.: Description of a singular appearance seen in the Sun on September 1, 1859. Mon. Not. R. Astron. Soc. **20**, 13 (1859)

Cliver, E.W., Dietrich, W.F.: The 1859 space weather event revisited: limits of extreme activity. J. Space Weather Space Clim. **3**, A31 (2013)

Crosley, M.K., Osten, R.A.: Constraining stellar coronal mass ejections through multi-wavelength analysis of the active M dwarf EQ Peg. Astrophys. J. **856**, 39 (2018)

Davenport, J.R.A.: The Kepler catalog of stellar flares. Astrophys. J. **829**, 23 (2016)

Dierckxsens, M., Tziotziou, K., Dalla, S., Patsou, I., Marsh, M.S., Crosby, N.B., Malandraki, O., Tsiropoula, G.: Relationship between solar energetic particles and properties of flares and CMEs: statistical analysis of solar cycle 23 events. Solar Phys. **290**, 841 (2015)

France, K., Loyd, R.O.P., Youngblood, A., Brown, A., Schneider, P.C., Hawley, S.L., Froning, C.S., Linsky, J.L., Roberge, A., et al.: The MUSCLES treasury survey I: motivation and overview. Astrophys. J. **820**, 89 (2016)

Gershberg, R.E.: Solar-Type Activity in Main-Sequence Stars. Springer, Berlin (2005)

Graham, D.R., Cauzzi, G.: Temporal evolution of multiple evaporating ribbon sources in a solar flare. Astrophys. J. Lett. **807**, L22 (2015)

Grenfell, J.L., Griessmeier, J.-M., von Paris, P., Patzer, A.B.C., Lammer, H., Stracke, B., Gebauer, S., Schreier, F., Rauer, H.: Response of atmospheric biomarkers to NOx-induced photochemistry generated by stellar cosmic rays for Earth-like planets in the habitable zone of M dwarf stars. Astrobiology **12**, 1109 (2012)

Haisch, B.M., Linsky, J.L., Bornmann, P.L., Stencel, R.E., Antiochos, S.K., Golub, L., Vaiana, G.S.: Coordinated Einstein and IUE observations of a disparitions brusques type flare event and quiescent emission from Proxima Centauri. Astrophys. J. **267**, 280 (1983)

Haisch, B., Strong, K.T., Rodonò, M.: Flares on the Sun and other stars. Ann. Rev. Astron. Astrophys. **29**, 275 (1991)

Hawley, S.L., Pettersen, B.R.: The great flare of 1985 April 12 on AD Leonis. Astrophys. J. **378**, 725 (1991)

Hawley, S.L., Allred, J.C., Johns-Krull, C.M., Fisher, G.H., Abbett, W.P., Alekseev, I., Avgoloupis, S.I., Deustua, S.E., et al.: Multiwavelength observations of flares on AD Leonis. Astrophys. J. **597**, 535 (2003)

Joy, A.H., Humason, M.L.: Observations of the faint dwarf star L 726-8. Publ. Astron. Soc. Pac. **61**, 133 (1949)

Karmakar, S., Pandey, J.C., Airpetian, V.S., Misra, K.: X-ray superflares on CC Eri. Astrophys. J. **840**, 102 (2017)

Knipp, D.J.: Understanding Space Weather and the Physics Behind It, 1st edn. McGraw-Hill Education, New York (2011)

Knipp, D.J., Ramsay, A.C., Beard, E.D., et al.: The May 1967 great storm and radio disruption event: extreme space weather and extraordinary responses. Space Weather **14**, 614 (2016)

Kööp, L., Heck, P.R., Busemann, H., Davis, A.M., Greer, J., Maden, C., Meier, M.M.M., Wieler, R.: High early solar activity inferred from helium and neon excesses in the oldest meteorite inclusions. Nat. Astron. **2**, 709 (2018)

Kowalski, A.F., Allred, J.C.: Parameterizations of chromospheric condensations in dG and dMe model flare atmospheres. Astrophys. J. **852**, 61 (2018)

Kowalski, A.F., Hawley, S.L., Carlsson, M., Allred, J.C., Uitenbroek, H., Osten, R.A., Holman, G.: New insights into white-light flare emission from radiative-hydrodynamic modeling of a chromospheric condensation. Solar Phys. **290**, 3487 (2015)

Loyd, R.O., France, K., Youngblood, A., Schneider, C. Brown, A., Hu, R., Segura, A., Linsky, J., Redfield, S., Tian, F., Rugheimer, S., Miguel, T., Froning, C.S.: The MUSCLES treasury survey. V. FUV flares on active and inactive M dwarfs. Astrophys. J. **876**, 71 (2018)

Mekhaldi, F., Muscheler, R., Adolphi, F., Aldahan, A., Beer, J., McConnell, J.R., Possnert, G., Sigl, M., Svensson, A., et al.: Multiradionuclide evidence for the solar origin of the cosmic-ray events of AD 774/5 and 993/4. Nat. Commun. **6**, 8611 (2015)

Miyake, F., Masuda, K., Nakamura, T.: Another rapid event in the carbon-14 content of tree rings. Nat. Commun. **4**, 1748 (2013)

Osten, R.: Solar explosive activity throughout the evolution of the solar system. In: Schrijver, C.J., Bagenal, F., Sojka, J.J. (eds.) Heliophysics Active Stars, Their Atmospheres, and Impacts on Plnetary Environments, p. 23. Cambridge University Press, Cambridge (2016)

Osten, R.A., Kowalski, A., Drake, S.A., Krimm, H., Page, K., Gazeas, K., Kennea, J., Oates, S., Page, M.: A very bright, very hot, and very long flaring event from the M dwarf binary system DG CVn. Astrophys. J. **832**, 1740 (2016)

Pye, J.P., Rosen, S., Fyfe, D., Schöder, A.C.: A survey of stellar X-ray flares from the XMM-Newton serendipitous source catalogue: HIPPARCOS-Tycho cool stars. Astron. Astrophys. **581**, A28 (2015)

Redfield, S., Linsky, J.L., Ake, T.B., Ayres, T.R., Dupree, A.K., Robinson, R.D., Wood, B.E., Young, P.R.: A far ultraviolet spectroscopic explorer survey of late-type dwarf stars. Astrophys. J. **581**, 626 (2002)

Schmidt, S.J., Shappee, B.J., van Saders, J.L., Stanek, K.Z., Brown, J.S., Kochanek, C.S., Dong, S., Drout, M.R., Frank, S., et al. The largest M dwarfs flares from ASAS-SN (2018). arXiv180904510S

Schrijver, C.J.: Socio-economic hazards and impacts of space weather: the important range between mild and extreme. Space Weather **13**, 524 (2015)

Segura, A., Walkowicz, L.M., Meadows, V., Kasting, J., Hawley, S.: The effect of a strong stellar flare on the atmospheric chemistry of an Earth-like planet orbiting an M dwarf. Astrobiology **10**, 751 (2010)

Shibata, K., Isobe, H., Hillier, A., Choudhuri, A.R., Maehara, H., Ishii, T.T., Shibayama, T., Notsu, S., Notsu, Y., et al.: Can superflares occur on our Sun? Publ. Astron. Soc. Jpn. **65**, 49 (2013)

Shibayama, T., Maehara, H., Notsu, S., Notsu, Y., Nagao, T., Honda, S., Ishii, T.T., Nogami, D., Shibata, K.: Superflares on solar-type stars observed with Kepler. I. Statistical properties of superflares. Astrophys. J. Suppl. Ser. **209**, 5 (2013)

Tabataba-Vakili, F., Grenfell, J.L., Griessmeier, J.-M., Rauer, H.: Atmospheric effects of stellar cosmic rays on Earth-like exoplanets orbiting M-dwarfs. Astron. Astrophys. **585**, A96 (2016)

Tilley, M.A., Segura, A., Meadows, V., Hawley, S., Davenport, J.: Modeling repeated M-dwarf flaring at an Earth-like planet in the habitable zone: I. Atmospheric effects for an unmagnetized planet. Astrobiology **19**, 6 (2019)

Tsurutani, B.T., Judge, D.L., Guarnieri, F.L., Gangopadhyay, P., Jones, A.R., Nuttall, J., Zambon, G.A., Didkovsky, L., Mannucci, A.J., et al.: The October 28, 2003 extreme EUV solar flare and resultant extreme ionospheric effects: comparison to other Halloween events and the Bastille Day event. J. Geophys. Res. Lett. **32**, L03S09 (2003)

Walkowicz, L.M., Basri, G., Batalha, N., Gilliland, R.L., Jenkins, J., Borucki, W.J., Koch, D., Caldwell, D.: White-light flares on cool stars in the Kepler quarter 1 data. Astron. J. **141**, 50 (2011)

Youngblood, A., France, K., Loyd, R.O.P., Brown, A., Mason, J.P., Schneider, P.C., Tilley, M.A., Berta-Thompson, Z.K., Buccino, A., Froning, C.S., et al.: The MUSCLES tresury survey IV. Scaling relations for ultraviolet Ca II, and energetic particle fluxes from M dwarfs. Astrophys. J. **843**, 31 (2017)

Chapter 13
The Effects of Heterogeneous Stellar Surfaces on the Analysis of Exoplanet Transit Light Curves and Spectra

The results of our ... temperature modeling efforts demonstrate that, in principle, a heterogeneous stellar photosphere can provide a transmission spectroscopy signal larger than that introduced by the exoplanetary atmosphere in the optical, adding another degeneracy to the modeling and interpretation of exoplanet spectra. (Rackham et al. 2017)

If stars were uniformly bright across their surfaces, then the analysis of transit observations would require only the subtraction of this bright background from the observed data in order to obtain the transit light curve and the absorption spectrum produced by atoms and molecules in the exoplanet's atmosphere. Even this unrealistically simple observing scenario is a data analysis challenge because the area of a transiting planet may be only 1% or smaller compared to that of the host star, and the area of a terrestrial exoplanet's atmosphere seen in absorption against the stellar disk emission may be only 0.01% that of the stellar disk area. Thus high signal/noise data are essential to minimize random noise.

However, there is a far larger source of noise produced the host star itself. This has two aspects. The first is that along the path that the exoplanet traverses across the stellar surface, the star is not uniformly bright. Instead, if the Sun is a reliable guide, the exoplanet will cross dark starspots, bright regions (faculae and network elements and perhaps even flares), and the more gradual center-to-limb brightness changes that are limb darkened in the optical and infrared but are limb brightened at UV and X-ray wavelengths and in the cores of chromospheric lines. The second aspect of stellar noise is the brightness changes in the stellar surface areas that are not occulted by the exoplanet. These brightness changes are both intrinsic as the localized heating rate changes with time in each portion of the surface and are also produced by bright and dark regions rotating into and out of view. The effects of stellar surface inhomogeneity are, therefore, to contaminate light curves with the noise of variable illumination along the transit path and from the unocculted star and to pollute exoplanet absorption spectra with the spectral shape of the background illumination by the host star with its spots and faculae. Also, to obtain the exoplanet's absorption spectrum, one must subtract the stellar

© Springer Nature Switzerland AG 2019
J. Linsky, *Host Stars and their Effects on Exoplanet Atmospheres*,
Lecture Notes in Physics 955, https://doi.org/10.1007/978-3-030-11452-7_13

spectrum obtained pre-ingress and post-egress, but the stellar spectrum can change appreciably over the several hours difference between these two calibration spectra. This chapter reviews the causes and magnitudes of stellar surface heterogeneity and their effects on measurements of exoplanet properties.

13.1   Heterogeneous Surfaces of the Sun and Sun-Like Stars: Spots and Faculae

At all wavelengths, the Sun is not uniformly bright across its apparent surface. Variations in the magnetic heating rate in the upper photosphere and lower chromosphere produce temperature variations at the optical depths where the emission originates. The Sun shows two main types of surface brightness structures that are both produced by magnetic fields: bright faculae and dark sunspots. The different surface coverage by faculae and sunspots as solar rotation moves these features into and out of the field of view dominate the changes in the Sun's total irradiance on short times scales of about 50 parts per million (Kopp et al. 2005) and about 0.1% over the course of the Sun's 11 year magnetic cycle. Figure 13.1 shows that the percent change in the solar spectral irradiance between minimum and maximum increases to shorter wavelengths (Fröhlich and Lean 2004). Young and more active stars and especially M dwarfs show more pronounced heterogeneity from such features particularly at shorter wavelengths. This heterogeneous and time variable surface structure provides a serious challenge to exoplanet observers who wish to understand transit spectra in terms of the exoplanet's atmosphere rather than being confused by the difficult to measure time and spatial variations in the background light source that constitute background noise that must be subtracted accurately. It is therefore important to investigate how changes in surface coverage of these features on solar-like stars could be inferred from the changes in magnetic activity.

Magnetic activity in the Sun and stars is commonly measured by the excess emission in the Ca II K line above that predicted from an atmosphere heated only by radiation and convection from below. The Ca II H and K lines are often used because they are the only lines in the optical spectrum that measure magnetic heating and are transitions from the ground state that can be simply modeled. Figure 13.2 illustrates the different ways of measuring the Ca II excess emission. The observed Ca II K line and Mg II k line fluxes for the Sun viewed as a star are compared with two models (Linsky and Ayres 1978). The dark solid lines are solar radiative/convective equilibrium models computed to fit the solar photosphere with no magnetic heating. The dashed lines are synthesized profiles computed from a chromosphere model with partial frequency redistribution spectral line formation. The synthesized profiles do not fit the shape of the emission peaks because Doppler motions (microtubulence) are not included, but the synthesized profiles do fit the integrated line core fluxes. The extra emission in the observed and synthesized profiles above the radiative/convective equilibrium model measures the magnetic

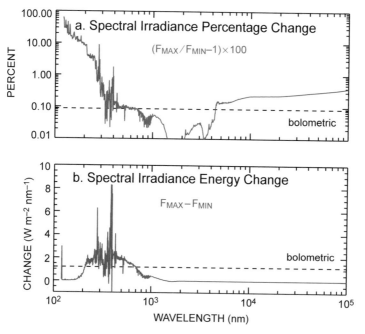

Fig. 13.1 *Top panel:* percent change in the solar spectral irradiance between maximum and minimum over the solar cycle. The mean bolometric change is 0.1% (dashed line). *Bottom panel:* energy change corresponding to the spectral irradiance change between maximum and minimum. The change in the total solar irradiance (dashed line) from minimum to maximum is about $1.3\,\mathrm{W\,m^{-2}}$ compared to the mean total solar irradiance of $1366\,\mathrm{W\,m^{-2}}$. Figure from Fröhlich and Lean (2004)

heating rate. The extra emission in the line core between the K_{1V} and K_{1R} features (secondary minima features on the violet and red sides of the Ca II K line emission peaks) and between the k_{1V} and k_{1R} features in the Mg II k line (similar terminology to the Ca II lines) measure the magnetic heating in the chromosphere. Outside of the K_1 and k_1 features the excess emission in the inner line wings measures magnetic heating in the upper photosphere that radiative/convective models do not include.

There are different ways of estimating the magnetic heating rate from the Ca II K line data. For example, Linsky and Ayres (1978) showed that for the Mg II k line the excess emission between the k_{1V} and k_{1R} features (the k_1 index) measures about 20% of the total chromospheric heating rate and the corresponding Mg II h line measures an additional 10%. They argue that the analysis of the Ca II K line is more complicated as the net flux between the K_{1V} and K_{1R} features (the K_1 index) measures magnetic heating but only 60% of the flux in the K_1 index measures magnetic heating in the chromosphere with the remaining magnetic heating in the upper photosphere. After this correction, the Ca II H and K lines together (60% of the K_1 index and 60% of the corresponding H_1 index) measure 10% of the total heating rate in the solar chromosphere.

Fig. 13.2 Observed flux of
the Ca II K and Mg II k lines
in integrated sunlight when
the Sun was relatively quiet
(filled triangles). The fluxes
of radiative equilibrium solar
models (solid lines) and
synthesized line profiles from
solar chromosphere models
with partial redistribution line
formation (dashed lines) are
included. Figure from Linsky
and Ayres (1978).
Reproduced by permission of
the AAS

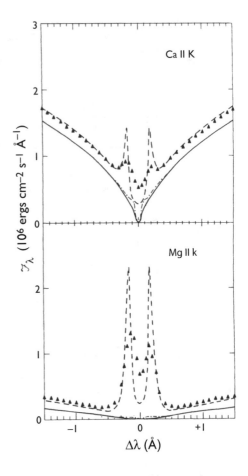

Noyes et al. (1984) described a different way of estimating the chromospheric heating rate from the R'_{HK} parameter. The Mt. Wilson HK program to study activity in cool dwarf stars (cf. Vaughan et al. 1978) operated between 1966 and 2002. The Mt. Wilson measurements are the flux in 0.109 nm triangular passbands centered on the H and K lines and the flux in 2.0 nm wide "continuum" windows centered at 390.07 and 400.107 nm on the blue and red sides of the Ca II lines. The S-index is the ratio of the fluxes in the H and K passbands to the continuum windows. Since the fluxes in these continuum windows depend on the spectral energy distribution of the star, Noyes et al. (1984) calibrated the S-index as a function of the star's B-V color to produce R_{HK}, the ratio of the total flux in the H and K passbands to the stellar bolometric flux (σT_{eff}^4). To correct for the photospheric contribution to the flux inside of the H and K passbands, they subtracted the flux on either side of the H_1 and K_1 minimum features to obtain R'_{HK}, which measures the magnetic heating rate in the chromosphere and upper photosphere (inside of the H_1 and K_1 features)

divided by the star's bolometric flux. The R'_{HK} parameter has been used in many studies of stellar activity cycles (e.g., Wilson 1978; Baliunas et al. 1995).

Several authors have considered alternative ways of correcting S-index measurements obtained by the Mt. Wilson, Lowell Observatory and other programs for photospheric emission. For example, Mittag et al. (2013) used the non-LTE PHOENIX grid of photospheric models to determine the absolute fluxes in the continuum windows and in the H and K line passbands. This corrected S-index should not contain any photospheric emission. From the more than 2500 stars with continuum-corrected S-indices, they could accurately measure the minimal Ca II fluxes for dwarfs and giants as a function of B-V color. They identified the minimal chromospheric heating rates as the basal chromospheric fluxes produced by non-magnetic heating first measured by Schrijver (1987) (see Fig. 2.3). Mittag et al. (2013) also suggested that a new parameter, $R^+_{HK} = R'_{HK} - R'_{basal}$ should be used to isolate the magnetic heating rate in stellar chromospheres.

In their study of photometric variability of stars in the spectral type range F7 V to K2 V, Radick et al. (1983) found that stars tend to become fainter as their variability at optical wavelengths increases. This was clearly seen in young stars in the Pleiades cluster, where Fang et al. (2016) later found that spot coverage can be as large as 40–50% on the basis of TiO band strengths. Radick et al. (1983) speculated that the variability is produced by the rotational modulation of dark starspots as they rotate into and out of view and that the decrease in mean brightness is due to a larger surface coverage by dark starspots than is seen on the Sun. Radick et al. (1998) extended this work to show that the less active stars including the Sun become brighter with increasing R'_{HK}, presumably because the brightness of faculae dominates over the darkness of spots for the less active stars, but the more active stars become fainter with increasing R'_{HK}, because the number and size of starspots dominates the brightness variation. The separation between these different behaviors occurs near log $R'_{HK} = -4.65$, corresponding to ages of $1-2\times10^9$ yr for $1M_\odot$ stars. With a mean log $R'_{HK} = -4.89$, the Sun lies in the facular dominated regime but near the threshold of spot dominance. Near the peak of solar activity, however, when the Zürich sunspot relative number exceeds about 150, the total solar irradiance begins to decrease with increasing activity indicating that sunspot darkening then exceeds facular brightening (Solanki and Fligge 1999). Radick et al. (1998) noted that stars otherwise similar to the Sun would be more facular dominated with increasing inclination from the line of sight as faculae become brighter and spots appear less dark near the stellar limb. For all stars in their sample, photometric variability at optical wavelengths and the total irradiance increase with magnetic heating in the chromosphere as measured by R'_{HK} (cf. Lockwood et al. 2007).

Based on 35 years of monitoring 72 Sun-like stars, Radick et al. (2018) confirmed their earlier results with a slight change in the dividing line between stars with correlated (facular dominated) and anticorrelated (spot dominated) behavior to log $R'_{HK} = -4.75$. For G-type stars, this occurs near $P_{rot} = 17$ days and an age of about 2.5 Gyr (Mamajek and Hillenbrand 2008) near the Vaughan–Preston gap (Vaughan and Preston 1980).

To better understand the photometric variability and mean brightness changes of solar-type stars, Shapiro et al. (2014) developed a model relating the surface area coverage of dark starspots and bright faculae to the optical variability and S-index. Over the course of the last three solar activity cycles, the area coverage by spots increased from a minimum of 0.003–0.28% and the area coverage of faculae increased from a minimum of 0.36–3%. Since the observed facular area coverage increases linearly with the S-index but the spot area coverage increases quadratically, the relative importance of spot darkening to facular brightening of the solar total solar irradiance and the flux at optical wavelengths increases with activity as measured by the S-index and R'_{HK} parameters. Their model for the optical brightness of solar-like stars is based on the functional forms of the solar spot and facula dependence on the S-index, which they extrapolated to stars more active than the Sun. Their model reproduces the change from facular brightening to spot darkening near $R'_{HK} = -4.8$ as was found by Radick et al. (1998) and predicts increased photometric variability with increasing inclination angle of the star's rotation axis with respect to the line of sight. The model also predicts how these properties would change if the spots and faculae are located near the rotation poles rather than at mid-latitudes as is the case for stars more active than the Sun. Thus the star's S-index and R'_{HK} parameters are good predictors of the heterogeneity of surface brightness for a solar-type star. What can be said about surface heterogeneity of stars much cooler than the Sun?

13.2 Heterogeneous Surfaces of Stars

Since high-resolution imaging of stellar surfaces is not feasible for stars and cooler stars may have qualitatively different surface brightness structures than the Sun, indirect techniques are needed to infer surface brightness heterogeneity, primarily the effect of starspots and faculae. For a comprehensive review of starspots and the diverse observing and analysis techniques used to infer their properties, see Berdyugina (2005).

The availability of long-duration high-precision photometry from instruments in space including *MOST*, *CoRoT*, and *Kepler* provides the raw material for the extracting spot parameters (e.g., spot coverage, locations on the stellar surface, and contrast relative to the nonspot photosphere) from photometric variations alone. Optical variability much larger than observed on the Sun increases to the cooler stars with *Kepler* photometry showing that 90% of early M dwarfs have higher variability than the Sun (Basri et al. 2013). From an analysis of the photometric variability of solar-like stars ($5600 < T_{eff} < 6300\,\mathrm{K}$), Maehara et al. (2017) showed that the dark spot area coverage increases from roughly 0.1% of the stellar disk for slowly rotating stars to as much as 10–30% of the stellar disk for stars with $P_{rot} < 3$ days. They also showed, not unexpectedly, that stars with superflares not unexpectedly tend to have large spots.

To better understand the causes of this photometric variability, Walkowicz et al. (2013) assessed the information content and degeneracies that are inherent in the photometric technique based on a set of numerical experiments including spots that represent more realistic spot groups but not including the presence of bright faculae. Problems with the photometric technique include errors in spot parameters that result from the unknown continuum brightness of the unspotted star (a consequence of spots located at the stellar pole or uniformly distributed in longitude), the degeneracy between stellar inclination angle and spot latitude, false rotation periods when spots are widely distributed in longitude, and the uncertainty in stellar inclination when noise is significant.

Despite these problems, there have been studies of the properties of spots and faculae using photometric techniques. An interesting example is the analysis of 142 days of uninterrupted photometry of the young (about 0.5 Gyr) active G7 V star CoRoT-2 observed by the *CoRoT* satellite. Lanza et al. (2009) found that the short term optical variability is about 20 times larger for the Sun and that spots rather than faculae dominate the optical variability with the area ratio of faculae to spots much smaller than for the Sun. Assuming that the brightness contrast of spots to the nonspotted photosphere is 0.665 like the Sun, they found that the mean fractional area covered by such spots is about 8% of the stellar photosphere compared to 0.2% when the Sun is most active. If the contrast is much smaller, for example 0.21 as is the case for very active stars, the fractional area coverage is reduced to about 3%, but still much larger than for the Sun.

Silva-Valio et al. (2010) analyzed 77 transit light curves in the same CoRoT-2 data set. These light curves showed that the path of CoRoT-2b across the star passed over between 1 and 9 spots. Figure 13.3 shows two of these light curves: one when the exoplanet's path traversed 9 spots and the other when only 1 spot. The effect of the exoplanet covering a dark spot is to make the star appear brighter and the light curve less deep at this time. The positive residuals seen in the figure result from the covering and uncovering of individual spots. Silva-Valio et al. (2010) found that along the transit path the average surface coverage by spots is $16 \pm 5\%$ and the equivalent blackbody temperature of the spots is $T_{spot} = 4700 \pm 300$ K, about 1000 K cooler than the star's $T_{eff} = 5625$ K. Thus $T_{spot}/T_{eff} = 0.84 \pm 0.05$. The inferred average spot size is 10 times larger than the largest sunspots, and spots are likely present on the stellar surface of CoRoT-2 at all times.

Zeeman Doppler images based on polarized spectral line profiles are sensitive to bright faculae but insensitive to faint spots and do not accurately show active regions with complex magnetic fields that cancel in circular polarization measurements (cf. Hébrard et al. 2016). However, Doppler images based on unpolarized light can identify both bright and dark regions across stellar surfaces. Barnes et al. (2015) obtained Doppler images of two rapidly rotating young M dwarf stars: GJ 791.2A (M4.5 V) and LP 944-20 (M9 V). They found for GJ 791.2A a mean spot filling factor of 3.2% with spots concentrated at high latitudes (59° and 75°). Their best fit is with spots 300 K cooler than the nonspot equivalent blackbody temperature (3000 K) and a spot brightness contrast relative to the unspotted photosphere of 0.41 in the wavelength band 644–1025 nm. For LP 944-20 the spots are concentrated at high latitudes but the mean spot filling factor is only 1.5%. The best fit is for spots 200 K cooler than the nonspot temperature of 2300 K and spot contrast of 0.48.

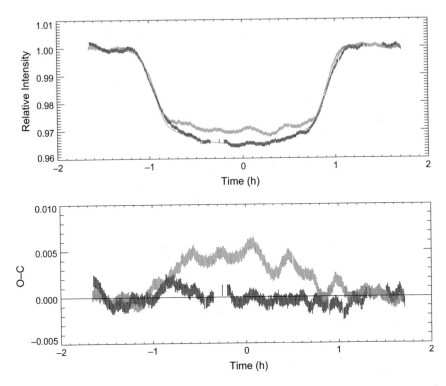

Fig. 13.3 *Top panel:* transit light curves of the active G7 V star CoRoT-2 observed by the *CoRoT* satellite. The red curve is for the time when the exoplanet's path crossed over 9 starspots and the black curve is for the time when there were a minimum number of spots along the transit path. *Bottom panel:* residuals after subtraction of the two light curves by the model of the star with no spots. Figure from Silva-Valio et al. (2010). Reproduced with permission of ESO

A new Doppler image of GJ 791.2A revealed a similar distribution of spots but with spots 400 K cooler than the nonspot photosphere and a spot brightness contrast of 0.32 rather than 0.41 (Barnes et al. 2017). For GJ 65A (M5.5 V) and GJ 65B (M6 V), also called UV Ceti, the best fits are for spots 400 K cooler than the nonspot photosphere, but the spot coverage factor is much larger for GJ 65B than GJ 6A. These results indicate that late-M dwarfs have spots with temperatures 85–90% that of the nonspotted photosphere compared to about 72% for the Sun. Figure 13.4 compares the difference between stellar photosphere and spot temperatures, $\Delta T = T_{photo} - T_{spot}$, as a function of photospheric temperature showing the trend of ΔT increasing from about 0.11 at $T_{photo} = 3500$ K to 0.29 at $T_{photo} = 5500$ K. Molecular spectra are important tools for assessing starspot magnetic fields and temperatures (Afram and Berdyugina 2015). The tentative identification of bright spots (faculae) from *Kepler* and *Spitzer* photometry of TRAPPIST-1 (M8 V) (Morris et al. 2018) argues that faculae should also be included in analyses of photometry of very cool stars.

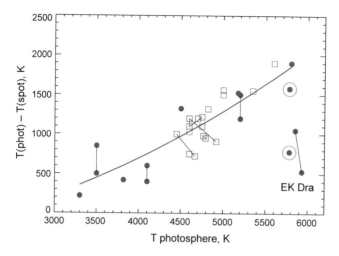

Fig. 13.4 Difference between stellar photosphere and spot temperatures for dwarf stars (filled circles) and active giants including RS CVn systems (squares). The vertical lines refer to the same star but different observations. The thick solid line is a polynomial fit to the stellar data excluding EK Dra. The circles with dots refer to the solar umbra (upper symbol) and solar penumbra (lower symbol). Figure from Berdyugina (2005)

In their analysis of optical and near-infrared absorption spectra of the sub-Neptune GJ 1214b during transits, Rackham et al. (2017) were not able to fit the light curves with any sensible combination of the exoplanet's chemical composition including the effects of clouds and hazes. Instead they developed a composite photosphere and atmospheric transmission (CPAT) model to simulate the effects of both stellar surface heterogeneity and exoplanet atmosphere properties in order to fit the observed optical and near-infrared spectra. A critical constraint was that the optical transit depths were shallower than the depths measured in the near-infrared. The best set of parameters to fit this constraint was that 3.2% of the stellar surface was covered by faculae with temperatures $\Delta T = 354 \pm 46\,\mathrm{K}$ warmer than the mean photosphere ($T_{\mathrm{eff}} = 3252\,\mathrm{K}$). The facular dominance over spots is consistent with the old age of the star, and the value of ΔT is similar to that of solar faculae (Topka et al. 1997).

Extending the previous analysis, Rackham et al. (2018) modeled the fractional coverage by spots and faculae in M0 V to M9 V stars consistent with a 1% variability full amplitude in the $\sim 1\,\mu\mathrm{m}$ band used by *Kepler, TESS, MEarth,* and *JWST*. Rackham et al. (2018) considered cases of many small Sun-like spots and a few giant spots with and without faculae. The coverage fraction by spots and faculae consistent with the 1% amplitude variability depends primarily on the assumed spot sizes. The rotational modulation of a few giant spots randomly distributed in longitude with small fractional coverage can explain the 1% mean amplitude variations, whereas a very large fractional coverage is required to explain the same amplitude variations from a large number of small spots because the spots are likely

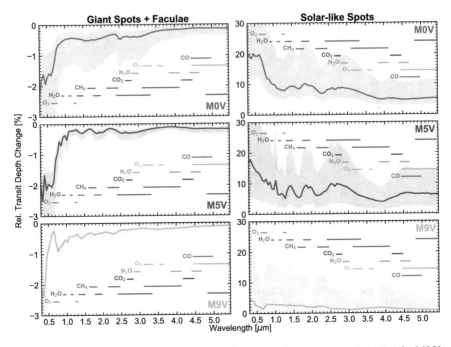

Fig. 13.5 Stellar contamination spectra (percent increase or decrease of transit depths) for M0 V, M5 V, and M9 V stars (solid lines). The range of variations in the contamination spectra is indicated by the blue shading. The six cases shown are for giant spots and faculae (left column) and for solar-like spots and faculae (right column). All examples are for 1% variability near 1 μm wavelength. The locations of important molecular are shown by horizontal lines. Figure from Rackham et al. (2018). Reproduced by permission of the AAS

distributed nearly uniformly in longitude. The inclusion of bright faculae in their simulations compensates for the darkening produced by spots and therefore requires a higher fractional coverage by spots to produce the same 1% amplitude variability. They computed stellar contamination spectra, the mean value and range of fractional changes in transit depths as a function of wavelength for each model of spots and faculae and host star spectral type. As shown in Fig. 13.5 the transit depths become less deep (negative transit depth change) for the case of giant spots and faculae when the brighter host star emission due to faculae dominates over the darkening by their few giant spots. The brighter host star emission makes the planet appear to be smaller and thus the transit depth shallower. On the other hand, transit depths become deeper (positive transit depth change), by as much as 20% in the visual and 10% in the near infrared when the host star's darkening by the many small spots dominates over the brightening by faculae. Since Newton et al. (2017) measured full amplitude variabilities of 1–8% in their large sample of M dwarf stars observed by *MEarth*, the above results concerning the covering factions of spot and faculae for 1% variability are lower limits for the very active stars.

13.3 Effects of Heterogeneous Surfaces on the Analysis of Exoplanet Transit Spectra

The previous section provides a basis for estimating the brightness heterogeneity across the surfaces of different types of stars. What are the effects of this heterogeneity on the properties of exoplanets inferred from transit light curves and spectra?

Radial velocity noise (RV jitter): There are many sources of RV jitter in stars as summarized by Dumusque (2016). For example, the rotation of spots and faculae onto and off of the visible stellar produces distortions in absorption line shapes that change the measured radial velocities from one observation to the next. Barnes et al. (2015) estimated that for the M4.5 V star GJ 791.2A the rms noise induced by starspots is $138\,\mathrm{m\,s^{-1}}$, but they estimate that this noise can be reduced by a factor of 2 by monitoring the presence of spots in contemporaneous Doppler images. Even so, the reduced $73\,\mathrm{m\,s^{-1}}$ jitter would severely compromise searches for exoplanets by the radial velocity technique. Magnetic activity indicators such as R'_{HK} can predict the RV jitter of active but not inactive stars (Oshagh et al. 2017). While the RV jitter due to spots and faculae can be greatly reduced by observing very inactive stars, there is a floor to the RV jitter set by flows generated by stellar magnetic fields. Magneto-convection produces upflows in granulation cells and downflows in intergranular lanes that change with the center-to-limb angle and magnetic field strength. Simulations of line profiles based on magneto-convection by Cegla et al. (2018) showed that this cause of jitter appears to be similar to the $9\,\mathrm{cm\,s^{-1}}$ level needed to discover Earth-like planets in the habitable zone, but that p-mode oscillations produce far larger RV jitter requiring sophisticated data analysis to minimize this component of the RV jitter. From observations of integrated sunlight, however, Haywood et al. (2016) found that the suppression of granular blueshifts in magnetized regions produces $2.4\,\mathrm{m\,s^{-1}}$ RV jitter. New observing and data analysis techniques are needed to minimize this jitter in order to discover an Earth-like exoplanet in the habitable zone.

Exoplanet radius and density: During the transit of a spotted star, the light curve will be deeper than for a nonspotted star and thus the exoplanet will appear to have a larger radius. For CoRoT-2a the effect is to make the planet appear to be 3% larger and thus have a density about 10% smaller than the true value (Lanza et al. 2009). For the facular-dominated case of TRAPPIST-1, Rackham et al. (2018) estimated that the transit depths are larger and inferred planetary radii are overestimated leading to a systematic density errors of $-8^{+7}_{-20}\%$ compared to their values obtained without including starspots (Grimm et al. 2018). The underestimated exoplanet densities mean that the volatile content of these exoplanets and the size of their atmospheres will be smaller when the effect of host star heterogeneity is taken into account.

Molecular column densities: Unocculted spots and faculae change the brightness of the host star and thus the depth of transit light curves. As a result, the relative depth of molecular absorption features in the exoplanet's atmosphere will differ from the case of illumination by an unspotted host star. The effect is larger at optical wavelengths but it is still important at near infrared wavelengths where many molecular bands are present (see Fig. 13.5). The effect can be 10% for early M dwarfs with 1% variability and more for larger variability. The effect of spots and faculae on transit spectra can be ten or more times larger than the absorption by molecules in an exoplanet's atmosphere and therefore must be taken into account in transit data analyses (Rackham et al. 2018). The contamination spectrum of spots and faculae can have spectral features especially for very cool stars. An example is the "inverted" water vapor features near $1.4\,\mu m$ seen in transit spectra of the TRAPPIST-1 exoplanets instead of water absorption in the exoplanet atmospheres. The "inversion" is produced by weaker illumination from the star near $1.4\,\mu m$ (Zhang et al. 2018). The effects of heterogeneous host star illumination can be very large at X-ray and ultraviolet wavelengths and for the analysis of Lyman-α spectra (Llama and Shkolnik 2015, 2016).

Analysis of the spectral slopes of transit spectra: McCullough et al. (2014) called attention to the effect of unocculted starspots making the host star appear both darker and redder and thus overestimates of the exoplanet's apparent radius by an amount that increases to shorter wavelengths. This apparent increase in radius of HD 189733b at wavelengths shorter than about $1\,\mu m$ has been interpreted by Pont et al. (2013) and others as Rayleigh scattering by small particles in the exoplanet's atmosphere, but the apparent radius increase can instead by explained by star spots covering 5.6% of the stellar surface. McCullough et al. (2014) noted that uncertainties in the available data are consistent with both explanations or a model in which unocculted starspots explain the increase in apparent radius in the 0.5–$1.0\,\mu m$ range and Rayleigh scattering provides most of the rise to shorter wavelengths. The effect of unocculted starspots on the slope of the extracted spectrum in the 1.1–$1.7\,\mu m$ region, however, is small (Zellem et al. 2017; Kreidberg et al. 2018).

Confusion of spot and exoplanet spectra: Host star spots and very cool photospheres with temperatures less than about 3000 K will have water absorption bands that make the host star darker than in adjacent wavelengths. As a result, transiting exoplanets will appear to have larger radii in the water vapor bands complicating the inference of water mixing ratios in exoplanet atmospheres. This may be the explanation for the "inverted" water feature near $1.4\,\mu m$ seen in the transit spectra of the exoplanets of TRAPPIST-1 (Zhang et al. 2018).

References

Afram, N., Berdyugina, S.V.: Molecules as magnetic probes of starspots. Astron. Astrophys. **576**, 34 (2015)

Baliunas, S.L., Donahue, R.A., Soon, W.H., Horne, J.H., Frazer, J., Woodard-Eklund, L., Bradford, M., Rao, L.M., Wilson, O.C., Zhang, Q., et al.: Chromospheric variations in main-sequence stars. Astrophys. J. **438**, 269 (1995)

Barnes, J.R., Jeffers, S.V., Jones, H.R.A., Pavlenko, Ya.V., Jenkins, J.S., Haswell, C.A., Lohr, M.E.: Starspot distributions on fully convective M dwarfs: implications for radial velocity planet searches. Astrophys. J. **812**, 42 (2015)

Barnes, J.R., Jeffers, S.V., Haswell, C.A., Jones, H.R.A., Shulyak, D., Pavlenko, Ya.V., Jenkins, J.S.: Surprisingly different star-spot distributions on the near equal-mass equal-rotation-rate stars in the M dwarf binary GJ 65 AB. Mon. Not. R. Astron. Soc. **471**, 811 (2017)

Basri, G., Walkowicz, L.M., Reiners, A.: Comparison of Kepler photometric variability with the Sun on different timescales. Astrophys. J. **769**, 37 (2013)

Berdyugina, S.: Starspots: a key to the stellar dynamo. Living Rev. Sol. Phys. **2**, 8 (2005)

Cegla, H.M., Watson, C.A., Shelyag, S., Chaplin, W.J., Davies, G.R., Mathioudakis, M., Palumbo, M.L., III, Saar, S.H., Haywood, R.D.: Stellar surface magneto-convection as a source of astrophysical noise II. Center-to-limb parameterisation of absorption line profiles and comparison to observations. Astrophys. J. **866**, 55 (2018)

Dumusque, X.: Radial velocity fitting challenge. I. Simulating the data set including realistic stellar radial-velocity signals. Astron. Astrophys. **593**, A5 (2016)

Fang, X.-S., Zhao, G., Zhao, J.-K., Chen, Y.-Q., Bharat Kumar, Y.: Stellar activity with LAMOST. I. Spot configuration in Pleiades. Mon. Not. R. Astron. Soc. **463**, 2494 (2016)

Fröhlich, C., Lean, J.: Solar radiative output and its variability: evidence and mechanisms. Astron. Astrophys. Rev. **12**, 273 (2004)

Grimm, S.L., Demory, B.-O., Gillon, M., Dorn, C., Agol, E., Burdanov, A., Delrez, L., Sestovic, M., Triaud, A.H.M.J., et al.: The nature of the TRAPPIST-1 exoplanets. Astron. Astrophys. **613**, 68 (2018)

Haywood, R.D., Collier Cameron, A., Unruh, Y.C., Lovis, C., Lanza, A.F., Llama, J., Deleuil, M., Fares, R., Gillon, M., Moutou, C., et al.: The Sun as a planet-host star: proxies from SDO images for HARPS radial-velocity variations. Mon. Not. R. Astron. Soc. **457**, 3637 (2016)

Hébrard, E.M., Donati, J.-F., Delfosse, X., Morin, J., Moutou, C., Boisse, I.: Modelling the RV jitter of early-M dwarfs using tomographic imaging. Mon. Not. R. Astron. Soc. **461**, 1465 (2016)

Kopp, G., Lawrence, G., Rottman, G.: The Total Irradiance Monitor (TIM): science results. Solar Phys. **230**, 129 (2005)

Kreidberg, L., Line, M.R., Thorngren, D., Morley, C.V., Stevenson, K.B.: Water, high-altitude condensates, and possible methane depletion in the atmosphere of the warm super-Neptune WASP-107b. Astrophys. J. **858**, 6 (2018)

Lanza, A.F., Pagano, I., Leto, G., Messina, S., Aigrain, S., Alonso, R., Auvergne, M., Baglin, A., Barge, P., Bonomo, A.S., et al.: Magnetic activity in the photosphere of CoRoT-Exo-2a. Active longitudes and short-term spot cycle in a young Sun-like star. Astron. Astrophys. **493**, 193 (2009)

Linsky, J.L., Ayres, T.R.: Stellar model chromospheres. VI - empirical estimates of the chromospheric radiative losses of late-type stars. Astrophys. J. **220**, 619 (1978)

Llama, J., Shkolnik, E.L.: Transiting the Sun: the impact of stellar activity on X-Ray and ultraviolet transits. Astrophys. J. **802**, 41 (2015)

Llama, J., Shkolnik, E.L.: Transiting the Sun II. The impact of stellar activity on Lyα transits. Astrophys. J. **817**, 81 (2016)

Lockwood, G.W., Skiff, B.A., Henry, G.W., Henry, S., Radick, R.R., Baliunas, S.L., Donahue, R.A., Soon, W.: Patterns of photometric and chromospheric variation among Sun-like stars: a 20 year perspective. Astrophys. J. **171**, 260 (2007)

Maehara, H., Notsu, Y., Notsu, S., Namekata, K., Honda, S. Ishii, T.T., Nogami, Daisaku, N., Shibata, K.: Starspot activity and superflares on solar-type stars. Publ. Astron. Soc. Jpn. **69**, 41 (2017)

Mamajek, E.E., Hillenbrand, L.A.: Improved age estimation for Solar-type dwarfs using activity-rotation diagnostics. Astrophys. J. **687**, 1264 (2008)

McCullough, P.R., Crouzet, N., Deming, D., Madhusudhan, N.: Water vapor in the spectrum of the extrasolar planet HD 189733b. I. The transit. Astrophys. J. **791**, 55 (2014)

Mittag, M., Schmitt, J.H.M.M., Schröder, K.-P.: Ca II H+K fluxes from S-indices of large samples: a reliable and consistent conversion based on PHOENIX model atmospheres. Astron. Astrophys. **549**, A117 (2013)

Morris, B.M., Agol, E., Davenport, J.R.A., Hawley, S.L.: Possible bright starspots on TRAPPIST-1. Astrophys. J. **857**, 39 (2018)

Newton, E.R., Irwin, J., Charbonneau, D., Berlind, P., Calkins, M.L., Mink, J.: The Hα emission of nearby M dwarfs and its relation to stellar rotation. Astrophys. J. **835**, 85 (2017)

Noyes, R.W., Hartmann, L.W., Baliunas, S.L., Duncan, D.K., Vaughan, A.H.: Rotation, convection, and magnetic activity in lower main-sequence stars. Astrophys. J. **279**, 763 (1984)

Oshagh, M., Santos, N.C., Figueira, P., Barros, S.C.C., Donati, J.-F., Adibekyan, V., Faria, J.P., Watson, C.A., Cegla, H.M., Dumusque, X., et al.: Understanding stellar activity-induced radial velocity jitter using simultaneous K2 photometry and HARPS RV measurements. Astron. Astrophys. **606**, A107 (2017)

Pont, F., Sing, D.K., Gibson, N.P., Aigrain, S., Henry, G., Husnoo, N.: The prevalence of dust on the exoplanet HD 189733b from Hubble and Spitzer observations. Mon. Not. R. Astron. Soc. **432**, 291 (2013)

Rackham, B., Espinoza, N., Apai, D., López-Morales, M., Jordán, A., Osip, D.J., Lewis, N.K., Rodler, F., Fraine, J.D., Morley, C.V., Fortney, J.J.: ACCESS I: an optical transmission spectrum of GJ 1214b reveals a heterogeneous stellar photosphere. Astrophys. J. **834**, 151 (2017)

Rackham, B.V., Apai, D., Giampapa, M.S.: The Transit light source effect: false spectral features and incorrect densities for M-dwarf transiting planets. Astrophys. J. **853**, 122 (2018)

Radick, R.R., Mihalas, D., Lockwood, G.W., Thompson, D.T., Warnock, A., III, Hartmann, L.W., Worden, S.P., Henry, G.W., Sherlin, J.M.: The photometric variability of solar-type stars. III. Results from 1981-82, including parallel observations of thirty-six Hyades stars. Publ. Astron. Soc. Pac. **95**, 621 (1983)

Radick, R.R., Lockwood, G.W., Skiff, B.A., Baliunas, S.L.: Patterns of variation among Sun-like stars. Astrophys. J. Supl. Ser. **118**, 239 (1998)

Radick, R.R., Lockwood, G.W., Henry, G.W., Hall, J.C., Pevtsov, A.A.: Patterns of variation for the Sun and Sun-like stars. Astrophys. J. **855**, 75 (2018)

Schrijver, C.J.: Magnetic structure in cool stars. XI - relations between radiative fluxes measuring stellar activity, and evidence for two components in stellar chromospheres. Astron. Astrophys. **172**, 111 (1987)

Shapiro, A.I., Solanki, S.K., Krivova, N.A., Schmutz, W.K., Ball, W.T., Knaack, R., Rozanov, E.V., Unruh, Y.C.: Variability of Sun-like stars: reproducing observed photometric trends. Astron. Astrophys. **569**, 38 (2014)

Silva-Valio, A., Lanza, A.F., Alonso, R., Barge, P.: Properties of starspots on CoRoT-2. Astron. Astrophys. **510**, A25 (2010)

Solanki, S.K., Fligge, M.: A reconstruction of total solar irradiance since 1700. J. Geophys. Res. **26**, 2465 (1999)

Topka, K.P., Tarbell, T.D., Title, A.M.: Properties of the smallest solar magnetic elements. II. Observations versus hot wall models of faculae. Astrophys. J. **484**, 479 (1997)

Vaughan, A.H., Preston, G.W.: A survey of chromospheric Ca II H and K emission in field stars of the solar neighborhood. Publ. Astron. Soc. Pac. **92**, 385 (1980)

Vaughan, A.H., Preston, G.W., Wilson, O.C.: Flux measurements of CA II H and K emission. Publ. Astron. Soc. Pac. **90**, 267 (1978)

Walkowicz, L.M., Basri, G., Valenti, J.A.: The information content in analytic spot models of broadband precision light curves. Astrophys. J. Suppl. Ser. **205**, 17 (2013)

Wilson, O.C.: Chromospheric variations in main-sequence stars. Astrophys. J. **226**, 379 (1978)

Zellem, R.T., Swain, M.R., Roudier, G., Shkolnik, E.L., Creech-Eakman, M.J., Ciardi, D.R., Line, M.R., Iyer, A.R., Bryden, G., Llama, J., Fahy, K.A.: Forecasting the impact of stellar activity on transiting exoplanet spectra. Astrophys. J. **844**, 27 (2017)

Zhang, Z., Zhou, Y., Rackham, B., Apai, D.: The near-infrared transmission spectra of TRAPPIST-1 planets b, c, d, e, f, and g and stellar contamination in multi-epoch transit spectra. Astron. J. **156**, 178 (2018)

Chapter 14
Star-Planet Interactions

In Chaps. 10 and 11, I surveyed the various ways in which a host star's radiation and wind can erode an exoplanet's atmosphere, change its chemistry, and thereby determine whether the exoplanet could be habitable. I now turn to the question of whether an exoplanet can change the properties of its host star, in particular, its rotation rate, UV and X-ray radiation, and the properties of its wind. The study of this feedback of an exoplanet on its host star is usually called star-planet interactions (SPI), although a more accurate term would be planet-star interactions. The various ways in which SPI could effect host stars including their radiative and wind emission and possible observational consequences have been reviewed by Guenther and Geier (2015) and Shkolnik and Llama (2018). Cuntz et al. (2000) were the first to explore possible tidal and magnetic interactions between host stars and close-in exoplanets. Subsequent theoretical and observational studies have developed their initial suggestions.

14.1 Tidal and Magnetic Interactions

A useful starting point for the discussion of SPI is the physical processes that occur in a close binary system when both stars have convective envelopes. Zahn (1977) showed that for stars with convective envelopes, turbulent viscosity is the primary force retarding the equilibrium tide to force synchronization of orbital and rotational periods with a time scale,

$$t_{sync} \sim 10^4((1+q)/2q)^2 P^4 \text{ years,} \qquad (14.1)$$

where P is the orbital period in days and $q = M_2/M$ is the mass ratio of the secondary star to the primary. The timescale is somewhat uncertain as the treatment of eddy-viscosity and its effects on tidal dissipation are uncertain, but the functional

© Springer Nature Switzerland AG 2019
J. Linsky, *Host Stars and their Effects on Exoplanet Atmospheres*,
Lecture Notes in Physics 955, https://doi.org/10.1007/978-3-030-11452-7_14

dependence on q and P should not change. Zahn (1977) derived a similar equation for the orbital circularization time and different equations for a warm primary star without a convective envelope. The fourth power of P in this equation requires that $t_{synch} \sim 1.6 \times 10^9$ years for binary stars of roughly equal mass with $P = 20$ days. Shorter period binaries will likely be synchronized and longer period binaries may not. The effect of orbital-rotational synchronization is to increase the rotation rate for short period binaries and especially for giant stars that would otherwise be slow rotators as a result of their increase in radius and their wind-driven angular momentum loss. Increased UV, X-ray, and gyrosynchrotron radiation are produced by the stronger magnetic fields that are produced as a result of faster rotation. RS CVn systems consisting of G and K somewhat evolved stars in short-period orbits, such as HR 1099 (G5 IV+K1 IV, P=2.84 days) and UX Ari (G5 V+K1 IV, P=6.44 days) are examples of such systems with high UV and X-ray luminosities. Short-period dwarf binary systems called BY Dra systems such as the prototype BY Dra (K4 V+K7.5 V, P=5.98 days) and YY Gem (dM1e+dM1e, P=0.81 days) are also very luminous UV and X-ray sources.

Now consider what happens when the secondary star in a close binary system is replaced by a Jupiter-mass planet. The same physical processes occur, but the dissipation that torques the equilibrium tide is weaker by the q^2 factor. For example, a $1 M_{Jup}$ planet in a 1 day orbit around a $1 M_\odot$ star would have $q = 0.001$ and $t_{sync} = 2.5 \times 10^9$ yr. A $10 M_{Jup}$ planet in a 3 day orbit would have a similar synchronization time, and the synchronizing times for lower mass stars will be shorter by the q^2 factor. These estimates indicate that the tidal effects of a Jupiter-mass planet in a short period orbit can spin-up its host star on timescales shorter than the time scales for the host star to evolve on the main sequence. Similar rough estimates have encouraged many observers to search for SPI effects in systems with Jupiter-mass planets and rotational periods shorter than a few days, in particular systems with large values of the parameter M_{planet}/P_{orb}. For a further discussion of tidal synchronization by close-in exoplanets see Lanza (2010) and references therein.

WASP-18 provides a perhaps rare example of how tidal interactions can supress a host star's magnetic field. WASP-18b is a $10.4 M_J$ exoplanet in a very tight 20 h orbit (a=0.02047 au) around its F6 V host star that is younger than 1 Gyr. Contrary to stars of similar spectral type and age, its far-UV emission lines, X-ray luminosity upper limit, and R'_{HK} chromospheric activity parameter are unusually low. The extremely weak level of chromospheric activity indicates a weak stellar magnetic field, which makes it difficult to detect possible SPI signals. Nevertheless, Fossati et al. (2018) pointed out that the tidal bulge produced by the massive hot Jupiter is about 500 km and the ratio of this tidal bulge to the pressure scale height in the upper layers of the host star's thin convective zone is 1.2, far larger than other F-type stars with hot Jupiters. Pillitteri et al. (2014) and Fossati et al. (2018) speculate that these huge tides can change the convective motions and meridional circulation patterns in the star to reduce or perhaps entirely disrupt magnetic dynamo processes, leaving this young star as inactive as an old star. They argued that this type of SPI does not "age" host stars with deep convective zones.

In addition to tidal forces, the interaction of magnetic fields of the host star and its exoplanet can change the activity of the host star and perhaps lead to observable effects. Significant electromagnetic interactions between a close-in Jupiter-mass exoplanet and its host star most likely occurs when the semi-major axis of a magnetized exoplanet is located inside of the host star's Alfvén radius, the distance from the star where the speed of the accelerating wind speed equals the local Alfvén speed. Inside of the Alfvén radius, the stellar wind is subsonic and the reconnection of oppositely directed stellar and planetary magnetic fields can release a plasma stream along field lines to impact the stellar chromosphere and corona. Using a simple magnetohydrodynamic model for an axisymmetric stellar magnetic field, Lanza (2008) showed that when the magnetic fields of the star and planet reconnect, accelerated electrons can flow down magnetic field lines to heat a small hot spot in the stellar chromosphere at an orbital phase well ahead of the exoplanet. In a more realistic simulation of the HD 189733 system consisting of a Jupiter-mass planet orbiting at only $8.8 R_{star}$ from its $0.82 M_\odot$ host star, Cohen et al. (2011) computed the interacting magnetic fields using a time-dependent 3D MHD code and the star's magnetic field obtained from a Zeeman Doppler image. They showed that short duration reconnection events can occur as the exoplanet traverses through the inhomogeneous stellar corona. The resulting downflow of plasma can produce both a "hot spot" of accreted plasma in the stellar atmosphere and a comet-like tail behind the exoplanet as shown in Fig. 14.1.

Matsakos et al. (2015) performed a systematic study of star-planet interactions for systems containing a hot Jupiter exoplanet with a range of parameters for the stellar EUV flux that drives the outflow from the exoplanet and the exoplanet's radius and mass. Both the stellar wind and the exoplanet's outflow are magnetic. They identified four types of flow patterns shown in Fig. 14.2 depending on the relative strength of the input parameters. The calculations used the PLUTO 3-D MHD code. When the planetary outflow is weak, corresponding to weak EUV emission from the host star, the stellar wind plasma is intercepted by the exoplanet's magnetic field leading to a bow shock and the formation of a thin planetary tails shown as Type I interactions in the figure. Type II interactions occur when the planetary outflow is strong producing a bow shock with all of the planetary outflow being swept back into a broad tail. Type III interactions differ from Type II because some of the planetary wind does not go into a tail but instead accretes onto the star. Type IV represents the case where the interaction between the stellar wind and relatively weak planetary wind occurs beyond the L_1 Lagrange point as in Roche lobe overflow, leading to accretion onto the stellar sub-planetary point. An interesting aspect of these calculations is that accretion can occur and in the Type III case the accretion occurs 70–90° ahead of the sub-planetary point as shown in Fig. 14.3. There are several examples of stellar bright spots ahead of the sub-planetary point seen in the Ca II K line (e.g., Shkolnik et al. 2008) and in X-ray and far-UV emission (e.g., Pillitteri et al. 2014). In addition to the high-temperature emission, accretion can also change the rotational period of the star.

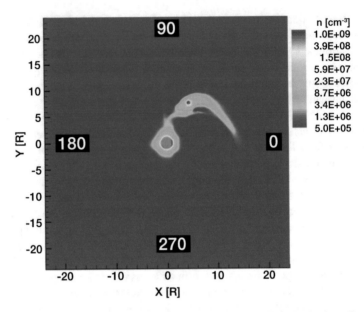

Fig. 14.1 Simulated plasma flows for the HD 189733 system computed with the three dimensional MHD code BATS-R-US. Densities near the host star (large red circle) and Jupiter-like exoplanet (small red circle) are color coded. The figure shows inflowing plasma onto the star ahead of the exoplanet's orbital position during a reconnection event and a plasma tail behind the exoplanet. Figure from Cohen et al. (2011). Reproduced by permission of the AAS

Are such phenomena observable? Can close-in Jupiter-like exoplanets have magnetic fields that can interact with the host star's magnetic field to produce observable SPI effects?

14.2   Observational Searches for SPI

There have been extensive searches for observational evidence of SPI as reviewed by Shkolnik and Llama (2018). One approach has been to search for orbital phase-resolved activity indicators of individual stars, and the second is to compare activity levels of a large sample of stars with and without known planets as a function of the parameter $M_{\text{planet}}/P_{\text{orb}}$.

14.2.1   Monitoring of Individual Stars

Shkolnik et al. (2003) pioneered the search for SPI using the orbital-phased technique in which one looks for enhanced emission in one or more activity

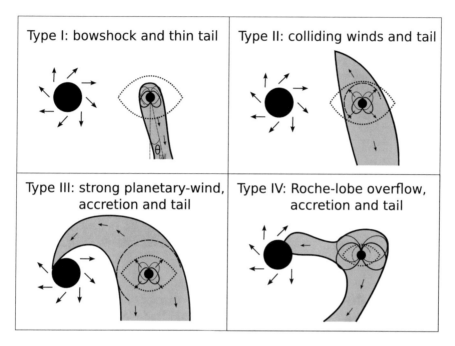

Fig. 14.2 Schematic of the types of star-planet interactions involving close-in hot Jupiter exoplanets. The host star is left of center, the exoplanet is right of center, and the shaded areas indicate material flowing out from the planet. The arrows indicate flow directions. The solid circular arcs indicate where magnetospheric pressure equals the ambient pressure, and the dashed circular arcs mark where the ambient pressure equals the ram pressure of the stellar wind. The dotted line is where the effective exoplanet gravity and centrifugal potential passes through the L_1 Lagrange point. Figure from Matsakos et al. (2015). Reproduced with permission of ESO

indicators phased to the orbital period of the exoplanet rather than to the rotational period of the host star. If the flux of an activity indicator, such the core of the Ca II K line, is seen to repeat with the exoplanet's orbital period, then the cause likely involves interactions of stellar and planetary magnetic fields, whereas if the activity indicator repeats with half of the exoplanet's orbital period, then the likely cause involves tidal forces as the host star's shape will be distorted both towards and away from the exoplanet. If an activity indicator repeats at the star's rotational period, then the cause is likely variable stellar activity unrelated to the exoplanet. Significantly different stellar rotation and exoplanet orbital periods are required for this technique to work.

In the first reported detection of SPI, Shkolnik et al. (2003) presented evidence that HD 179949, an F8 V star with a rotational period of about 7 days and a Jupiter-like exoplanet in a 3.092 day orbit, showed enhanced Ca II flux with a peak near orbital phase $\phi = 0.8$ ahead of the planet's inferior conjunction ($\phi = 0$). Since the enhanced Ca II flux is phased to the orbital period observed over 100 orbits, they interpreted the observations as evidence for chromospheric heating due to magnetic

Fig. 14.3 Logarithmic density (color coded) and flow vectors (arrows) for Type III interactions. The figure shows the fragmented accretion and tail flows. Figure from Matsakos et al. (2015). Reproduced with permission of ESO

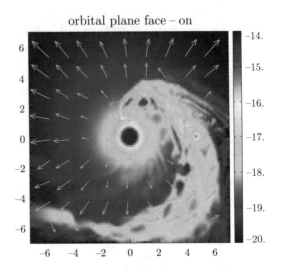

interactions of the close-in exoplanet (a=0.045 au corresponding to about $8R_*$) with the stellar magnetic field. Subsequent observations (see Fig. 14.4) confirmed the reality of enhanced Ca II flux and, therefore, excess chromospheric activity for this star repeating near $\phi = 0.75$.

Several other stars have been monitored in different activity parameters to search of SPI effects. For example, Walker et al. (2008) found evidence that an active region on τ Boo A (F7 V) is present near $\phi = 0.8$ of a Jupiter-like planet ($P_{orb} = 3.3125$ days) on the basis of photometric variations observed by the *MOST* satellite and Ca K line flux variations. They concluded that the active region is possibly induced by the host star's exoplanet. Wood et al. (2018), however, found no evidence for the exoplanet affecting the host star's coronal temperature distribution, but they concluded that the exoplanet somehow altered the star's coronal abundances. Two other stars, υ And (F7 V) and HD 189733 (K1 V), show evidence for SPI (Shkolnik et al. 2008), although Poppenhaeger et al. (2010, 2011) could find no correlation of X-ray or optical emission with orbital period for υ And or other stars. In an analysis of six archival Ca II K-line data sets for HD 189733, Cauley et al. (2018) identified a strong SPI signal for one data set (August 2013) spanning 8 orbits in which the residual K-line flux peaked near $\phi \approx 0.9$ corresponding to a phase lead of 40° ahead of the planet. They argued that the SPI signal was strong at this time because the stellar magnetic power near the planet was larger than for the other data sets.

Monitoring of HD 171156, a G0 V star with a hot Jupiter in a highly eccentric 21.2 day orbit showed an increase in R'_{HK} and X-ray detection at periastron but not at apastron, indicating that close proximity of the exoplanet to its host star is essential for the detection of an SPI signal (Maggio et al. 2015).

There are two unanticipated aspects of these SPI detections that need to be understood in terms of the relevant physics. The first is that the enhanced activity does not peak at the sub-planet point ($\phi = 0$) when the planet is closest to the star

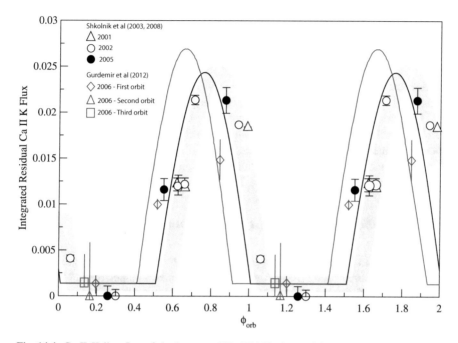

Fig. 14.4 Ca II K line flux of the host star HD 179949 observed between 2001 and 2006 by Shkolnik et al. (2003, 2008) and Gurdemir et al. (2012). The different symbols represent the different observing times. ϕ_{orb} is the orbital phase of the exoplanet relative to inferior conjunction. The solid line is a model of the relative number of starspots. The peaks of the observed Ca II K line fluxes and the spot model are near orbital phase 0.75 ahead of inferior conjunction. Figure from Shkolnik and Llama (2018)

but rather near $\phi = 0.8$, a location far from the planet. This can be explained by simulations of interacting stellar and planetary magnetic fields (Lanza 2008; Cohen et al. 2011) in which the close-in exoplanet orbits inside of the Alfvén point in the stellar wind allowing the downflow of plasma released by the reconnection of stellar and planetary magnetic fields to the chromosphere and corona. One important result of these calculations for the case of HD 189733, where the exoplanet's orbit has a semi-major axis of only $8.8R_*$, is that the plasma's splashdown point is ahead of the planet at an orbital phase that depends on the strength and orientation of the planet's magnetic field.

A second aspect of these SPI observations is that the detections are sporadic with positive detections occurring only about 75% of the time. This can be explained by the need for reconnection events to release plasma from the exoplanet to the host star, and such events occur only when the fields are oppositely directed and interacting. Since the star's magnetic field is inhomogeneous and rotating with the star's rotation period, which differs from the orbital period, there will be orbits when reconnection events can and cannot occur resulting in the on/off nature of SPI events (cf. Shkolnik et al. 2008). An illustrative example is the large changes in dynamic pressure seen by the exoplanet orbiting Proxima Centauri that produces variable compression of the exoplanet's magnetic field as an exoplanet crosses different magnetic field regions and current sheets in the wind of its host star (Garraffo et al. 2016). The triggering of magnetic energy release could result from the planet's motion through the star's magnetic field that changes the field topology, thereby increasing the helicity dissipation of the two magnetic fields (Lanza 2009).

SPI can also feed back on the planet's atmosphere. In a model for magnetic field interactions, Lanza (2013) finds that accelerated electrons penetrate deep into a close-in planet's atmosphere near the magnetic poles with energy comparable to the star's EUV radiation. Thus SPI can contribute to the loss of a planet's atmosphere.

14.2.2   Comparing Stars with and Without Planets

A number of studies have searched for SPI effects by comparing large samples of stars hosting planets with stars that are not known to host a planet, the nonhost stars. These large statistical studies have so far provided inconclusive results probably because the two star samples are very different. Planet-hosting stars that are bright enough to have UV and X-ray observations, which are very sensitive to SPI, are mostly discovered by radial velocity searches that select for low-activity and thus older stars to minimize the star's radial velocity noise. Non-hosting stars are generally more active and younger and thus have not been targets for radial-velocity studies. Since activity is correlated with age, the two stellar samples have very different ages. This sample bias makes it difficult to obtain unambiguous results concerning whether higher activity is an age or a SPI effect (Canto Martins et al. 2011; Poppenhaeger and Schmitt 2011; Shkolnik 2013). Host stars discovered from transits are generally too distant for UV and X-ray studies.

With the *HST* MUSCLES survey of seven M dwarfs with exoplanets, France et al. (2016) searched for a correlation of UV emission line fluxes with the SPI parameter, M_{planet}/P_{orb}. They found that the Pearson correlation coefficient was near zero for the Mg II chromospheric lines (indicating no correlation with M_{planet}/P_{orb}) but near 0.8 for the transition region lines (Si III, Si IV, C IV, and N V). To test whether this tentative SPI correlation for transition region lines is supported by a much larger star sample, France et al. (2018) compared the FUV fluxes observed by STIS and COS of 71 known planet-hosting stars with 48 stars without known planets. The UV emission lines of the planet-hosting stars are systematically 5–10 times fainter than the nonhosting stars (see Fig. 14.5), which they ascribe to a low activity bias in the selection of targets for radial velocity monitoring and the subsequent bias of selecting planet-hosting stars for follow-up *HST* observations. Thus the systematically low UV fluxes of the planet-hosting stars are not useful identifiers for SPI. To compensate for this bias, they selected small groups of stars with similar spectral types and rotational periods and found similar UV emission line fluxes for the planet-hosting and nonhosting stars, providing no evidence for SPI. For the full sample of stars, France et al. (2018) found a statistically significant

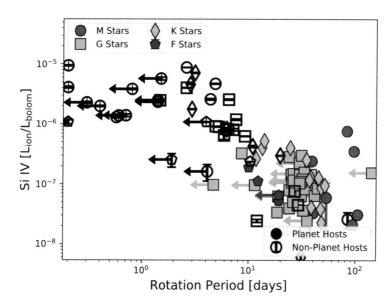

Fig. 14.5 The luminosity ratio of the Si IV 139.3 and 140.2 nm doublet to L_{bol} for 71 host stars and 48 nonhost stars with color coded spectral types vs stellar rotation period. Arrows indicate upper limits. The data illustrate the bimodal nature of the two stellar data sets. Figure from France et al. (2018). Reproduced by permission of the AAS

correlation with the SPI parameter, M_{planet}/P_{orb}, but they found no correlation of UV fluxes with the SPI parameter in a principal component analysis. Thus the statistical approach of searching for correlations of enhanced UV fluxes with the SPI parameter is severely compromised by the much larger UV activity levels of the biased sample of stars with different ages, rotational periods, and other properties. A further complication is that most if not all of the stars identified as non-planet-hosting may actually have planets that have not yet been detected or even been searched for.

Poppenhaeger and Wolk (2014) tried a different approach to the search for evidence of SPI. They obtained X-ray observations of five widely separated binaries in which one star has a known exoplanet and the other does not. For two of the systems, HD 189733 (K0 V+M4 V) and CoRoT-2 (G7 V+K9 V), the primary star has a much larger X-ray luminosity for its age, which is assumed to be the same as that of the secondary star, indicating that tides produced by the close-in exoplanet have forced the primary star to spin faster and, therefore, be more active and X-ray luminous than predicted for its age.

In summary, there is now considerable evidence for SPI effects when there is a high-mass exoplanet in close proximity to its host star, but corrections for sampling bias remains a challenge. Additional observations and simulations are need to further test the importance of SPI effects. In particular, there could be observable differences in the large scale magnetic fields of stars with and without exoplanets, but such differences have not yet been detected (Fares et al. 2013). The very low activity level of WASP-18, an F6 IV-V star with a hot Jupiter in a 20 h orbit, suggests that the massive planet may disrupt the stellar magnetic dynamo leading to very weak X-ray emission (Pillitteri et al. 2014).

14.3   Can Exoplanet Magnetic Fields Be Detected?

The likely detection of SPI phenomena resulting from the interactions of stellar and planetary magnetic fields requires that the exoplanets have magnetic fields. Unfortunately, there are no direct measurements of magnetic fields from planets outside of the solar system, although there are several lines of evidence that many and perhaps most exoplanets should have magnetic fields.

Planetary magnetic fields are created by internal dynamos driven by thermal and compositional convection in their electrically conducting interiors. Olson and Christensen (2006) found that the Earth, Jupiter, Saturn, Uranus, Neptune, and Jupiter's moon Ganymede, which all have measured magnetic fields, have internal convective velocities above the critical magnetic Reynolds number required for magnetic dynamo action. Furthermore, the similarity of the magnetic parameters for these planets suggests that they all can be described by the same convective dynamo family. In their analysis of 125 published magnetic dynamo models, they found scaling laws that reasonably well predict the observed magnetic fields of these planets and can be used to predict magnetic properties of exoplanets. Christensen et al. (2009) extended this study to obtain scaling laws that fit both planets and stars.

Figure 14.6 plots the measured magnetic moments of solar system planets as a function of M_{planet}/P_{rot}. The magnetic moment is the product of an electrical current (measured in Amperes) and the square of the diameter of the current loop (measured in m^2) in the core of a planet. The ratio of the dipole magnetic moments of Jupiter to Earth is more than a factor of 10^5, although the ratio of planet radii squared is only 125. Thus the magnetic properties of a planet do not just scale with the planet's radius but rather depend on the strength of the convective flows that are driven either by heat from the planet's core or distributed sources throughout the planet's volume. These quantities are unknown for exoplanets. Fortunately, the relation shown in Fig. 14.6 and the similarity of the magnetic parameters of the terrestrial planets (Olson and Christensen 2006) provide some confidence that one can predict the magnetic moments of exoplanets based only on their masses and rotational periods. If so, then close-in exoplanets such as hot Jupiters with their synchronous rotation periods of one to several days will likely have magnetic moments smaller than Jupiter which has a rotation period less than 10 h. Predicting magnetic moments for rocky Earth-like planets is more uncertain because both Venus and Mars do not presently have global magnetic fields perhaps because these two planets do not have tectonic plates.

The discovery of near-UV absorption prior to ingress, the beginning of a transit at optical wavelengths, led Vidotto et al. (2010) to propose that the pre-ingress

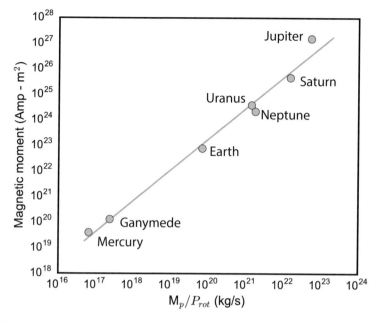

Fig. 14.6 Magnetic moments of solar system planets and one moon of Jupiter. Figure from Shkolnik and Llama (2018)

absorption is due to a magnetic bow shock ahead of the exoplanet in its orbit. For the case of WASP-12, a G0 V star with a Jupiter-mass exoplanet in a 1.09 day orbit (a=0.023 au), the planet has a supersonic orbital speed of 230 km s^{-1} that can produce a bow shock in front of the planet if the stellar magnetic field is aligned with the orbital motion or at an angle if the magnetic field is misaligned. Pressure balance between the total pressure (gas and magnetic) in the corona at the location of the planet and the planet's hot gas and magnetic pressure results in an upper limit of 24 G for the mean dipolar magnetic field strength at its surface. Since this is an upper limit that depends on many assumed parameters, this analysis of the transit timing of WASP-12b provides some support for the existence of of magnetic fields on this hot Jupiter exoplanet.

Low-mass brown dwarfs have many similarities to exoplanets in that they are fully convective, do not have a nuclear source of heat in their cores, and have similar radii as most of their volume is degenerate matter. Kao et al. (2016) detected pulsed radio emission from three of four T-class brown dwarfs with T_{eff} in the range 800–1100 K and masses in the range 20 to >30 M_J. They interpreted the 100% circularly polarized pulses as electron cyclotron maser (ECM) emission (cf. Hallinan et al. 2015; Pineda et al. 2017). Assuming that this emission is at the fundamental frequency for ECM, the implied magnetic field is >2.5 kG. In a subsequent survey of radio emission from L7 to T6.5 brown dwarfs with T_{eff} in the 800–1800 K range, Kao et al. (2018) derived minimum surface magnetic field strengths of 2.3–4.4 kG from radio observations of pulsed ECM auroral emission. Although these brown dwarfs rotate faster and have masses 12 to >27 times that of Jupiter, the presence of kilogauss magnetic fields on the low-mass brown dwarfs suggests that Jupiter-like planets could also have strong magnetic fields amplified by dynamo processes. The internal structure of low-mass brown dwarfs may not the same as for cooler exoplanets, in particular the much higher luminosity of brown dwarfs suggests that they have higher convective velocities and thus larger magnetic fields than exoplanets. Nevertheless, the measurement of strong magnetic fields in low-mass brown dwarfs provides another argument in support of the presence of magnetic fields on exoplanets. Searches for ECM emission from exoplanets are underway with no detections yet.

Finally, there is a technique that has produced a realistic estimate of the magnetic moment of one particular hot Jupiter. Kislyakova et al. (2014) proposed that the magnetic moment of an exoplanet can be inferred from the size of the neutral hydrogen cloud obstacle formed by the interaction of the stellar wind and the exoplanet's magnetosphere. They modeled the interaction between the wind of the G2 V solar-like host star HD 209458 with its hot Jupiter exoplanet and determined the best parameters to fit the observed broad Lyman-α absorption observed during transits. Since the host star is similar to the Sun, they could assume solar parameters for its wind. They found that the magnetic moment of the Jupiter-like exoplanet HD 2019458b has a magnetic moment 10% as large as Jupiter. This is consistent with a weaker magnetic dynamo resulting from the slower 3.52 day rotation period if synchronous with the orbital period compared to Jupiter's 10 h rotation period. Although this method requires knowledge of the often uncertain wind parameters of the host star, it is applicable to other types of exoplanets including Earth-like exoplanets with broadened Lyman-α absorption lines produced by hydrogen in their magnetospheres.

References

Canto Martins, B.L., Das Chagas, M.L., Alves, S., Leão, I.C., de Souza Neto, L.P., de Medeiros, J.R.: Chromospheric activity of stars with planets. Astron. Astrophys. **530**, 73 (2011)

Cauley, P.W., Shkolnik, E.L., Llama, J., Bourrier, V., Moutou, C.: Evidence of star-planet interactions in the HD 189733 system from orbitally-phased Ca K variations (2018). arXiv181005253C

Christensen, U.R., Holzwarth, V., Reiners, A.: Energy flux determines magnetic field strength of planets and stars. Nature **457**, 167 (2009)

Cohen, O., Kashyap, V.L., Drake, J.J., Sokolov, I.V., Garraffo, C., Gombosi, T.I.: The dynamics of stellar coronae harboring hot Jupiters. I. A time-dependent magnetohydrodynamic simulation of the interplanetary environment in the HD 189733 planetary system. Astrophys. J. **733**, 67 (2011)

Cuntz, M., Saar, S.H., Musielak, Z.E.: On stellar activity enhancement due to interactions with extrasolar giant planets. Astrophys. J. Lett. **533**, L151 (2000)

Fares, R., Moutou, C., Donati, J.-F., Catala, C., Shkolnik, E.L., Jardine, M.M., Cameron, A.C., Deleuil, M.: A small survey of the magnetic fields of planet-host stars. Mon. Not. R. Astron. Soc. **435**, 1451 (2013)

Fossati, L., Koskinen, T., France, K., Cubillos, P.E., Haswell, C.A., Lanza, A.F., Pillitteri, I.: Suppressed far-UV stellar activity and low planetary mass loss in the WASP-18 System. Astron. J. **155**, 113 (2018)

France, K., Loyd, R.O.P., Youngblood, A., Brown, A., Schneider, P.C., Hawley, S.L., Froning, C.S., Linsky, J.L., Roberge, A., et al.: The MUSCLES treasury survey I: motivation and overview. Astrophys. J. **820**, 89 (2016)

France, K., Arulanantham, N., Fossati, L., Lanza, A.F., et al.: Far-ultraviolet activity levels in F, G, K, and M dwarf exoplanet host stars (2018). arXiv180907342F

Garraffo, C., Drake, J.J., Cohen, O.: The space weather of Proxima Centauri b. Astrophys. J. Lett. **833**, L4 (2016)

Guenther, E.W., Geier, S.: The effects of close-in exoplanets on their host stars. In: Characterizing Stellar and Exoplanetary Environments. Astrophysics and Space Science Library, vol. 411, p. 169. Springer, Berlin (2015)

Gurdemir, L., Redfield, S., Cuntz, M.: Planet-induced emission enhancements in HD 179949: results from McDonald observations. Publ. Astron. Soc. Aust. **29**, 141 (2012)

Hallinan, G., Littlefair, S.P., Cotter, G., Bourke, S., Harding, L.K., Pineda, J.S., Butler, R.P., Golden, A., Basri, G., Doyle, J.G., et al.: Magnetospherically driven optical and radio aurorae at the end of the stellar main sequence. Nature **523**, 568 (2015)

Kao, M.M., Hallinan, G., Pineda, J.S., Escala, I., Burgasser, A., Bourke, S., Stevenson, D.: Auroral radio emission from late L and T dwarfs: a new constraint on dynamo theory in the substellar regime. Astrophys. J. **818**, 24 (2016)

Kao, M.M., Hallinan, G., Pineda, J.S., Stevenson, D., Burgasser, A.: The strongest magnetic fields on the coolest brown dwarfs. Astrophys. J. Suppl. Ser. **237**, 25 (2018)

Kislyakova, K.G., Holmström, M., Lammer, H., Odert, P., Khodachenko, M.L.: Magnetic moment and plasma environment of HD 209458b as determined from Lyα observations. Science **346**, 981 (2014)

Lanza, A.F.: Hot Jupiters and stellar magnetic activity. Astron. Astrophys. **487**, 1163 (2008)

Lanza, A.F.: Stellar coronal magnetic fields and star-planet interaction. Astron. Astrophys. **505**, 339 (2009)

Lanza, A.F.: Hot Jupiters and the evolution of stellar angular momentum. Astron. Astrophys. **512**, A77 (2010)

Lanza, A.F.: Star-planet magnetic interaction and evaporation of planetary atmospheres. Astron. Astrophys. **557**, A31 (2013)

Maggio, A., Pillitteri, I., Scandariato, G., Lanza, A.F., Sciortino, S., Borsa, F., Bonomo, A.S., Claudi, R., Covino, E., Desidera, S., et al.: Coordinated x-ray and optical observations of star-planet interaction in HD 17156. Astrophys. J. Lett. **811**, L2 (2015)

Matsakos, T., Uribe, A., Königl, A.: Classification of magnetized star-planet interactions: bow shocks, tails, and inspiraling flows. Astron. Astrophys. **578**, 6 (2015)

Olson, P., Christensen, U.R.: Dipole moment scaling for convection-driven dynamos. Earth Planet. Sci. Lett. **250**, 561 (2006)

Pillitteri, I., Wolk, S.J., Sciortino, S., Antoci, V.: No X-rays from WASP-18. Implications for its age, activity, and the influence of its massive hot Jupiter. Astron. Astrophys. **567**, 128 (2014)

Pineda, J.S., Hallinan, G., Kao, M.M.: A panchromatic view of brown dwarf aurorae. Astrophys. J. **846**, 75 (2017)

Poppenhaeger, K, Schmitt, J.H.M.M.: A correlation between host star activity and planet mass for close-in extrasolar planets? Astrophys. J. **735**, 59 (2011)

Poppenhaeger, K., Wolk, S.J.: Indications for an influence of hot Jupiters on the rotation and activity of their host stars. Astron. Astrophys. **565**, L1 (2014)

Poppenhaeger, K., Robrade, J., Schmitt, J.H.M.M.: Coronal properties of planet-bearing stars. Astron. Astrophys. **515**, A98 (2010)

Poppenhaeger, K., Lenz, L.F., Reiners, A., Schmitt, J.H.M.M., Shkolnik, E.: A search for star-planet interactions in the υ Andromedae system at x-ray and optical wavelengths. Astron. Astrophys. **528**, A58 (2011)

Shkolnik, E.L.: An ultraviolet investigation of activity on exoplanet host stars. Astrophys. J. **766**, 9 (2013)

Shkolnik, E.L., Llama, J.: Signatures of star-planet interactions. In: Handbook of Exoplanets, p. 1734 (2018)

Shkolnik, E., Walker, G.A.H., Bohlender, D.A.: Evidence for planet-induced chromospheric activity on HD 179949. Astrophys. J. **597**, 1092 (2003)

Shkolnik, E., Bohlender, D.A., Walker, G.A.H., Cameron, A.C.: The on/off nature of star-planet interactions. Astrophys. J. **676**, 628 (2008)

Vidotto, A.A., Jardine, M., Helling, C.: Early UV ingress in WASP-12b: measuring planetary magnetic fields. Astrophys. J. Lett. **722**, L168 (2010)

Walker, G.A.H., Croll, B., Matthews, J.M., Kuschnig, R., Huber, D., Weiss, W.W., Shkolnik, E., Rucinski, S.M., Guenther, D.B., Moffat, A.F.J., Sasselov, D.: MOST detects variability on τ Bootis A possibly induced by its planetary companion. Astron. Astrophys. **482**, 691 (2008)

Wood, B.E., Laming, J.M., Warren, H.P., Poppenhaeger, K.: A Chandra/LETGS survey of main-sequence stars. Astrophys. J. **862**, 66 (2018)

Zahn, J.-P.: Tidal friction in close binary stars. Astron. Astrophys. **57**, 383 (1977)

Chapter 15
Summary and Final Comments

Now that exoplanets are being discovered almost daily through transit, radial velocity, and other techniques, the direction of exoplanet research is changing from discovery to characterization of exoplanet properties. Of particular importance is the existence and composition of exoplanet atmospheres and whether these atmospheres are supportive of life forms on the surfaces of rocky planets. It is now recognized that the emissions of host stars, both their radiation and wind properties, determine whether their exoplanets retain their initial atmospheres and any secondary atmospheres that may later emerge. While there is only one present example of an inhabited planet, there are now more than a handful of possibly habitable rocky exoplanets and the number of such planets will surely increase rapidly in the near future. Although the presence of surface water was the original criterion for the term habitable zone, or more precisely "liquid water habitable zone" (LWHZ), there are additional factors that also determine habitability and many of these factors involve the present and past properties of the host star.

The spectral energy distribution emitted by a host star depends critically on the star's activity. Young rapidly rotating stars show far higher X-ray and UV emission than slowly rotating old stars of the same spectral type. The chemical composition of an explanet's atmosphere responds to the stellar UV emission, especially the very strong Lyman-α radiation that can be as energetic as the entire UV radiation of an M dwarf and the strongest UV emission component of a G-type star. Far-UV (FUV) emission photo-dissociates H_2O, CO_2, CH_4, and other molecules leading to the production of oxygen molecules including O_3. Near-UV (NUV) radiation, on the other hand, dissociates O_3. Thus the relative strength of FUV to NUV radiation controls the oxygen chemistry, and the FUV/NUV flux ratio depends on stellar activity and thus on the stellar magnetic field. Since these photochemical reactions occur relatively high in an exoplanet's atmosphere, typically at the 0.1–1 mbar level, vertical mixing through flows and diffusion control the timescale over which photochemical reactions can determine the bulk chemistry of an exoplanet's atmosphere. The rate of volcanic outgassing and chemical interactions with an ocean

© Springer Nature Switzerland AG 2019

J. Linsky, *Host Stars and their Effects on Exoplanet Atmospheres*,
Lecture Notes in Physics 955, https://doi.org/10.1007/978-3-030-11452-7_15

must also be considered in assessing the long term evolution of an exoplanet's atmosphere.

What should be done to evaluate the observations of the many nearby exoplanets that will be discovered with the new space observatories *TESS*, *JWST*, and *PLATO* and the increasing powerful ground-based spectrometers and direct imaging instruments. First, it is essential to obtain simultaneous or near-simultaneous observations of the host star across its electromagnetic spectrum at the same time as the exoplanet is observed. This is especially important for young M dwarfs that are highly variable. The MUSCLES (France et al. 2016) and MEGA-MUSCLES (PI: C. Froning) surveys of M dwarfs are examples of such panchromatic spectra. With such data one can estimate the extreme-UV emission, reconstruct the Lyman-α profile, and estimate the stellar mass-loss rate. With a panchromatic spectrum and stellar wind estimate, one can infer the exoplanet's present mass-loss rate and atmospheric chemistry. A sensible assessment of habitability, however, requires a realistic evaluation of these properties back in time at least to when a young star enters the main-sequence. This involves characterizing the output of host stars with similar T_{eff} but very different ages. An example is the study of solar-type stars by Guinan et al. (2003), but similar studies of different mass stars is essential.

To better understand the processes occurring in the atmospheres of stars, we need detailed physical models that include as much of the essential physics as feasible. Models are of two types: (1) semi-empirical models that include non-LTE radiative transfer and multispecie excitation and ionization statistical equilibria with a thermal structure that leads to good fits to the observed panchromatic spectrum, and (2) theoretical models that compute the thermal structure from first principles and energy balance. Both types of models must adequately describe the ionization and excitation equilibria of many atoms and ions, compute the profiles and fluxes of optically thick emission lines in non-LTE with partial frequency redistribution, include all important processes including diffusion of electrons and neutrals in the thermally steep transition region, and include molecular spectra that are especially important for M stars but are also important in some wavelength regions for G and K stars. Since active stars are spatially inhomogeneous and their radiation is time variable, either multi-component or multi-dimensional models are needed. Computing such models will be a severe challenge.

A deeper understanding of habitability will require more data or more realistic simulations concerning host star winds, flares, coronal mass ejections, and high energy proton emission for stars with different masses, ages, and rotation rates. Exoplanet magnetic fields have not yet been measured but are needed to properly estimate exoplanet mass-loss rates. These tasks are underway by many groups and are required if we are to understand host stars and their effects on exoplanet atmospheres.

I conclude with Table 15.1 that summarizes the many unfavorable factors that can prevent or restrict habitability after a host star reaches the main sequence and its exoplanets have lost their initial gas envelopes. The factors that involve the host star are highlighted in bold face. These factors, which are discussed at length in the preceding chapters, identify which types of stars are more favorable for

Table 15.1 Unfavorable factors for habitability

Factor (stellar factors **bold face type**)	Effect on the exoplanet
Major changes in luminosity	**Exoplanet moves in or out of LWHZ**
Many superflares	**Loss of the explanet's atmosphere**
Many high-energy proton events	**O_3 destruction, surface sterilization**
Very strong UV radiation	**Photodissociation of H_2O, CO_2, CH_4**
Very strong EUV radiation	**Hydrodynamic loss of atmosphere**
Very strong wind	**Nonthermal loss of atmosphere**
Exoplanet mass $\geq 1.5 M_{Earth}$	Retains initial gas envelope
Low initial or accreted H_2O	Very dry surface
Highly eccentric orbit	Variable surface temperatures
Orbital instabilities	Exoplanet moves in or out of LWHZ
Weak exoplanet magnetic field	Nonthermal loss of atmosphere

supporting habitable exoplanets. Although they are the most common type of star in the Galaxy, M dwarfs have many negative factors which are made more severe by the multi-billion years that such stars remain very active and the close proximity of the LWHZ to the host star. Solar-type stars, likely including the whole G and F spectral classes, are relatively inactive after 1 Gyr or less and are radiatively stable while they are on the main sequence for several billion years. These stars are good candidates for hosting habitable exoplanets, but such stars are far less common in the Galaxy than M dwarfs. The intermediate temperature K-type stars should also be good candidates for hosting habitable exoplanets as these stars are more common, they become relatively inactive after 1 or 2 Gyr, and their LWHZs are located at intermediate distances compared to solar-type and M stars. Fortunately, the question of which types of stars best support habitability will be better understood in the next few years as stellar activity phenomena are monitored. Extensive monitoring of the most favorable host stars by the new generation of space and ground-based observatories could guide us from the present stage of habitability assessments to the discovery of inhabited exoplanets.

References

France, K., Loyd, R.O.P., Youngblood, A., Brown, A., Schneider, P.C., Hawley, S.L., Froning, C.S., Linsky, J.L., Roberge, A., et al.: The MUSCLES treasury survey I: motivation and overview. Astrophys. J. **820**, 89 (2016)

Guinan, E.F., Ribas, I., Harper, G.M.: Far-ultraviolet emissions of the Sun in time: probing solar magnetic activity and effects on evolution of paleoplanetary atmospheres. Astrophys. J. **594**, 561 (2003)

Printed in the United States
By Bookmasters